Biology

A Family-Style Science Program

Michelle Copher and Karen Loutzenhiser

Published by HooDoo Publishing
United States of America

ISBN 979-8-9874658-3-7

Available at Layers-of-Learning.com

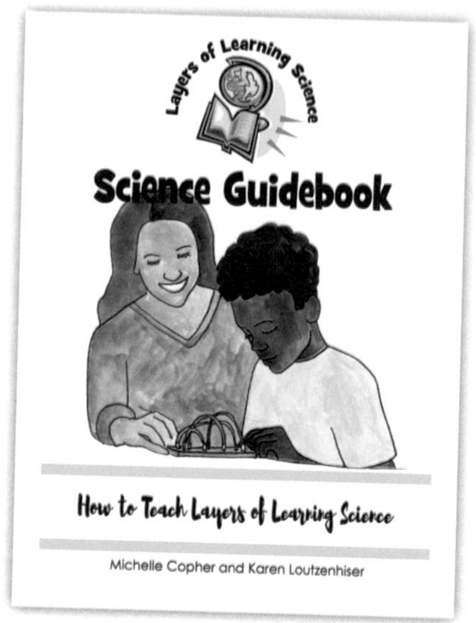

Science Guidebook: How To Teach Layers of Learning Science

Learn the philosophy behind Layers of Learning Science, how to plan a unit, how to schedule your learning, and how to assess and grade.

This is a valuable and inexpensive PDF guide to using the curriculum effectively.

Science Notebook

The Science Notebook is modeled after the notebooks that real scientists keep. Students record data and sketches from experiments and observations. But they also add foldables, take notes, and record their learning. The science courses include instructions on filling the Science Notebook.

Printable Packs

If you purchased this book directly from Layers of Learning, your Printable Packs were included on your receipt. If you purchased it elsewhere you can retrieve your free Printable Packs with the coupon code: FREEPACK.

Scan the QR code to visit the Biology Resources page.

How To Use This Course

Biology is a family-style program, which means your whole family, from ages six to eighteen and beyond, can use the program together. The activities are meant to be a family affair, with individual expectations being tailored to the ages and abilities of each child. We also encourage you to share completed individual work, like reports, posters, and projects.

Like all of Layers of Learning, Biology is a pick-and-choose curriculum; you don't need to complete everything in the book. Instead, browse through and choose the library books, Explorations, and sidebars that appeal to you and are appropriate for your kids. Generally you will choose one or two Explorations to do together each week.

Scheduling

Each unit within this book is designed to last about a month and then be repeated in subsequent years, but the exact schedule and timing are completely up to you. If your kids are engaged and enthusiastic about a topic, feel free to carry on for a little while longer.

In Biology each Exploration is one complete lesson plan. The exact length of each one varies and depends on your needs, the ages of the students, and how absorbed in a lesson you get. However, you can generally plan on one to two hours per lesson. Science should be a weekly part of your educational plan.

Sidebars

You will find sidebars throughout Biology. The sidebars are little snippets of information you can read aloud while children work, things you can touch on during a Morning Meeting, or springboards for lessons you create on your own. Do as many or as few as you like.

Printable Packs

This curriculum includes printables. For printing convenience and to keep your costs down, the printables are found in digital Printable Packs that you can retrieve from the Layers of Learning catalog. Find the product page for this book and scroll down to the link to download the Printable Pack. If you purchased this book directly from Layers of Learning, you will see a link to download the printables on your receipt. Otherwise, use the coupon code FREEPACK at checkout to get the Printable Packs for free.

Resources

Every unit comes with its own YouTube playlist of videos to use as enrichment or as video lectures to supplement lessons. The videos can be played during a lesson. Frequently Explorations include instructions to watch a video. The videos are curated in these playlists.

On the Layers of Learning website, you can also find links to websites and resources that are especially useful when teaching each unit. The web links are located under "Resources" on the main menu.

At Layers of Learning, we believe learning is about exploring and we invite you to joyfully explore with us. In the words of Robert Louis Stevenson, "The world is so full of a number of things, I'm sure we should all be as happy as kings."

Table of Contents

LIVING THINGS

Biology is the science of life. Biologists are scientists who study living organisms like plants and animals. Living things have unique characteristics—they must grow, use energy, reproduce, respond to the environment, and be made of cells. Some living things are made of only one cell, while others have hundreds, thousands, millions, or billions of cells. There is an enormous number of varieties of living things on Earth, and biology is the science of all of those things.

In this unit, we'll learn about cells. They are microscopic membranes with plasma and DNA inside. There are many different kinds of cells that can do many different things. They can move, create proteins, generate energy, store products, transmit information, and so much more.

To organize our knowledge of living things, biologists have put the different kinds of life into categories, a process we call classification. Over time, the categories have been refined and changed as our knowledge has increased. For the most part, the categories a living thing fits into has to do with how it looks. For example, things with hair are in a different category from things with feathers. The spider, wasp, and flower, below, are all living things, but they have their own classifications within the category of living things.

As we learn about living things, the cells they all have in common, and how they are classified into groups, you can think of yourself as a biologist. You may only be beginning, but you are learning all about the science of life.

Unit Overview

Key Concepts:

- Biologists use the scientific method, keep careful notes, and use microscopes.
- Cells are the smallest unit of life and all cells come from other cells.
- Microorganisms are too small to see with the naked eye. They include single-celled bacteria, protists, algae, fungi, archaea, and tiny multicellular protists and animals.
- Biologists classify living things based on common characteristics. Living things are organized into domains, kingdoms, phyla, classes, orders, families, genera, and species.

Vocabulary:

- Biology
- Scientific method
- Microscope
- Cell
- Life
- Cell membrane
- Cytosol
- Organelle
- Prokaryotic cells
- Eukaryotic cells
- DNA
- Multicellular organism
- Stem cell
- Sexual reproduction
- Asexual reproduction
- Microorganism
- Bacteria
- Protists
- Classification
- Species

Scientific Skills:

- Scientific method
- Scientific sketching
- Using a microscope

Theories, Laws, & Hypotheses:

- Cell theory

Step I: Library List

Choose books from your library that go with this topic. Here's a list of some favorites and also a list of search terms so you can utilize what your library offers. Read the books with your kids and/or assign them some to read independently. It is from these books your kids will learn most of the facts they need from this unit.

Search for: microscope, living things, cells, microbes, microorganisms, bacteria, classification

☺ ☺ ☻ *Encyclopedia of Science* from DK. Read "What Is Life?" on page 306; "Cells" on pages 338-339; "Classifying Living Things," "Viruses," "Bacteria," and "Single-Celled Organisms" on pages 310-314; "Growth and Development" on pages 362-363; and "Asexual Reproduction" on page 366.

☺ ☺ ☻ *The Usborne Science Encyclopedia*. Read "Plant Cells" on page 250, "Classifying Plants" on pg. 254, "Animal Cells" on pg. 298, and "Classification" on pg. 340.

☺ ☺ ☻ *The Kingfisher Science Encyclopedia*. Read pages 49-52 about life and classification.

☺ ☺ ☻ *The Usborne Complete Book of the Microscope* by Kirsteen Rogers. Get this book if you are new to microscopes and want to really utilize yours to the utmost. It explains how a microscope works, teaches you how to use one, and gives lots of ideas of things to look at through the microscope.

☺ ☺ ☻ *The Animal Book: A Visual Encyclopedia of Life on Earth* by David Burnie. This is a big, colorful encyclopedia of living things from bacteria and mushrooms to elephants and more. Use this as a browsing book and reference book through most of biology this year.

☺ *Living Things and Nonliving Things* by Kevin Kurtz. A thought-provoking book for young kids about what makes something alive.

☺ *Enjoy Your Cells* by Fran Balkwill. Simple picture book for young kids about the cells that make up our bodies.

☺ *Tiny Creatures: The World of Microbes* by Nicola Davies. Microbes live everywhere all around us and in us. This book tells you what single-celled creatures do and what they are.

☺ *What Kind of Living Thing Is It?* by Bobbie Kalman. Living things are alike in many ways, but it is the differences we use to sort them. The reader looks at characteristics of living things and sorts them into groups.

Family School Levels

The colored smilies in this unit help you choose the correct levels of books and activities for your child.

☺ = Ages 6-9
☺ = Ages 10-13
☻ = Ages 14-18

On the Web

For videos, web pages, games, and more to add to this unit, visit the Biology Resources at Layers-of-Learning.com.

You will find a link to video playlists, web links, and more.

Bookworms

If you're looking for a family read-aloud, we'd like to suggest this one.

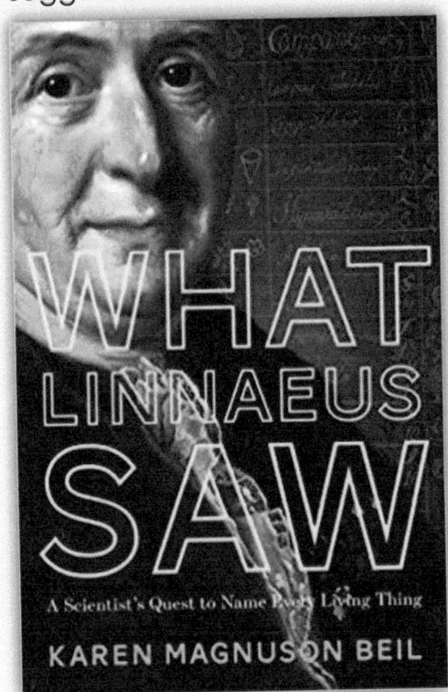

What Linnaeus Saw by Karen Magnuson Beil tells the story of Linnaeus and his quest to classify living things, beginning with an odd animal from America.

Deep Thoughts

What is life?

Living things:

- Have cells with DNA
- Use energy
- Respond to their surroundings
- Grow and develop
- Reproduce

The definition above is a pretty good one and is a accepted by most biologists, but it has problems too.

Is a virus alive? It doesn't have cells, and it doesn't have DNA, but it does have RNA, a sub-unit of DNA, it uses energy, and it reproduces, though not without a living cell to use parasitically.

You would think that the definition of life would be pretty easy, but scientists debate over it.

How can you tell if something is alive?

Bookworms

Who Goes First?: The Story of Self-Experimentation in Medicine by Lawrence K. Altman is a page turner about scientists and patients who were pioneers in risky endeavors.

For ages 14 and up.

☺ *It's Catching: The Infectious World of Germs and Microbes* by Jennifer Gardy. A fun book (really) all about the microorganisms and diseases they cause.

☺ *The Vast Wonder of the World* by Melina Mangal. A lovely picture book about the pioneering cell biologist Ernest Everett Just, an African American who researched egg cells in the early 1900s.

☺ ☺ *How To Replicate A Wood* by Tali Lavy. The entire story about a forest, surrounded by a stone wall, is a metaphor for cell replication.

☺ ☺ *Proteins* by Tali Lavy. Proteins make up many of the parts of cells, including DNA. This book is an illustrated, fun, and accurate introduction to this super important topic.

☺ *The Bacteria Book* by Steve Mould. Besides bacteria, this books also covers cells and their functions. Lots of pictures and just the right amount of text.

☺ *The Basics of Cell Life With Max Axiom* by Amber J. Keyser. This is a graphic novel science book that teaches about prokaryotic and eukaryotic cells, their parts, and how they make up all living things.

☺ *The Surprising World of Bacteria with Max Axiom* by Agnieszka Biskup. Graphic novel about bacteria. A fun and painless way to learn about the microscopic world.

☻ *The Cell: A Visual Tour of the Building Blocks of Life* by Jack Challoner. Quite a bit of detail, but told with easy to understand language and lots of full-color photos and illustrations.

☻ *March of the Microbes: Sighting the Unseen* by John L. Ingraham. Discusses all the single-celled life and how it has a huge effect on the world. Easy and entertaining.

☻ *A Tour of the Cell, What Is DNA, Mitosis, Meiosis, Classification of Life, Protists,* and *Bacteria* from Bozeman Science on YouTube. Use these videos as lectures for your high schooler. Have your student take notes.

Step 2: Explore

Choose hands-on explorations from this section to work on as a family. They should be appealing activities that will create mental hooks so your kids remember the information in the unit. Save the rest of the explorations for the next time you do this unit in four years when your kids are older. You can also read the sidebars together and explore some little rabbit trails.

This unit includes printables. See the introduction for instructions on retrieving your Printable Pack.

Biology Skills

Biology is the study of living things. Biologists work outdoors in the field studying plants, animals, ecosystems, and populations. Other biologists work in labs studying bacteria, medicine, viruses, and disease. All biologists use the scientific method and careful note taking during their work.

Teaching Tip

Every child needs his or her own Science Notebook. You can buy one from Layers of Learning or make your own with a spiral bound notebook. We prefer either an art sketchpad or a book of graph paper for science. If you make your own, be sure to reserve a few pages at the beginning for a table of contents that you'll add to throughout the course.

EXPLORATION: The Scientific Method & Osmosis

For this activity, you will need:

- "Scientific Method" from the Printable Pack
- Video about the scientific method from the unit's YouTube Playlist
- Colored pencils and a pen for labeling
- Scissors
- Glue stick
- Gummy bears
- 250 ml beaker
- Electronic balance
- 2 small bowls
- Salt
- Water
- Ruler
- Science Notebook

Scientists conduct experiments to learn about the natural world. They design experiments that are controlled so they can observe the results of just one variable within the experiment. They hypothesize, experiment, record their observations and data along the way, and continually look for new information. This process of methodically searching for answers about the natural world is called the **scientific method**.

1. Watch a video about the scientific method. Take notes

Memorization Station

Biology: the study of living things

"Bio" is Greek for life and "logy" is Greek for branch of study

Scientific method: the systematic observation, experimentation, and measurement of the natural world to find out why and how things happen

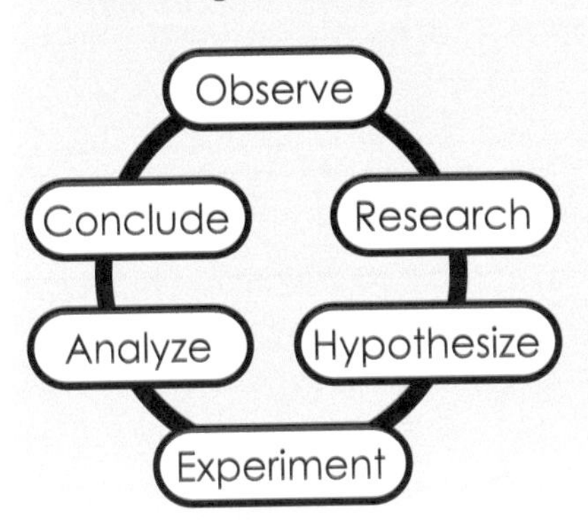

on the "Scientific Method" anchor chart, cut it out around the solid lines, and add it and paste your notes into your Science Notebook. Add it to your table of contents too.

2. Head a new page in your Science Notebook "Gummy Bear Osmosis." Write down a hypothesis of what you think will happen if you soak a gummy bear in water overnight. Then, write a hypothesis of what you think will happen if you soak a gummy bear in saltwater overnight.
3. Measure 150 ml of water into each of two bowls. Add 10 g of salt to one of the bowls. Label the bowl with the salt so you don't get them mixed up.
4. Drop a gummy bear into each bowl and let it sit overnight.
5. Draw a labeled diagram of your procedure in your Science Notebook.
6. Observe the gummy bears. Measure the gummy bear in plain water, the gummy bear in saltwater, and a gummy bear that was left dry. Record your results in your Science Notebook.

   The gummy bear in plain water absorbed a lot of liquid because it needed to balance the solutes in the gummy bear with the solute in the water. This is called osmosis, the natural balancing out of dissolved things in water.

   The gummy bear in saltwater absorbed some liquid, but not as much as the one in plain water. It needed to absorb less to balance out the solutes, because the saltwater had more dissolved things in it.

   Did you hypothesize that the gummy bear in saltwater would absorb less water than the one in plain water?
7. Discuss what you learned together and identify these parts of the scientific method within the experiment:
   - Purpose (often a question)
   - Hypothesis (expected results based on observations)
   - Procedure (the steps of an experiment)
   - Results (along with all data collected along the way)
   - Conclusion

EXPLORATION: Using A Microscope

For this activity, you will need:

- "Parts of a Microscope" and "Parts of a Microscope Instructions" from the Printable Pack

Famous Folks

Anton van Leeuwenhoek was a Dutch scientist who invented the microscope, making it possible for the first time to see individual cells.

Dutch means that Leeuwenhoek was from the Netherlands. Find the Netherlands on a map of the world.

On the Web

An organization called Nature's Notebook is actively seeking people who live in the United States, U.S. territories, and Canada to observe nature and help them collect data on birds, insects, plants, and other species. They train you in what to look for and how to report data and then help you know when and where to look.

Search for them online to find out more and get involved in real scientific data collection.

Memorization Station

Microscope: an instrument used to magnify and view very small objects

- Colored pencils and a pen for labeling
- A light microscope (If you decide to buy a microscope, look for a compound microscope that plugs into the wall and has 3 objective lenses. A good high school microscope runs around $130. You don't *need* a microscope, but for high school, especially, it is nice.)
- Empty microscope slides with slide covers
- Eye dropper
- Water
- Methylene blue stain (optional - makes it easier to see)
- Prepared microscope slides (highly recommended if you have a microscope - look for a "biology slide set" from a science supplier)

Microscopes are a basic tool of biologists. They allow the scientist to see things that are too small for the human eye.

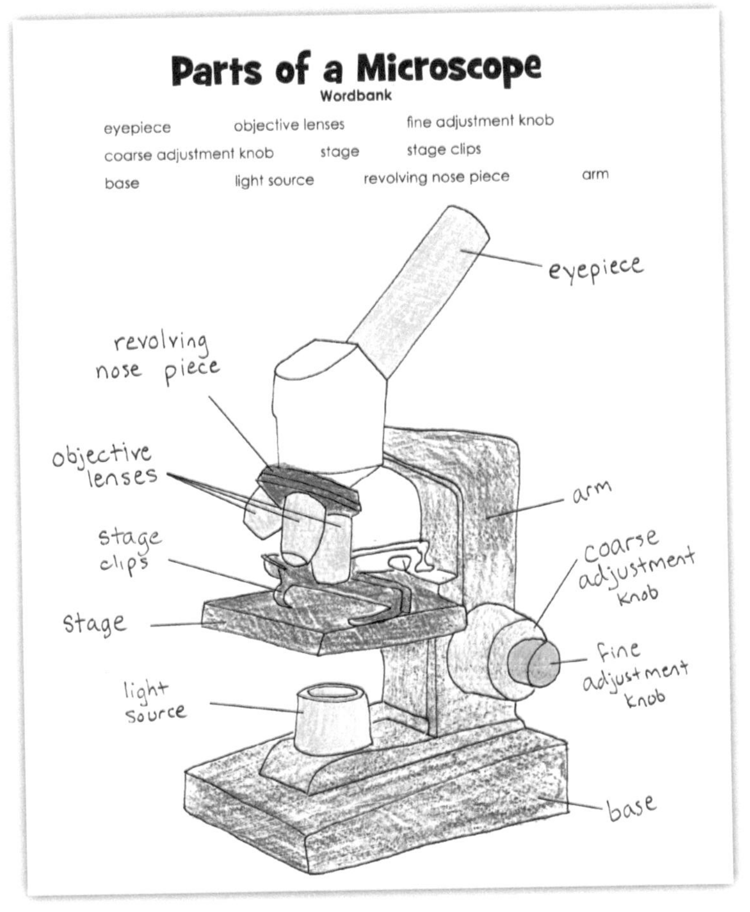

1. Start with "Parts of a Microscope" labeling and coloring the worksheet using the image above and the directions from "Parts of a Microscope Instructions."

2. Get out your microscope and find all the parts from "Parts of a Microscope" on your instrument. Practice carrying it safely, storing it where it will stay clean, and carefully moving the parts, like the revolving nose piece and coarse and fine adjustments.

Deep Thoughts

Most of the early advances in cell biology happened in Europe, especially Germany, the Netherlands, and England.

What was happening in 1700s to 1800s Europe to make this possible?

Famous Folks

Englishman Robert Hooke used an early microscope to peer at tiny creatures and cells. He meticulously drew what he saw and published his *Micrographia* in 1665.

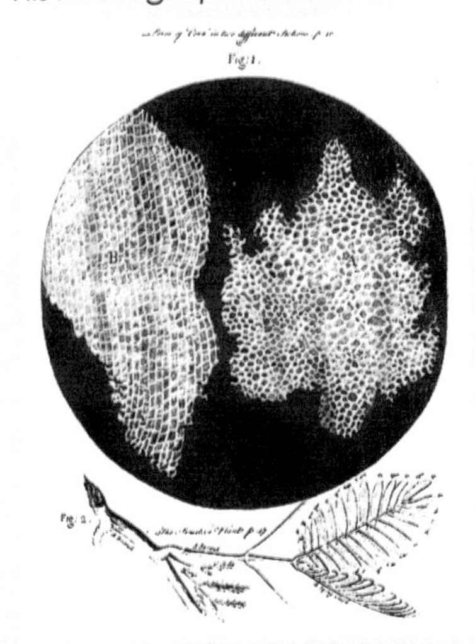

Hooke was the first to use the word "cell" to describe what he was seeing when he studied cork under a microscope.

Find England (part of Great Britain) on a map of the world. This is where Robert Hooke lived and did his work.

Additional Layer

Take the "Microscope Match-up" quiz from the Printable Pack to see how much you remember. An answer sheet is also included.

Additional Layer

Cells are generally considered the smallest form of life, but some scientists believe viruses are also living.

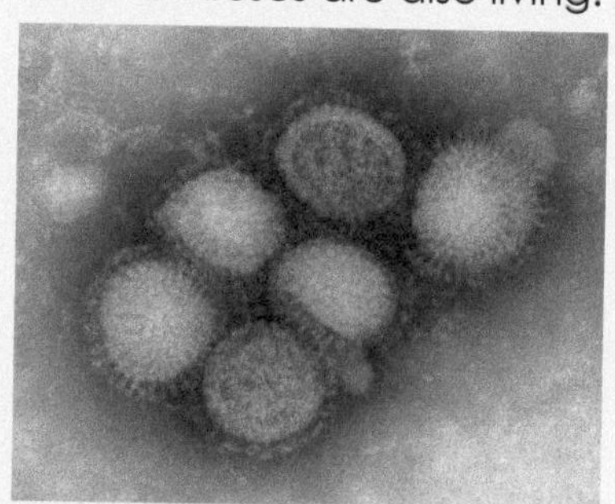

H1N1 influenza virus

A virus is a long strand of DNA or RNA inside a capsid or protein coat. They are much smaller than bacteria (think of a watermelon compared to a bb). They don't have normal cell structure and can only replicate by using a living cell.

Learn more about what a virus is and how it works.

Writer's Workshop

1.74 million species have been cataloged and described so far. We know there are many more that have not been discovered. Estimates of the number of living species ranges from 8 million to a trillion.

Around 18,000 new species are described by scientists every year.

Imagine a biologist searching for and finding a new species. It could be in a jungle, in a sample of dirt from his backyard, or in her lab with a microscope.

Write a series of field notebook entries that describe the find.

3. Put a prepared slide from a slide set onto the stage. Revolve the nose piece until the lowest magnification (4x) is pointing straight down at the slide. Turn on the light. Look through the eyepiece, adjusting with the coarse adjustment until the specimen comes into focus. Get it just right with the fine adjustment knob.
4. Move the revolving nose piece to the next level of magnification (10x). Use only the fine adjustment knob this time to get the specimen into focus. How much more can you see?
5. Finally, move the revolving nose piece to the highest level of magnification (40x). Use only the fine adjustment knob to bring the specimen into focus. If you can't find the specimen, move back down to 10x magnification, get it in focus, then try again at the higher magnification.
6. You can make your own slides of anything you want to look at. Take some time to play around. Gently scrape a few skin cells from inside your cheek with a toothpick or carefully scrape a thin layer off of an onion. Place the specimen on a clean glass slide. Drop 1-2 drops of water onto the specimen, add a drop of Methylene blue stain to make the specimen easier to see, then carefully cover the specimen with a slide cover. To keep bubbles out, tilt the slide cover on at an angle.

   Try looking at carpet fibers, dirt you sweep off the floor, a fingernail clipping, mold from your fridge, an insect, or whatever you can think of. Larger, dryer, or stiffer things like a fingernail or an insect don't need a drop of water and a slide cover.

EXPERIMENT: Scientific Sketching

For this activity, you will need:

- "Scientific Sketching" from the Printable Pack
- Pencil
- Science Notebook
- Colored pencils
- Natural specimens from outside (various options)
- Bean seed
- Plastic bag
- Paper towel

Scientific sketching isn't about making pretty pictures, it's about noticing and recording details. It is one method of recording data. It also forces you to really look carefully, much more so than taking a photograph. Scientific sketching allows a scientist to draw attention to important details that might be lost in a photograph.

1. Use the "Scientific Sketching" anchor chart from the

Printable Pack to discuss what makes a good scientific sketch.

- Make the sketch as accurate as you can. Don't make a stylized picture of a flower; draw the flower as you actually see it from one perspective.
- Make the sketch big, filling most of the page, so you can draw and see details. This is true whether you are sketching a bird or a microbe under a microscope.
- Color the picture to show how it looks in real life.
- Make the picture detailed, with all the parts in all the right places. Include the little things like stamens or stripes or knee joints.
- Explain the picture with labels, paragraphs, date and location, and other details.

Add illustrations to the anchor chart to show what each term means. For example, draw a stylized flower next to a realistic looking flower by the word "accurate."

Cut out the anchor chart and paste it into your Science Notebook.

2. Go outside and find a living specimen to sketch. It could be a leaf, an insect, a worm, or a snail, and so on. Look at the specimen carefully and record as many details in your Science Notebook in picture form as you can.
3. Somewhere next to your sketch, record your observations in word form, including the date and location of the specimen. If you can identify the specimen, that is great too. Label the parts of the specimen.
4. Sketch a natural process. Get a bean seed. Place it into a plastic bag with a wet paper towel. Sketch the bean seed on day one and add labels. Sketch the bean seed every three days, for at least three sketches total, as new developments happen, adding new notes and labels each time.

EXPLORATION: Biology Lab Safety

For this activity, you will need:

- Paper
- Colored markers

Biology includes laboratory work that can be dangerous if done sloppily. It's important to be orderly and neat, have a clear, clean space to work, always wear proper safety equipment, and wash up afterward. You also need to be aware of allergies. If you or another member of the group is allergic to a plant or animal that is recommend in an

Famous Folks

In 1985, Dr. Barry Marshall of Australia purposely drank a beaker full of the bacteria *Heliobacter pylori*. At the time, scientists said that bacteria couldn't survive in stomach acid and couldn't cause diseases of the digestive tract. Marshall thought they were wrong.

Dr. Marshall quickly became deathly ill and did tests to prove that the bacteria had colonized in his stomach. He took antibiotics, which saved his life, and he also proved that bacteria could cause symptoms such as ulcers, vomiting, and bad breath.

Millions of people every single year are cured of gastric ulcers because of Dr. Marshall's crazy self-experiment. Marshall received a Nobel Prize in medicine and probably was scolded by his mother.

Don't drink bacteria, kids.

Writer's Workshop

Write a story about a series of horrible lab accidents all caused by lab safety violations.

Take your character from one disaster to the next!

Deep Thoughts

Where did life come from in the first place?

For almost 2,000 years, beginning with Aristotle in around 350 BC, many people thought life sprang spontaneously from inorganic matter. This is the theory of spontaneous generation. For example, maggots seem to spring from rotting meat, leeches seem to ooze out of the scum at the bottom of ponds, and fruit flies seem to appear magically on decaying fruit.

Then, in the 1600s, a scientist named Francesco Redi said that to really prove spontaneous generation you would have to isolate the materials from the surroundings. So he put meat in a jar and screwed a lid on. No maggots appeared. Later, Louis Pasteur proved that even microorganisms come from other microorganisms.

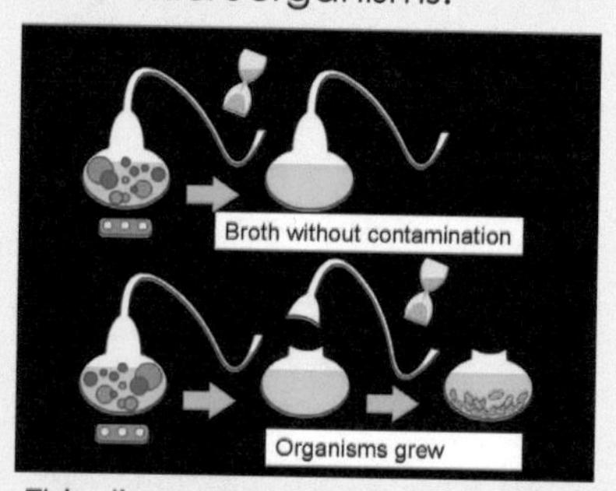

This diagram shows how Pasteur heated broth then kept it in a flask with a curved top to let in air but keep microbes out.

Spontaneous generation was finally debunked. We now believe in cell theory, which says that cells must come from other cells.

But where then did the first cell ever on Earth come from? What do you think?

experiment, either use a substitution or skip it.

1. Create an anchor chart showing safe practices in a biology lab. You can use the one above as inspiration.

   Include these concepts:

   - Have a clean, clear space to work.
   - Read the entire experiment before beginning.
   - Pay attention to and follow all warnings given in the experiment. Warnings look like this:

WARNING: This lab uses live cultures of bacteria. These can be harmful. Always wear gloves. Never eat or drink during this lab.

   - Handle sharp instruments with care. Dissections will be recommended in this course. Sharp scalpels and probes are a necessary tool. Mentors should demonstrate how to use them safely before kids are allowed to use them.
   - Chemicals may be recommended for some labs. Always handle and dispose of chemicals properly.
   - Always wear gloves, goggles, and other safety equipment, as recommended. If you don't have a supply of these safety items on hand, get some before you begin experiments.
   - Clean hands with soap. Clean surfaces with soap and/or disinfectant. Clean any bowls, utensils, or lab equipment with soap and/or disinfectant after every lab.
   - Never eat or drink during a biology lab unless the lab specifically instructs you to do so. This is especially important during labs with microorganisms or dissections.

Cells

Cells are the most basic building block of **life**. All living things are made of cells and knowing about cell functions is essential to understanding everything else in biology.

EXPLORATION: Simplified Cell Model

For this activity, you will need:

- Plastic wrap
- Small cardboard box (like a gelatin box)
- Green dessert gelatin
- Refrigerator
- Raisins
- Grapes
- Fruit leather
- Small nuts like peanuts or chopped walnuts
- Science Notebook

A cell is the smallest unit of life. Cells all have a **membrane**, which is like a sack, holding the cell together. Inside the membrane is a thick liquid called **cytosol**. In the cytosol are **organelles**, tiny parts of the cells that do all the jobs the cell needs to live and reproduce. All cells come from other cells.

1. Prepare the desert gelatin according to the directions on the box.
2. Cut off one large side of the box, then line it with plastic wrap. The box represents the cell wall, as found in plant cells. The plastic wrap represents the cell membrane. Set the box on a small plate.
3. Pour the gelatin into the lined box while it is still warm, until it is about three quarters full. The gelatin represents the cytosol, the liquid that fills all cells.

Memorization Station

Cell: smallest living unit of an organism, consist of a cell wall or membrane surrounding cytoplasm and various organelles

Life: a living thing can grow, reproduce, use energy, and respond to its environment

Cell membrane: a layer of lipids and proteins that surrounds a cell

Cytosol: liquid inside a cell

Organelle: specialized structures inside a cell

Add the "Living Things" definition box from the Printable Pack to your Science Notebook.

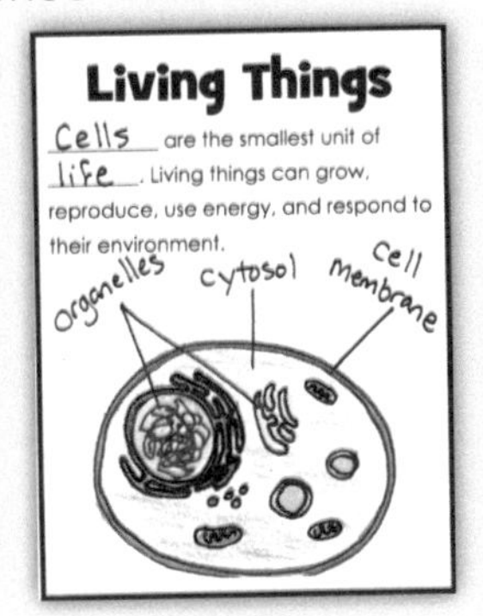

Additional Layer

This is an artist's rendering of the structure of a bacterial flagellum.

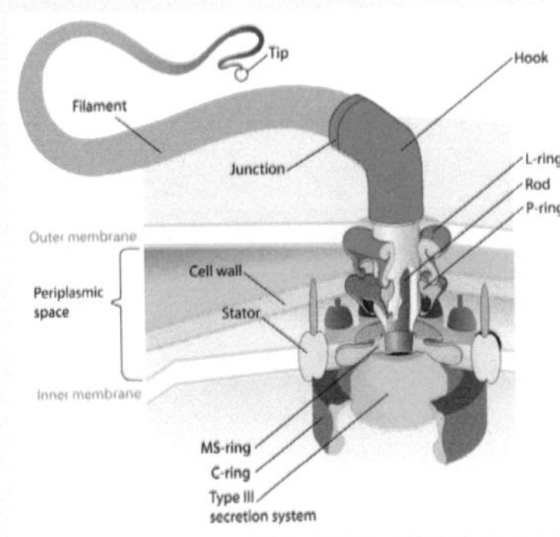

Even simple seeming structures of life are surprisingly complex. Learn how a flagellum works.

Fabulous Fact

Proteins are made of molecules called amino acids. There are 21 different amino acids available to make all the proteins in the living world. They are manufactured inside of cells by ribosomes on instructions from RNA, made by copying snippets of DNA.

This is an amino acid called Glutamine:

O O
H_2N OH
H_2N H

Glutamine can be linked on long chains of dozens or even hundreds of other amino acids to make specific proteins that do specific jobs.

Enzymes, hormones, structural pieces, mucus, hemoglobin in blood, and thousands of other cell products are made of proteins. For example, Insulin, a hormone that regulates blood sugar in humans, is made of a chain of 51 amino acids which are manufactured only in pancreas cells.

This is a 3-D diagram of a complex insulin protein. The curves show the way the protein actually shapes itself in reality.

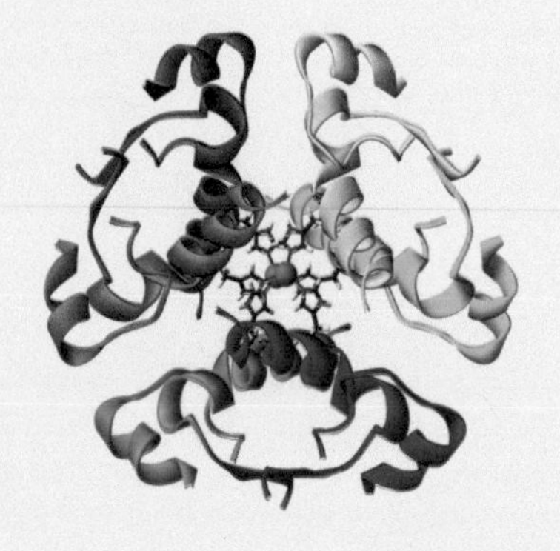

4. Put a grape in the box to represent the nucleus of the cell where the DNA is.
5. Then, accordion fold some fruit leather and place it next to the nucleus to represent the endoplasmic reticulum, where the proteins are made as directed by the DNA.
6. Next, add some raisins to the cell. These represent vesicles, tiny sacks that can be used to store materials until the cell is ready to either use them or get rid of them.
7. Finally, poke several small nuts around inside the cytosol. These represent mitochondria, which generate chemical energy for the cell.
8. Put your model in the refrigerator and wait an hour to see the finished cell model. Name each of the parts you learned and what it does.
9. Draw a picture of your cell model in your Science Notebook. Write down each of the parts: cell wall, cell membrane, cytosol, nucleus, endoplasmic reticulum, vesicle, and mitochondria. Also, write their definitions.

EXPLORATION: Cell Diagrams

For this activity, you will need:

- "Prokaryotic and Eukaryotic Cells" from the Printable Pack
- Colored pencils and a pen for labeling
- Microscope (optional)
- Prepared slides of a bacteria and a euglena (Optional - If you don't have a microscope, you can look up microscope images of bacteria and euglena online to view the structures.)
- Pencil
- Science Notebook

There are two main types of cells: prokaryotic and eukaryotic. **Prokaryotic cells** are simpler and smaller than eukaryotic cells. They do not have a nucleus and have only one strand of DNA instead of two. They are single-celled organisms like bacteria.

Eukaryotic cells are the cells of animals, plants, protists, and fungi. They have membrane-bound organelles, a nucleus with a membrane, double-stranded DNA, and are much larger than prokaryotic cells.

1. Color and label the "Prokaryotic and Eukaryotic Cells" sheet. First, label the cell membrane on each diagram. Color the cell membrane light blue. The cell membrane

surrounds the cell and protects it from the outside environment. It's like the "skin" of the cell.

Prokaryotic and Eukaryotic Cells

Prokaryotic Cells

capsule DNA cell wall cell membrane pilus flagellum
ribosomes cytosol

cytosol
ribosomes
DNA
cell membrane
cell wall
capsule
pilus
flagellum

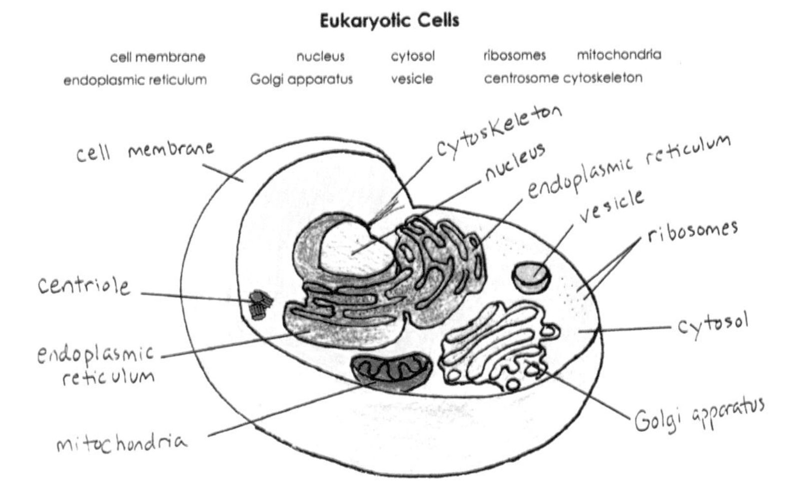

2. Label the cytosol on each diagram and color it yellow. Cytosol is the fluid that fills the cell. The cytosol and all the things floating in it are the cytoplasm.

3. Label the ribosomes. They are shown as tiny dots on these diagrams and we won't color them. Ribosomes make proteins by stitching amino acids together at the direction of RNA, which is replicated sections of DNA.

4. Label the nucleus on the eukaryotic cell and the DNA on the prokaryotic cell. Color them purple. The nucleus of the eukaryotic cell contains the DNA inside a membrane. In the prokaryotic cell, there is no membrane around the DNA.

5. Label the cell wall and the capsule of the prokaryotic cell. Color them two different shades of blue. The cell wall protects the cell and gives it shape. Some eukaryotic cells also have cell walls. Plants, for example, have stiff cells walls to give the plant structure. Some

Memorization Station

Prokaryotic cells: simple, small cells without a nucleus and one strand of DNA

Eukaryotic cells: large, complex cells with a nucleus, membrane-bound organelles, and two strands of DNA

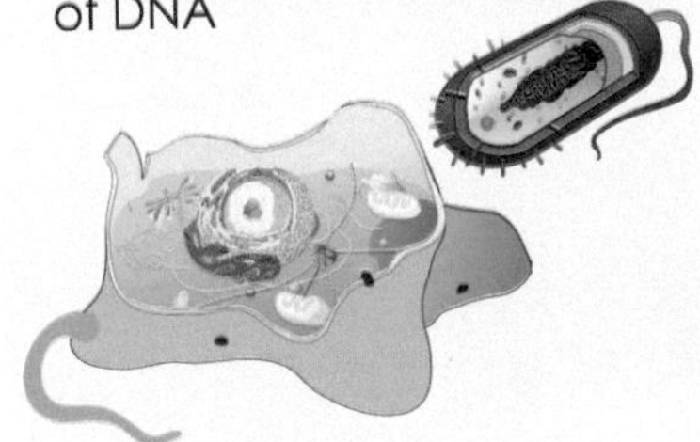

Memorization Station

Cell theory says that:

1. Cells are the smallest unit of life.
2. All living things are made up of cells.
3. Cells always arise from other cells.

Memorize cell theory and its three parts.

This theory was first proposed by Germans Matthias Jakob Schleiden and Theodor Schwann in 1839.

Cell theory is the unifying theory of biology, on which the rest of the study of life rests.

Without cell theory, we can't understand disease, genetics, how medicines work, reproduction, how pollutants affect organisms, or most other aspects of biology.

Additional Layer

Cut out and glue the "Functions of a Cell" pieces from the Printable Pack to a page in your Science Notebook.

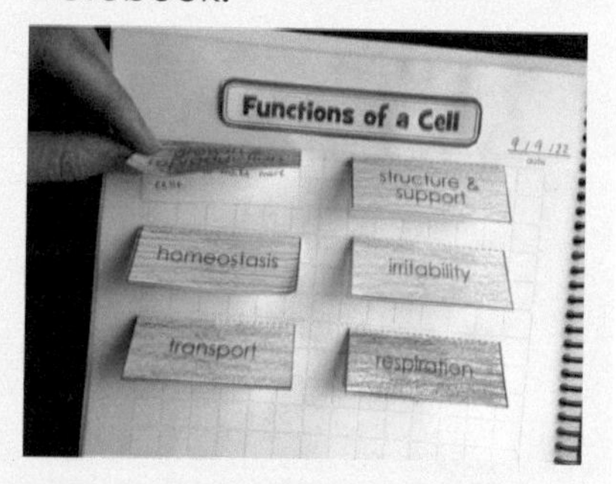

Look up each term listed on the flaps and write the definitions inside each flap.

Every cell does these things to stay alive and do its job.

Bookworms

The Immortal Life of Henrietta Lacks by Rebecca Skloot is about a woman whose cancer cells became the most studied cells in all of science.

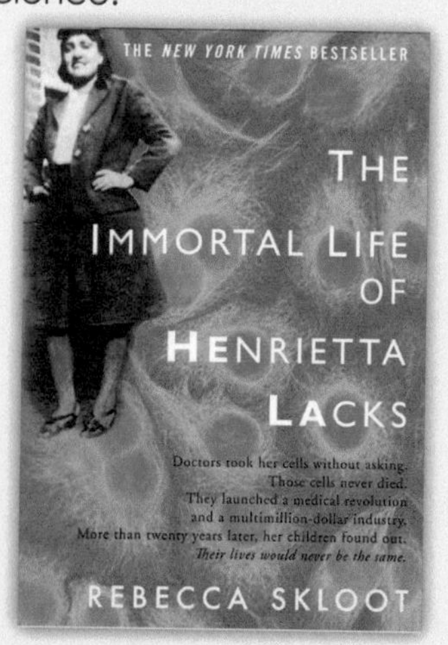

It's also about the ethics of medicine, the abuse of poor people, and the racism that led to Henrietta's cells being worth billions of dollars while her family remained in poverty.

For ages 14 and up.

types of bacteria also have a capsule which protects the cell even further.

6. Label the pilus on the prokaryotic cell. Pili are hairlike appendages used to pass DNA between bacteria for reproduction. They can also be used for locomotion or to grasp a surface.
7. Label the flagellum on the prokaryotic cell and color it pink. The flagellum is a long whip-like tail used for locomotion. It is present in some bacteria and occasionally in eukaryotic cells.
8. Label the cytoskeleton of the eukaryotic cell. The cytoskeleton is made of protein filaments that connect the cell membrane to different organelles in the cell. It gives the cell structure and is a pathway for cell signals.
9. Label the centrosome and color it orange. Centrosomes are only in protozoa and animal cells. They organize the cytoskeleton and are especially important during mitosis (cell division).
10. Label the mitochondria and color it red. Mitochondria are where chemicals are assembled for energy storage. The cell can use this energy later, as needed.
11. Label the endoplasmic reticulum and color it pink. The endoplasmic reticulum is connected to the nucleus and is the site of most of the protein, lipid, and hormone synthesis of the cell.
12. Label the vesicle and color it green. Vesicles are membranes that contain materials the cell is importing or exporting. They are used to move or store substances.
13. Label the Golgi apparatus and color it light green. The Golgi apparatus is used to put together proteins that come from the endoplasmic reticulum. The Golgi apparatus makes vesicles that carry the proteins to the cell wall. The proteins will later be secreted from the cell.
14. Look at a bacteria and a euglena under a microscope and compare them. A bacteria is a prokaryote and a euglena is a eukaryote. Most of the structures of the cells are too small to see with a normal light microscope, but you should still be able to notice some major differences. Sketch each of them in your Science Notebook.

EXPERIMENT: Cell Membranes

For this activity, you will need:

- Plastic sandwich bag
- Water

- Corn starch
- 250 ml beaker
- Electronic balance
- Iodine
- Small bowl

A cell membrane protects the cell and holds it all together. Things like carbohydrates, proteins, vitamins, and fats still need to be able to get in and out of the cell though. The cell membrane has pores and is semi-permeable. Some things can get through, but not everything.

1. Head a page in your Science Notebook "Cell Membranes" and add it to your table of contents.
2. Measure 10 g of corn starch with an electronic balance and 200 ml of water with a beaker. Pour them both into a plastic sandwich bag, seal the bag, and mix it up.
3. Fill a small bowl half full of water. Add 20 drops of iodine to the water in the bowl.
4. Place the sandwich bag with corn starch into the bowl of iodine water. Let it sit for 30 minutes, then check the starch water.

   In the presence of iodine, starch turns dark blue-purple. The iodine is able to travel through the bag and into the starch. This is like the semi-permeable membrane of a cell. Some things can travel through the cell wall, but not everything.
5. Draw a labeled picture of your experiment in your Science Notebook and write an explanation of how cell membranes are semi-permeable.

Additional Layer

Cell membranes are made of a double thick layer of lipids, or fats. This is called the phospholipid bilayer. A phospholipid molecule is made of a water-loving, or hydrophilic, head with a phosphorus group and a water-hating, or hydrophobic, tail made of lipids.

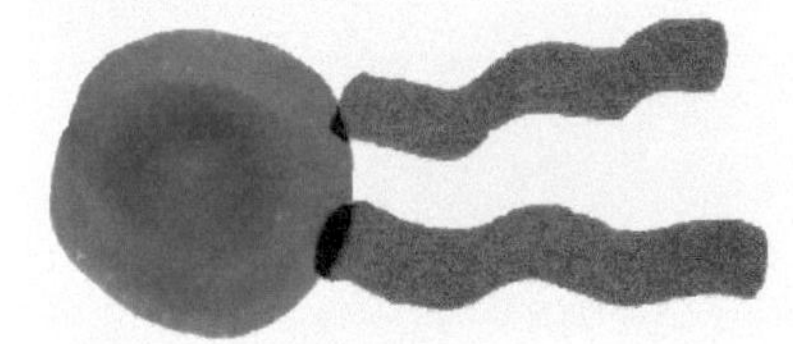

Because the tails do not like water, they all arrange themselves in a sheet with the tails facing inward, away from the fluid inside and outside the cell.

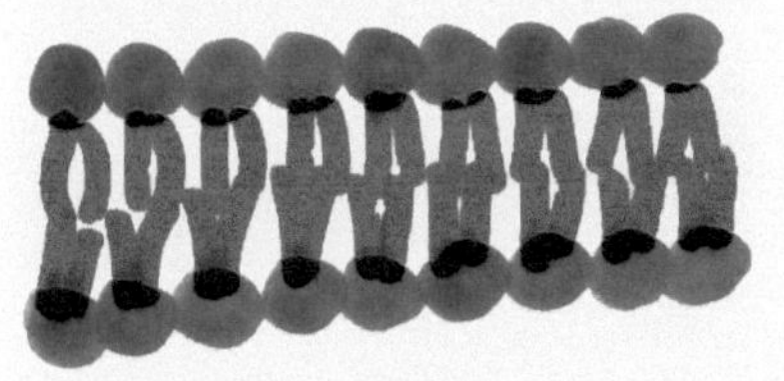

There are no edges, the phospholipid bilayer automatically arranges itself into a sphere. The only things that can get though this layer are oxygen and carbon dioxide.

Piercing the phospholipid bilayer are proteins that allow water, glucose, and nutrients in and out of the cell.

Larger molecules or larger quantities of molecules can be delivered in and out of the cell by a piece of cell wall indenting and then snipping itself off to create a vesicle, a little membrane container.

Memorization Station

DNA: short for deoxyribonucleic acid, this is the code inside every living cell that directs the cell's activities

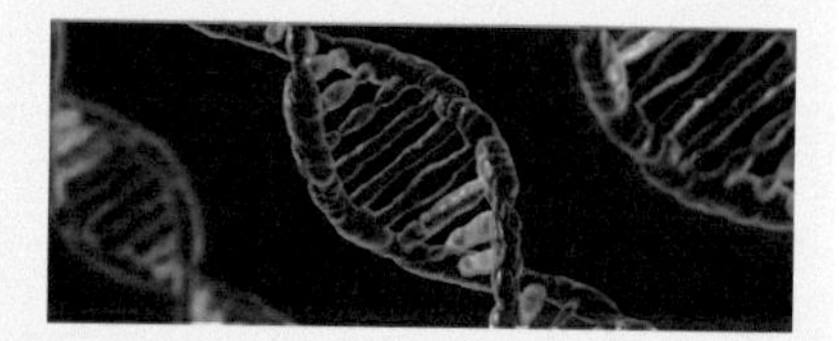

Additional Layer

Not all traits can be explained purely through DNA. Living things are so incredibly complex that we are still only at the beginning of understanding how traits are passed on and expressed. We do know that methyl carbon groups can be added to DNA, suppressing, activating, or altering a gene without changing the DNA of the individual. We also know that proteins can change shapes, changing genetic expression without changing the DNA code. There is currently a hypothesis that cells can communicate with one another in response to a changing environment, so that the cells can change and help the organism to survive. The study of ways that gene expression can be altered is called epigenetics and it is a new and expanding area of genetics.

This branch of research suggests that the ability to change and survive is an inborn trait of life, built into the structure of the DNA, as opposed to changes in living things being purely the result of random mutations.

EXPERIMENT: DNA Extraction

For this activity, you will need:

- Dry split peas
- Blender
- Salt
- Water
- Electronic balance
- Dish washing liquid
- Microscope and slide (optional)
- Strainer
- Medium bowl
- Meat tenderizer
- Isopropyl alcohol
- Graduated cylinder

DNA stands for deoxyribonucleaic acid. It is a single, huge molecule, made of billions of atoms, shaped like two chains coiled around one another, nicknamed the double helix. A helix is a corkscrew shape and this is the shape of each of the chains in the DNA. DNA is inside the nucleus of each cell of a living thing. The DNA controls what the cell does. It tells the cell to make proteins that are then combined in different ways to do different jobs like regulating temperature, breaking down nutrients from food, or building new cells. Only about 3% of human DNA is used for coding for proteins, but DNA has other functions besides protein regulation, which are just being discovered now.

1. In a blender, mix 70 g of dry split peas, 230 ml of water, and 1 g of salt. Blend for 15 seconds. The pea cells are now separated into their individual cells. If you would like, you can put a thin smear of the pea juice on a microscope slide and inspect the cells.
2. Pour the pea mixture through a strainer and into a large bowl. Toss the chunks and keep the liquid. Add 14 ml of dish washing liquid to the pea liquid and stir.

3. The dish washing liquid will dissolve the fatty lipids that make up the plasma membrane and the nuclear membrane, leaving the DNA free floating.
4. Let it sit for 10 minutes.
5. Add 8 g of meat tenderizer to the bowl. Meat tenderizer is an enzyme that will help unravel the DNA.
6. Carefully pour isopropyl alcohol into the pea juice until you have doubled the volume of the liquid. Observe the place where the alcohol and pea juice meet. White stringy stuff should have formed. This white stuff is DNA.

EXPERIMENT: Cell Specialization

For this activity, you will need:

- Pencil or pen for drawing
- Colored pencils
- Science Notebook
- Internet connection or book about cells

Many living things are made up of more than one cell. We call these **multicellular organisms**. Multicellular organisms usually have many specialized cells. In your body, some of your cells are muscle cells, others are skin cells, others are blood cells, and still others are brain cells and so on. The cells you color and label in a "diagram of a cell" are generic. But cells in real life have specific shapes, sizes, and functions. An organism starts out as a single cell which can then, as the living thing grows, divide and replicate into the many different types of cells the organism needs to live.

Cells that have the ability to differentiate into different types of cells are called **stem cells**. There are more than 200 different kinds of cells in a human body, all of which originally came from stem cells in a developing baby.

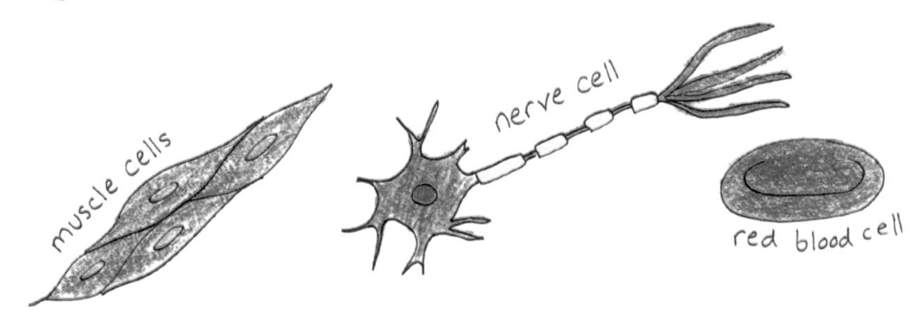

1. Draw a circular stem cell in the middle of a page in your Science Notebook. Don't forget a title for your page and to add it to the table of contents.
2. Around the basic cell, draw different types of cells found in your body—skin cells, blood cells, nerve cells, muscle

Additional Layer

The DNA strands are held together with four different nucleotides: cytosine, guanine, adenine, and thymine. These are called C, G, A, and T for short.

Cytosine always pairs with guanine and adenine always pairs with thymine.

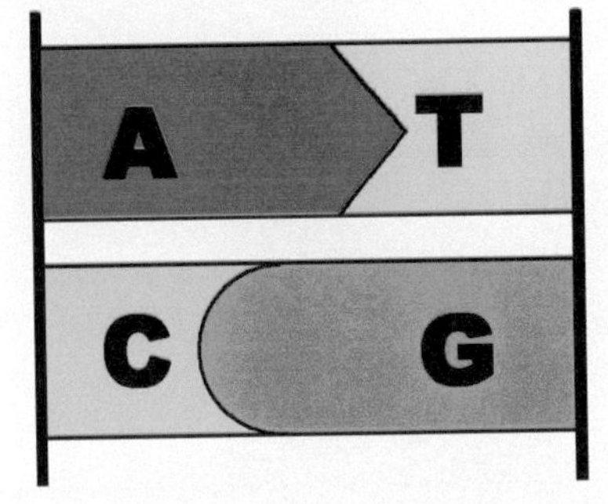

These nucleotide bonds can come apart or "unzip" in sections. Then the C, G, A, and T chemicals can be copied. Small sections of the DNA are copied like this all the time. The copies are called RNA, they're just short sections of the DNA, but single-stranded instead of double. The RNA leaves the nucleus through the tiny pores in the membrane and then moves to ribosomes. Ribosomes read the RNA code and then use it to build proteins.

Memorization Station

Multicellular organism: a living thing that is made up of more than one cell

Stem cell: has the ability to become any type of cell

Memorization Station

Memorize the steps of mitosis, or cell division.

1. Prophase - DNA copy

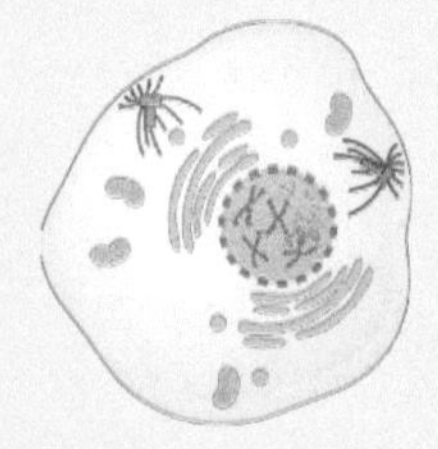

2. Prometaphase - microtubules attach

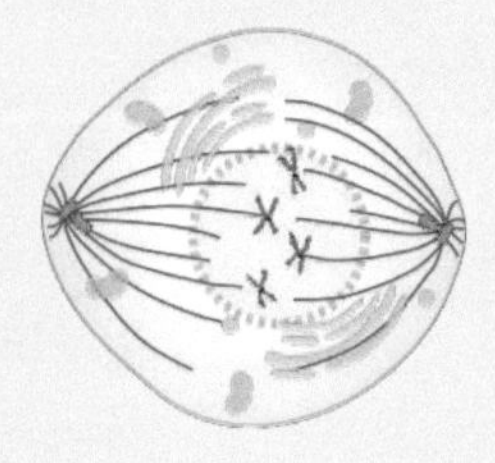

3. Metaphase - chromosomes line up

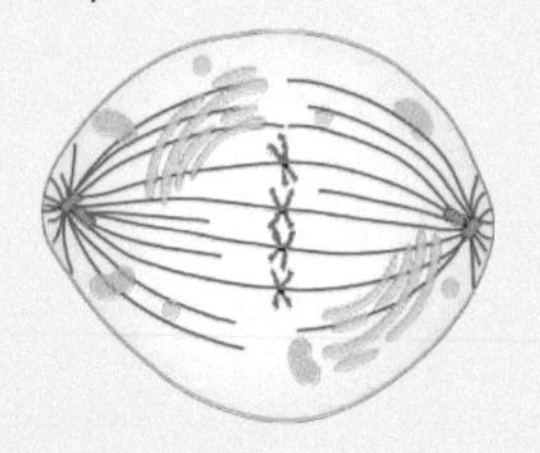

4. Anaphase - chromosomes pulled apart

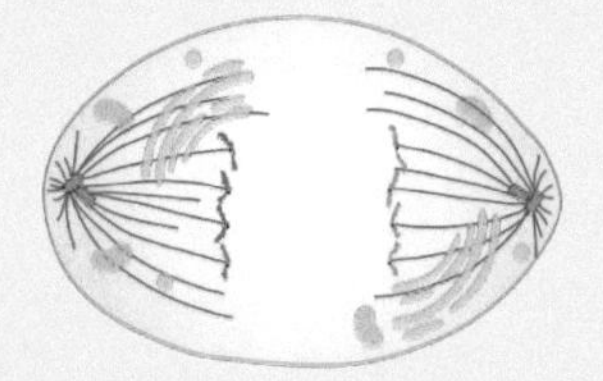

5. Telophase - nuclear membranes reform

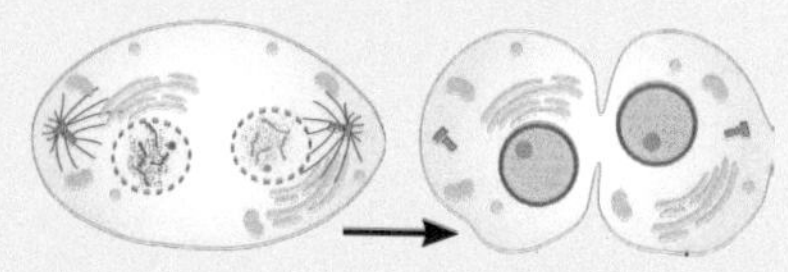

These steps of cell division are followed by the normal functioning of the cell called interphase.

cells, bone cells, and others you think of. You can look up images of cells online to help in your drawings.

3. Draw arrows out from the stem cell to the different types of cells. Color the cells.

☺ ☻EXPLORATION: Cell Division

For this activity, you will need:

- "Cell Division" from the Printable Pack
- Video about mitosis from the YouTube playlist for this unit
- Colored pencils
- Paper
- Glue stick
- Microscope (optional)
- Mitosis slide from a microscope slide set (optional)

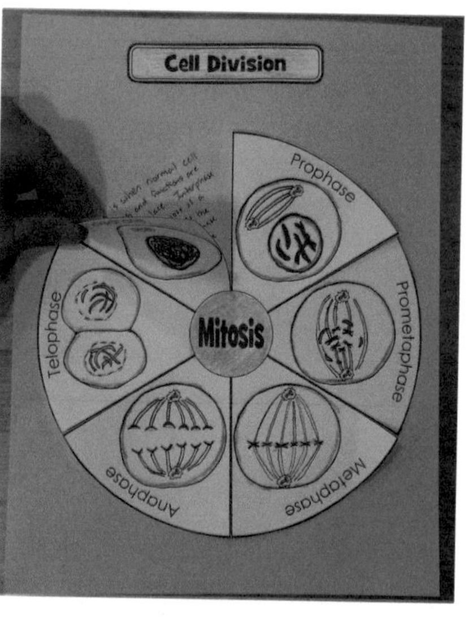

Cells divide to make new cells. This process is called mitosis in eukaryotes. There are five steps to mitosis plus a period called interphase.

- Interphase - This is when the cell is functioning as it normally does. The DNA is unraveled in the nucleus so it can be used to make the proteins the cell needs.
- Prophase - The DNA makes a complete copy of itself. Then, the DNA coils in on itself and forms into distinct chromosomes.

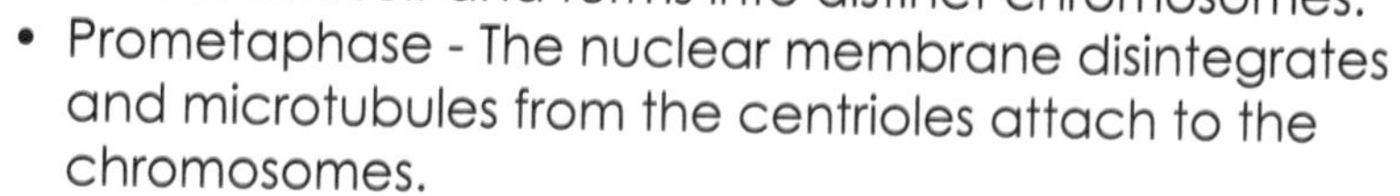

- Prometaphase - The nuclear membrane disintegrates and microtubules from the centrioles attach to the chromosomes.
- Metaphase - The chromosomes line up in the center of the cell with microtubules attached to each half of each chromosome.
- Anaphase - The chromosomes are pulled apart to opposite sides of the cell by the microtubules.
- Telophase - Each set of chromosomes is put back inside a nuclear membrane and the cell splits into two identical cells.

1. Color the "Cell Division" sheet. Cut out the circle on the heavy lines, leaving the center of the circle in tact, but cutting up each of the heavy radius lines. Cut out the title.

2. Watch a video about mitosis.

3. Glue the title and the center of the circle to a page in your Science Notebook, leaving the flaps loose. Under each flap, write a short description of each phase of mitosis.
4. If you have a microscope, look at a mitosis slide under a microscope to see what the phases look like in real life.

☻EXPLORATION: Meiosis and Making Sex Cells

For this activity, you will need:

- "Meiosis" from the Printable Pack (2 pages)
- Video about meiosis from the YouTube playlist for this unit
- Glue stick
- Two colors of clay or play dough

Normal cells have two copies of all the chromosomes. They are known as diploid. "Di" for two. But gametes, egg and sperm in animals, are haploid, they have only one copy of each of the chromosomes. The goal of these haploid cells is to create variation in the genetic code of the population.

Meiosis is a special type of cell division used for making gametes, or sex cells. First, the cell copies its DNA, exchanges fragments of the DNA, and then divides into two daughter cells. Then each of the two daughter cells divides again, forming two new daughter cells but with only half of the normal genetic information. Later, when the sperm and egg meet, they each have half of the genetic information and this combines to make the new complete copy of the genome for the new offspring. This is how genes are mixed up in populations of animals, plants, protists, and fungi to produce more variation.

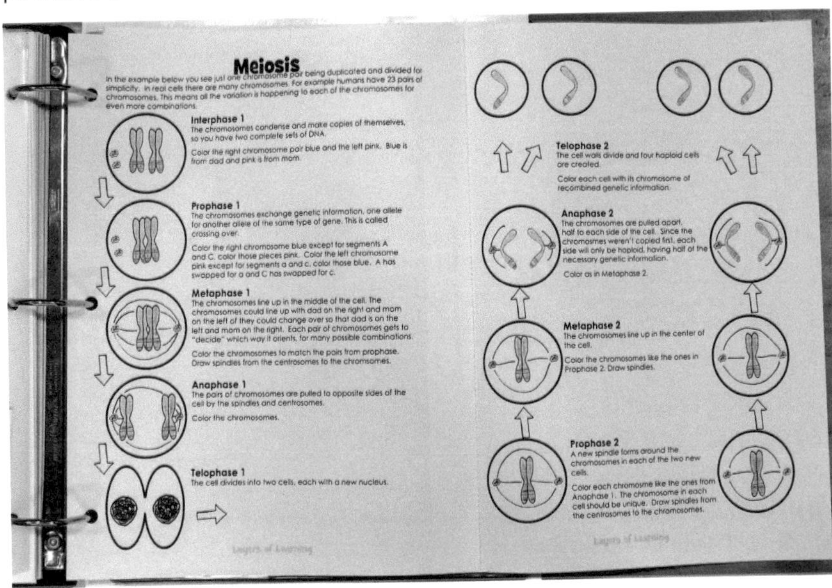

Meiosis

In the example below you see just one chromosome pair being duplicated and divided for simplicity. In real cells there are many chromosomes. For example humans have 23 pairs of chromosomes. This means all the variation is happening to each of the chromosomes for even more combinations.

Interphase 1
The chromosomes condense and make copies of themselves, so you have two complete sets of DNA.
Color the right chromosome pair blue and the left pink. Blue is from dad and pink is from mom.

Prophase 1
The chromosomes exchange genetic information, one allele for another allele of the same type of gene. This is called crossing over.
Color the right chromosome blue except for segments A and C, color those pieces pink. Color the left chromosome pink except for segments a and c, color those blue. A has swapped for a and C has swapped for c.

Metaphase 1
The chromosomes line up in the middle of the cell. The chromosomes could line up with dad on the right and mom on the left of they could change over so that dad is on the left and mom on the right. Each pair of chromosomes gets to "decide" which way it orients, for many possible combinations.
Color the chromosomes to match the pairs from prophase. Draw spindles from the centrosomes to the chromsomes.

Anaphase 1
The pairs of chromosomes are pulled to opposite sides of the cell by the spindles and centrosomes.
Color the chromosomes.

Telophase 1
The cell divides into two cells, each with a new nucleus.

Prophase 2
A new spindle forms around the chromosomes in each of the two new cells.
Color each chromosme like the ones from Anaphase 1. The chromosome in each cell should be unique. Draw spindles from the centrosomes to the chromosomes.

Metaphase 2
The chromosomes line up in the center of the cell.
Color the chromosomes like the ones in Prophase 2. Draw spindles.

Anaphase 2
The chromosomes are pulled apart, half to each side of the cell. Since the chromosmes weren't copied first, each side will only be haploid, having half of the necessary genetic information.
Color as in Metaphase 2.

Telophase 2
The cell walls divide and four haploid cells are created.
Color each cell with its chromosome of recombined genetic information.

1. Watch a video about meiosis.

Fabulous Fact

Your genes are part of your chromosomes, which are made of exactly one molecule of DNA each. Chromosomes come in pairs. Humans have 23 pairs of chromosomes. Other organisms have different numbers of chromosomes. Mosquitoes have 3 pairs, peas have 7 pairs, female honeybees have 16 pairs, and the red king crab has 104 pairs. The number of chromosomes is not necessarily related to the complexity of the organism.

Fabulous Fact

When mistakes are made in meiosis, the offspring either dies before birth or is born with developmental disabilities and often dies soon after birth.

Additional Layer

Even though bacteria reproduce asexually through binary fission, they do sometimes also exchange genetic material.

Bacteria will grasp one another with cilia, replicate a short portion of DNA, and then send their sections to one another, swapping like sections.

This is called conjugation.

Learn more about conjugation on Khan Academy or in a YouTube video.

Deep Thoughts

Genes do much to determine an organism's looks and behaviors, but there are other factors involved as well.

Environment and choices play a big role. A person who has alcoholism running in her family will not necessarily become an alcoholic, for example. Big debates in biology and psychology rage over how much of a person is determined by genetics versus environment. This is the "nature versus nurture" debate.

Talk about it. How much of your personality, looks, and behavior are determined by your genes and how much do you think is determined by your environment? How much comes from the choices you make?

Memorization Station

Sexual reproduction: gametes, or sex cells, from a male and a female parent combine to create a new, unique offspring

Asexual reproduction: a single parent divides to produce offspring which are an exact copy of the parent

Microorganism: living things that are too small to see with the naked eye

Bacteria: single-celled organism lacking a nucleus

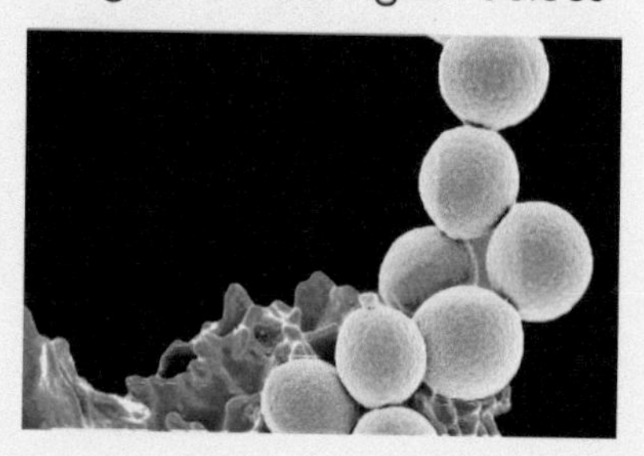

2. Color the "Meiosis" sheet from the Printable Pack, following the directions. The sheet explains the basics of meiosis. Glue the two sheets together along the long edge where the paper says "glue." The paper can then be folded and put in your Layers of Learning Notebook in the Science section.
3. Roll a dozen or so little balls of clay in one color that you will assign to be the "dad" color. Then roll another dozen balls of clay in a second color that you will assign to be the "mom" color. Each ball represents one gene.
4. Form two chromosomes out of six of the balls of clay, one from "dad" and one from "mom." Now, use the rest of the balls of clay to retell the story of meiosis. The balls of clay can be swapped between the chromosomes as in Prophase 1, to get variation.
5. Try telling the story again, but with two sets of chromosomes, so you can see how more chromosomes means more variation. Now you can see how the right and left sides of Metaphase 1 begin to have significance as even more variation is introduced.
6. Try again with three sets of chromosomes. You should notice as the number of chromosomes increases, the possible variations also increase.

EXPLORATION: Asexual Reproduction

For this activity, you will need:

- "Asexual Reproduction" from the Printable Pack
- Video about asexual reproduction from the YouTube playlist for this unit
- Pencil
- Colored pencils
- Internet connection
- Spider plant in a pot
- Soil
- Plant pot or container that can hold a plant

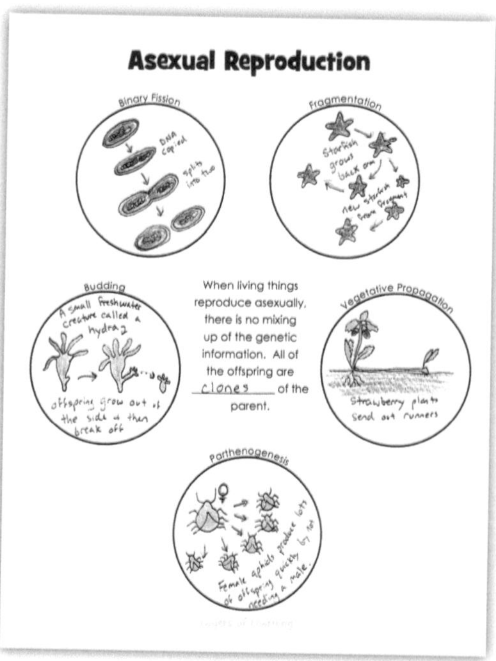

There are two broad categories of reproduction: sexual and asexual. In **sexual reproduction,** a male and female each contribute half of the genetic information needed to create the offspring. In **asexual reproduction,** only one parent is needed. This means that the offspring are all genetically identical to the parent; they are clones of the parent.

1. Watch a video about asexual reproduction.
2. Complete the "Asexual Reproduction" sheet. Fill in the blank with the word "clones." Draw each type of asexual reproduction in the circles.
    - Binary fission: a single-celled bacteria splits into two.
    - Fragmentation: an organism, like a starfish, breaks a piece off and both pieces form new individuals.
    - Budding: a new individual begins to grow off the side of the parent and then breaks free.
    - Vegetative propagation: a plant sends out runners and roots a new plant.
    - Parthenogenesis: an embryo develops from an unfertilized egg cell. Some amphibians, fish, reptiles, aphids, bees, nematodes, and others can do this.

    Many species, like strawberry plants, can reproduce sexually or asexually, depending on the conditions.
3. Grow a spider plant and propagate the plantlets. The spider plant usually produces plantlets during the summer months. It's fine to begin growing the spider plant any time of year and then watch the asexual reproduction whenever the plant is ready. Grow the plants in a well-drained soil mixture. Water them every few days when they begin to get dry. Spider plants like even moisture, not too wet and not too dry.

    When the plant produces plantlets, inspect the bottom of the plantlet for tiny root beginnings. Once it has roots started, prepare a second pot of soil and set it next to your mature plant. Set the plantlet on top of the soil and be careful not to disturb it while the roots are burrowing into the soil. Once the roots are established, snip the runner off the parent plant. You have a clone!

Microorganisms

Microorganisms, or microbes, are living things that are too small to see with the naked eye. They include **bacteria**, archaea, protists, protozoans, algae, micro-animals, and fungi. A microorganism can be a single cell, colonies of cells, or multicellular. They live everywhere from the deepest parts of the ocean and high in the atmosphere to inside your body, in hot thermal pools, and in dry deserts. All other forms of life depend on microorganisms for survival.

Fabulous Fact

All living things need energy.

Some living things can produce their own energy from sunlight and a few can produce energy from chemicals. These are autotrophs. "Auto" means "self" and "troph" means "energy."

Other living things must get their energy by eating other living things. These are heterotrophs. "Hetero" means "other."

Fabulous Fact

The smallest, simplest bacteria we know of is the *Candidatus Pelagibacter communis*.

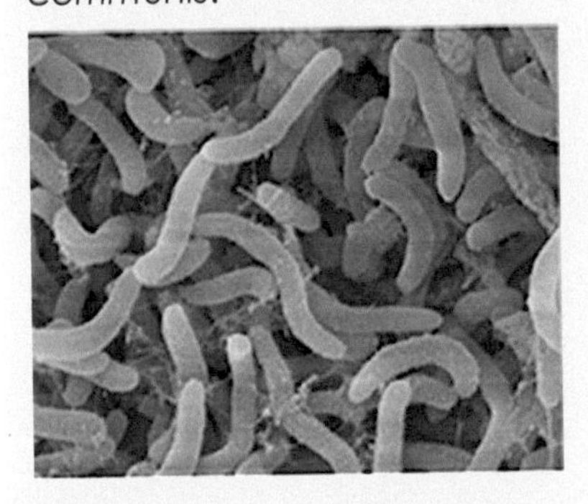

It also has the smallest genome of any living thing that we know of. This tiny bacterium is so prolific in our oceans that the combined mass of this one species outweighs the combined mass of all the fish in the sea.

In spite of its tiny size and relative simplicity, it still has enough genetic information to fill 20 books or so, contains hundreds of specialized proteins without any one of which it would die, and is made up of around 100 billion atoms.

Living things are mind bogglingly complex.

Fabulous Fact

Bacteria can be separated into two big groups:

- Gram positive
- Gram negative

Bacteria samples are stained with a "gram stain."

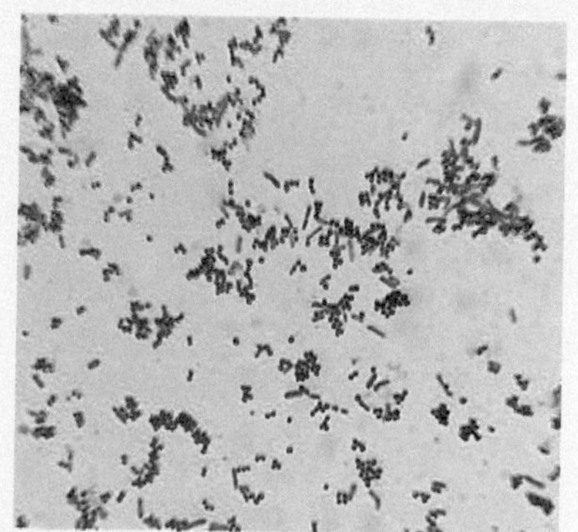

Photo by Microrao, CC by SA 4.0 Wikimedia

If the stain turns the cell wall pink, they are gram-negative. Gram-negative bacteria have thinner cell walls.

If the stain turns the cell walls purple, then they are gram-positive. These bacteria have thicker cell walls.

Fabulous Fact

Bacterial cell walls are made of a polymer called peptidoglycan, a very tough and resilient coating. Penicillin works by attacking the links in peptidoglycan, making the cell wall disintegrate.

Plant cell walls are made of pectin and then a layer of cellulose.

Fungi have cell walls made of chitin, the same stuff that forms a crab's shell or an insect's exoskeleton.

Animals do not have cell walls at all and rely only on the membrane for protection.

EXPERIMENT: Culturing Bacteria

WARNING: Colonized bacteria can be harmful and should be handled with caution. Wear gloves and wash hands thoroughly before and after. Never eat or drink anything while doing the lab work. Sterilize countertops, tools, and any implements used while doing the lab with bleach or alcohol. When finished, seal bacteria colonies and petri dishes in plastic bags and dispose of them in the trash.

For this activity, you will need:

- Petri dishes with agar, pre-prepared
- Sterile cotton swabs
- Disposable gloves
- Grease pencil or permanent marker
- Microscope
- Microscope slides
- Needle
- Candle
- Distilled water
- Science Notebook
- Zipper plastic bags - for disposal of the colonies
- Science Notebook

Bacteria are single-celled organisms that live individually or in colonies. They can be classified by their shapes. Spherical bacteria are called cocci. Rod-shaped bacteria are known as bacilli. And helical bacteria are called spirilla.

1. Divide a petri dish into thirds with a grease pencil on the outside of the plastic bottom. Label each section A, B, and C.
2. Use a sterile cotton swab to obtain a bacteria sample. You can swab your mouth, a doorknob, a toilet seat, a countertop, or any other surface you would like to test.

Swab across your test location, then gently swab the agar in the petri dish in a zig zag motion in one of the sections. Repeat for each section. Record the origin of each section's bacteria in your Science Notebook.

3. Let the dish sit in a warm place, like on top of your refrigerator, for 3-7 days. Observe the petri dish each day to see if there is growth. Though you can't see individual bacteria without a microscope, the colony will grow large enough to see. Each species of bacteria forms different shapes and colors of colonies. Draw sketches of your observations in your Science Notebook.
4. After you have seen growth in your colonies, sterilize a needle by passing it through a candle flame for 30 seconds. Let the needle cool, then gently touch the needle to the colony. Smear the needle across a microscope slide. Place a couple of drops of distilled water onto the slide, cover with a cover slip.
5. Observe your bacteria under a microscope. Draw what you see in your Science Notebook. Can you identify the shape: cocci, bacilli, or spirilla?
6. Take the lab further by designing your own experiments. Can you test the effectiveness of hand washing? Can you test the effectiveness of cleaning products? Can you test the ideal conditions for bacterial growth? Can you test whether bacteria are in the air or drinking water?

EXPLORATION: Archaea

For this activity, you will need:

- Video about archaea from the YouTube playlist
- Poster board or piece of card stock (for a mini poster)
- Colored pens
- Glue stick
- Colored paper

Archaea are small and simple, like bacteria, but they function more like eukaryotes. They are totally unique from both major groups in that they can use odd sources of energy like ammonia, metal, and hydrogen gas. The first archaea discovered were extremophiles living in hot volcanic pools and salt lakes, but since then, thousands of species have been discovered living everywhere you can imagine, including on your skin and in your gut. No parasitic or pathogenic archaea have been discovered so far.

1. Watch a video about archaea.
2. Choose a specific archaea to learn more about. Some

Writer's Workshop

Your cousin thinks popcorn kernels are not alive. You think they are. Write about an experiment you could design to test your hypothesis.

Additional Layer

Most cells need oxygen to survive, These are called aerobic. But some cells can survive anaerobically, or without oxygen. Nearly all anaerobic cells are bacteria or archaea.

Learn more about cellular respiration.

Additional Layer

Look up a specific type of bacteria and learn about it. Make a model of it out of household items like bottles, clay, chenille stems, toothpicks, or whatever other things you can find.

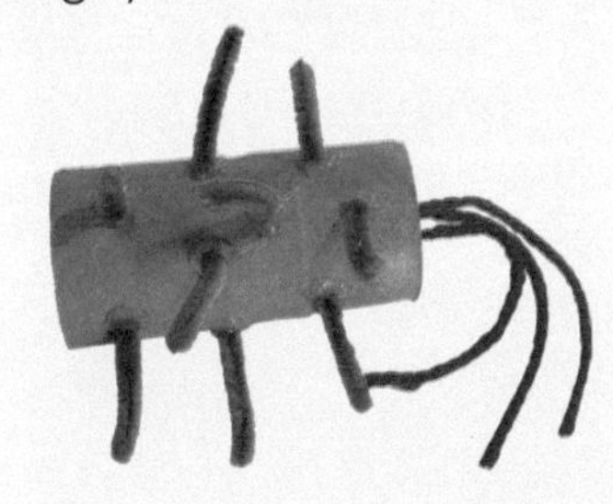

Fabulous Fact

Some species of microbes are used in fermenting foods like cheese, breaking down garbage or sewage, breaking down food in animal guts, or producing fuel. Many of them are useful for the environment and for humans.

Microorganisms that cause disease are called pathogens.

Bookworms

A Garden in Your Belly by Masha D'yans is a picture book about the bacteria that live inside and on the human body. Without them, we would die. For ages 6 to 11.

I Contain Multitudes: The Microbes Within Us and a Grander View of Life by Ed Yong is also about the microbes that live inside and on animals. This one is for ages 14 and up.

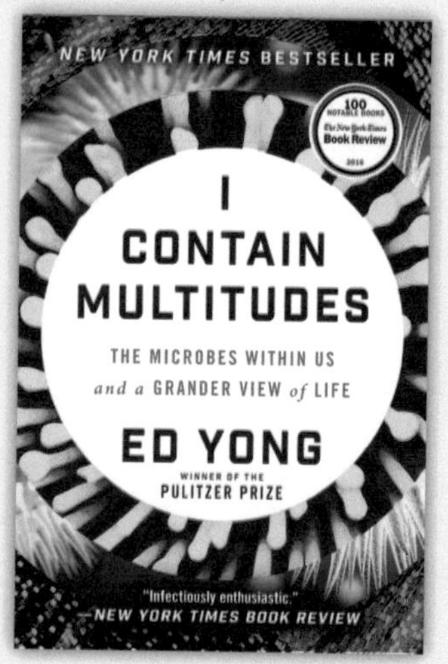

Memorization Station

Protists: Single-celled eukaryotic creatures, also called protozoa

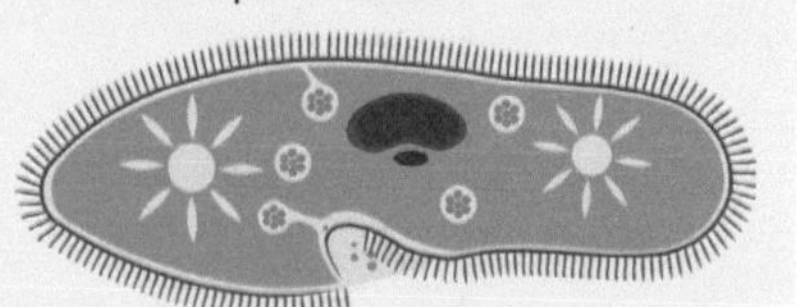

This is a single-celled paramecium.

possibilities: Halobacterium salinarum, Methanobacteria, Haloquadratum, or Sulfolobus.

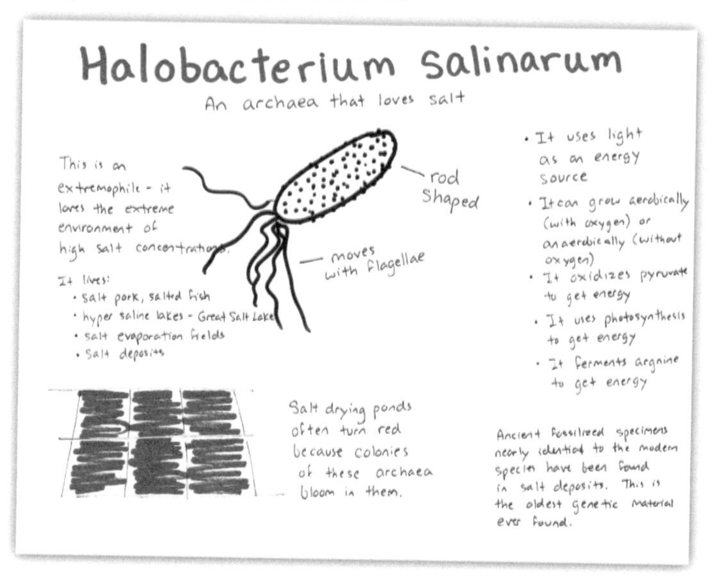

3. Make a poster about your archaea. Print pictures from the internet or draw illustrations. Add short paragraphs, lists, and captions. Include information on life cycle, special cell structures, and the environment your organism lives in.

4. Share your poster with a group, like your family.

EXPERIMENT: Protists in Pond Water

WARNING: Some microorganisms in pond water cause disease. Wash your hands following this lab and carefully clean up any spills. Do not eat or drink anything during this lab. Dispose of the pond water down the drain and thoroughly clean sinks, dishes, and utensils used.

For this activity, you will need:

- "Life in Pond Water" from the Printable Pack
- Jar
- Pond water or water from a freshwater aquarium
- Dropper
- Microscope
- Microscope slides and slipcovers
- Science Notebook

Protists, or protozoa, are single-celled eukaryotic creatures. Paramecium, amoeba, and euglena are three common species. They feed on other microorganisms or the dead parts of microorganisms. Some of them are parasitic.

1. Take a jar to a pond. Fill it three quarters full with pond water. Get the water from the bottom of the pond for best results.

2. Put a few drops of water from the pond onto a microscope slide using a dropper. Carefully place a slipcover on top of the water drops.

3. Starting with lowest magnification, observe the slide and look for microorganisms. Pond water contains many different types of creatures so you may see bacteria, protozoa, algae, and small multicellular creatures like tardigrades.
4. Use the "Life in Pond Water" sheet to identify life in your sample. Draw sketches of things you observe in your Science Notebook.
5. Take the time to research one or two of the protozoa to find out about their life cycles, how they move, what they eat, and where they live.

Classification

Classification is a system of organizing life for human convenience. Often, real living things defy our attempts to categorize them, so it is an imperfect system. Still, understanding how biologists organize life is important to scientific communication and basic understanding.

EXPLORATION: Three Domains and Four Kingdoms

For this activity, you will need:

- Video on classification from the YouTube playlist
- "Domains and Kingdoms of Life" from the Printable Pack (2 pages)
- Scissors

Fabulous Fact

Algae are plant-like protists that use chlorophyll to produce energy from sunlight. Some of them can move on their own.

Their cells are different from other life, both plants and animals, fungi, and bacteria.

They are eukaryotes but after that, their classification is murky and unsettled. Some species are in the plant kingdom, others are in the fungi kingdom, and some are not in any kingdom at all.

Bookworms

Fuzzy Mud by Louis Sachar is about a boy who discovers something microbial growing in the mud in the woods near his home. It seems to come from the off-limits research facility nearby.

Ages 10 and up.

Memorization Station

Classification: organizing things into groups based on shared characteristics

In biology, this is also sometimes called taxonomy.

- Glue stick
- Pencil
- Colored pencils

There are three domains of life: archaea, bacteria, and eukaryote. These divisions are based on the structures of the cells of the organisms in the groups. Archaea have a single lipid layer on their membranes and no nucleus. Bacteria have a double lipid layer membrane and no nucleus. All archaea and bacteria are single-celled, though some live in colonies. Eukaryotes have a nucleus. Some eukaryotes are single-celled while others are multicellular.

Eukaryotes can be further divided into four kingdoms: protista, plantae, fungi, and animalia. Protista are single-celled eukaryotic organisms. Plantae are multicellular plants, which contain chloroplasts and produce energy from sunlight. Fungi are multicellular with chitin in their cell walls and do not use sunlight to produce energy. Animalia are multicellular animals that must eat to get energy.

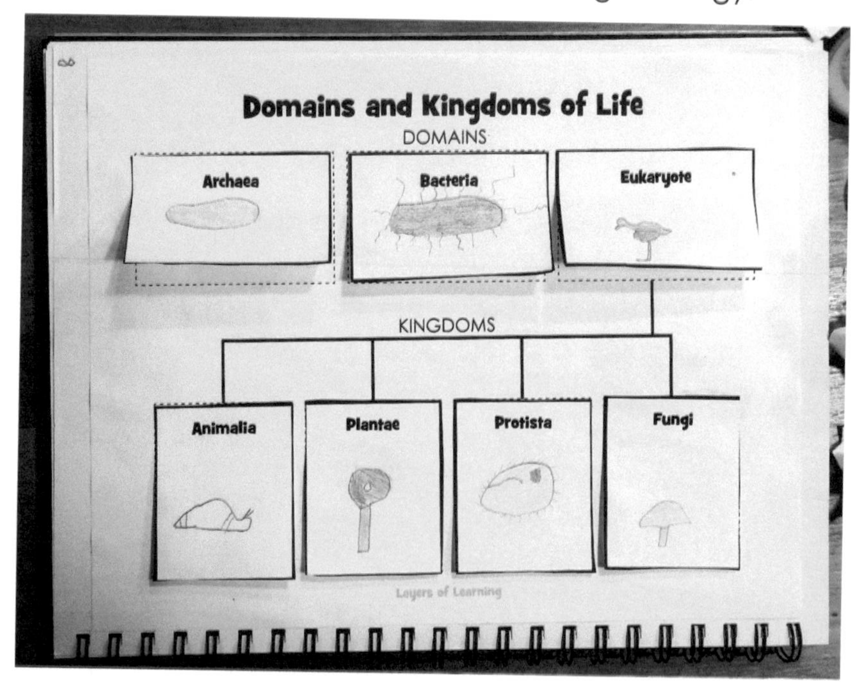

1. Watch a video about biological classification.
2. Cut out the rectangles with the heavy black lines and glue them onto the dashed boxes of the "Domains and Kingdoms of Life" chart. Glue just the top edge of each rectangle so you create a flap.
3. On top of the flap, beneath the name of the group, draw and color a picture of an organism from the group.
4. Under each flap, write a description of the group, using the information above.

Fabulous Fact

The two-part system of naming organisms is called binomial nomenclature. You use the genus and the species name for all organisms. This prevents confusion.

People are *Homo sapiens*. *Homo* is the genus and *sapiens* is the species.

House cats are in the genus *Felis* and the species *catus*. *Felis catus*.

Fabulous Fact

Sometimes the levels of organization are not specific enough and sub-levels are inserted. For example, there could be a sub-phylum or a sub-species.

Fabulous Fact

Many people have different ideas about how living things should be classified. This is because, in reality, nature isn't organized this way at all and organisms cross lines all the time. But people really like neat and tidy groups.

You may see a different type of classification in another text book or on a site online. None of them are wrong because there isn't just one way to do it. But throughout Layers of Learning we will stick to the style of classification introduced here with three domains and four kingdoms of Eukaryotes.

EXPLORATION: Classification

For this activity, you will need:

- "Living Things" from the Printable Pack (3 pages)
- "Modern Classification" from the Printable Pack
- Scissors
- Book or video about classification

Classification is the study of naming, describing, and organizing living things into categories. Over time, many different scientists have had ideas about how living things should be sorted. There isn't a right or wrong answer, just different approaches. But it is important to create a system that includes all living things, even those that have not yet been described and named.

1. Cut apart the 3 pages of "Living Things" cards. You can work as a group or give each child her own set of cards.
2. Sort the cards into groups however you'd like. Be able to explain why each living thing fits in its group. There are no right or wrong answers, but the groups should be logical.
3. Watch a video or read a book about modern biological classification to understand how living things are being classified by biologists right now.
4. Use the "Modern Classification" sheet, cut apart into rectangles, and organize the "Living Things" cards into the current groups.
   - **Bacteria: Salmonella Bacteria, E. coli Bacteria, streptococcus Bacteria**

Memorization Station

Memorize the levels for classifying living things from most general to most specific:

Domain

Kingdom

Phylum

Class

Order

Family

Genus

Species

You can use a mnemonic to help you remember the correct order:

Dear **K**ing **P**hillip **C**ame **O**ver **F**or **G**ood **S**oup

Or make up your own.

Famous Folks

Carl Linnaeus was a Swedish botanist and zoologist who developed the system scientists use for classifying living things.

He also invented the 3x5 index card. He needed a way to add items to the middle of a list or move them around as new information emerged.

- **Archaea: Haloquadratum Archaea, Halobacterium Archaea, Colony of Methanosarcina Archaea**
- **Protista: Single-celled Algae, Single-celled Paramecium, Single-celled Volvox, Amoeba**
- **Fungi: White Mushrooms, Bread Mold, Bread Yeast, Shelf Mushroom**
- **Plantae: Fern, Gerbera Daisy, Douglas Fir Tree, Grass, Hosta Plant, Lily, Moss, Oak Tree, Palm Trees, Sea Weed, Liverwort**
- **Animalia: Angelfish, Tree Frog, Boa Constrictor, Deer, Robin, Grasshopper, Earthworm, Ant, Bat, Cat, Centipede, Cod Fish, Cow, Dolphin, Elk Coral, Fox, Red Tailed Hawk, Honeybee, Leech, Black Widow Spider, Lobster, Mouse, Owl, Penguin, Rabbit, Salamander, Sea Horse, Snail, Sting Ray, Tapeworm, Microscopic Tardigrade, Tortoise, Vulture, Gray Wolf**

☻ ☻EXPLORATION: Species

For this activity, you will need:

- "Living Things" from the Printable Pack (You will be using one card each so you only need the first page, but you can print all three if you'd like)
- "Species" from the Printable Pack
- Pencil
- Encyclopedia, physical or online
- Scissors
- Glue stick

A **species** is a specific type of living thing. You can tell if something is a species because it can only reproduce with others of its kind. Species is the most specific of the levels of organization. You can see all the levels of organization on the "Species" printable.

1. Choose one of the "Living Things" cards. Cut out your organism card and paste it to the upper left of a page in your Science Notebook. Cut out the "Species" title and glue it the top center of the page.
2. Look your organism up in a physical or online encyclopedia. Find out its levels of classification. Write each group your organism belongs to on the large rectangle on the "Species" printable.
3. Cut out the large rectangle on the solid lines, including the lines that divide the rectangle into smaller parts. Glue it down on the right side of the sheet of paper along the edge that is not cut.
4. Find out the characteristics of each group in an encyclopedia. Under each flap, write what types of

Teaching Tip

Caring for a living thing is a valuable life lesson. A living thing has specific needs that must be met or it will die.

There is a great difference between a pet fish and a pet rock. During this unit, discuss the needs of living things and how that makes them unique.

If you don't already have a pet or a plant for your children, get one during this unit. Discuss the specific needs of the living thing in your home. Assign each of your children, in turn, the job of caring for the living thing for a set period of time (one day, one week, for the lifetime of the living thing).

Writer's Workshop

Write a how-to on caring for a pet or plant that you own.

You may want to use the *Writer's Workshop Descriptions & Instructions* unit for help.

organisms are included in each group. As you move from domain down to species you will notice that the groups become more and more specific until the species level includes just one kind of organism.

5. All living things are named with their genus and species, the two most specific levels of classification. The genus part of the name is always capitalized and the species part of the name never is. Both parts are written in italics. The names are also either Latin or Latinized, which means they are made to sound like a Latin word. So a horse's scientific name is *Equus ferus*. *Equus* is the genus, which includes donkeys and zebras as well as horses. *Ferus* is specific to horses only. Each species has only one member, though there can be subspecies.

   Write your organism's scientific name under its picture. Write in italics by slanting your letters to the right and adding extra curves to the tops and bottoms of straight letters like t and i.

Step 3: Show What You Know

During this unit, choose one of the assignments below to show what you have learned during the unit. Add this work to your Layers of Learning Notebook. You can also use this assignment to show your supervising teacher or your charter school as a sample of what you've been working on in your homeschool, if needed.

There are more ideas for writing assignments in the "Writer's Workshop" sidebars.

Coloring or Narration Page

For this activity, you will need:

- "Living Things" from the Printable Pack
- Writing or drawing utensils

Living Things

Cells are the smallest unit of life. A cell is a microscopic membrane filled with tiny organelles, each doing a different job to keep the cell alive. Some living things are just one single cell, while others are made up of billions of cells. A living thing can respond to heat, light, and food. It can also reproduce to make more living things like itself.

Layers of Learning

1. Depending on the age and ability of the child, choose either the "Living Things" coloring sheet or the "Living Things" narration page from the Printable Pack.
2. Younger kids can color the coloring sheet as you review some of the things you learned about during this unit. On the bottom of the coloring page, kids can write a sentence about what they learned. Very young children

Memorization Station

Species: a specific kind of living thing that can exchange genes or reproduce with others of its kind

Fabulous Fact

Donkeys and horses can be mated to produce mules, but donkeys and horses are still different species because the mules are always sterile, they cannot reproduce at all. The offspring of the mating also has to be able to reproduce for something to be a species.

Also, donkeys and horses never reproduce with one another in the wild, only when humans are manipulating things.

Writer's Workshop

Write a blog post from a bacteria, telling about an everyday experience or an adventure the bacteria cell had. Include at least one image.

This is Yersinia pestis, the bacteria that causes bubonic plague.

Unit Trivia Questions

1. A scientist who studies living things is called a/an:

   A. Astronomer

   B. Anthropologist

   C. Chemist

   D. Biologist

2. Name at least three things living things must do to be considered alive.

   - **Have cells with DNA**
   - **Use energy**
   - **Respond to their surroundings**
   - **Grow and develop**
   - **Reproduce**

3. Spontaneous generation is the idea that living things spring from non-living matter. The opposite of spontaneous generation is ______________ .

   Cell Theory

4. Which of these is not a lab safety rule?

   A. Eat during labs as long as you are careful.

   B. Be careful with sharp tools.

   C. Follow all of the directions.

   D. Heed warnings.

5. There are two types of cells: prokaryotic and eukaryotic. Describe the differences.

   Prokaryotic cells are smaller, have fewer organelles, and the DNA is not inside a membrane.

   Eukaryotic cells are more complex, have membranes around the organelles, and have double-stranded DNA.

can explain their ideas orally while a parent writes for them.

3. Older kids can write about some of the concepts you learned on the narration page and color the picture as well.
4. Add this to the Science section of your Layers of Learning Notebook.

Living Things Scavenger Hunt

For this activity, you will need:

- "Living Things Scavenger Hunt" from the Printable Pack

1. Give each person a copy of the "Living Things Scavenger Hunt" and have them race around your home, a store, a park, or another location and locate things that are living and nonliving. The first one to fill their scavenger hunt wins. You can write or draw each item.
2. Review and discuss together some of the things you've learned during this unit about living things.

Science Experiment Write-Up

For this activity, you will need:

- The "Experiment" write-up or "Experiment Report Template" from the Printable Pack

1. Choose one of the experiments you completed during this unit and create a careful and complete experiment write-up for it. Make sure you have included every specific detail of each step so your experiment could be repeated by anyone who wanted to try it.
2. Do a careful revision and edit of your write-up, taking it through the writing process, before you turn it in for grading.

Writer's Workshop

For this activity, you will need:

- A computer or a piece of paper and a writing utensil

Choose from one of the ideas below or write about something else you learned during this unit. Each of these prompts corresponds with one of the units from the Layers of Learning Writer's Workshop curriculum, so you may choose to coordinate the assignment with the monthly unit you are learning about in Writer's Workshop.

- **Sentences, Paragraphs, & Narrations:** Write your own well-crafted definition of life.
- **Descriptions & Instructions:** Describe what makes something alive.
- **Fanciful Stories**: Make a character sketch of a biologist who could be the hero of a story.
- **Poetry:** Write a limerick about bacteria.
- **True Stories:** Write a one-paragraph biography of Anton van Leeuwenhoek, the inventor of the microscope.
- **Reports & Essays:** Write a report about one type of human cell. Describe its functions and shape. Why is the cell shaped the way it is?
- **Letters:** Most experimental science is done with funding from universities, governments, and private donors. Scientists have to apply for the money. Write a letter to a potential donor asking for your research into cells to be funded. Be convincing and professional.
- **Persuasive Writing:** Why are cells small? Write a convincing argument for why they must be small.

Big Book of Knowledge

For this activity, you will need:

- "Big Book of Knowledge: Living Things" printable from the Printable Pack, printed on card stock
- Writing or drawing utensils
- Big Book of Knowledge

1. Color, draw on, or write on the Big Book of Knowledge page. Record concepts, definitions, and facts you learned during this unit. It's a record of the things you learned and hope to remember. Add the page to your Big Book of Knowledge.
2. Use your Big Book of Knowledge regularly to help you review, quiz, or create games that will help you commit the things you've learned to memory.

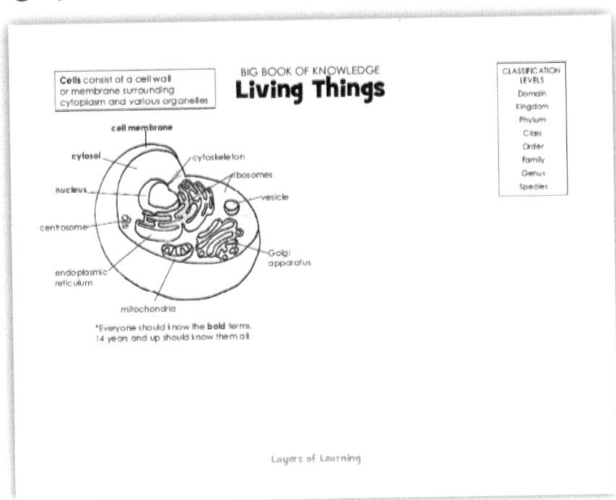

Big Book of Knowledge

The Big Book of Knowledge is a book for you, the mentor, to use as a constant review of all of the things you're learning about. You can use it to quiz your kids or prepare tests or review games. Whenever you learn something in Layers of Learning that you want your kids to remember, add it to your Big Book of Knowledge.

Assemble your Big Book of Knowledge in a binder or with binder rings. Divide it into sections for each subject.

In the Printable Pack for this unit you will find a "Big Book of Knowledge" sheet. You can add this sheet to others you collect or create yourself as you progress through the Layers of Learning curriculum. Customize the Big Book of Knowledge to your family by adding facts and topics that you enjoyed exploring as you were learning.

Visit Layers of Learning online to find more information on how to assemble and use your own Big Book of Knowledge.

You will also find cover and section pages to print along with creative games to play with your Big Book of Knowledge to keep school, even the tests, fun!

BIODIVERSITY

Unit Overview

Key Concepts:
- The variety of life on Earth comes from the genetic code in DNA.
- The genetic code is passed on from parent to child and given variation and unique expressions with sexual reproduction or asexual recombination.
- The genetic code can change so that living things can survive in a changing environment.
- Changes in genetic codes over time have made possible all the variety of life we see on earth.

Vocabulary:
- DNA
- Trait
- Genes
- Dominant
- Recessive
- Chromosome
- Chromatid
- Centromere
- Genotype
- Phenotype
- Locus
- Allele
- Homozygous
- Heterozygous
- Genetic mutation
- Genetic variation
- Species fitness
- Adaptation
- Natural selection
- Evolution
- Epigenetics
- Speciation
- Species
- Hybridization
- Allopatric speciation
- Sympatric speciation

Theories, Laws, & Hypotheses:
- Laws of Mendelian Inheritance
- Theory of evolution

Biodiversity is the vast variety of living things on Earth. There are tiny bacteria, huge squid, unique insects, and plants that hunt and eat meat. All of these living things add to the variety of life, they increase the biodiversity of the planet. How did all these different kinds of life come about, and how do they continue to survive even in changing environments?

All living things have DNA, a set of instructions that tells the organism how to stay alive and reproduce. The DNA contains genes that instruct cells on their jobs and make a living thing look and behave the way it does. Studying the DNA and the genes is called genetics. Everything a living thing does is coded into DNA, which is a strict set of instructions with incredible flexibility built in. Each species has its own code and, within each species, each individual has a set of variations that make the individual unique.

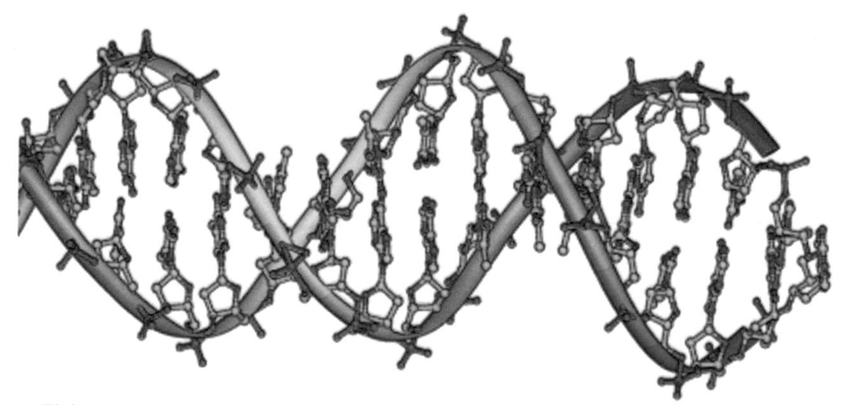

This is an artist's idea of what DNA looks like. The chemicals in DNA are a code, exactly like a computer code, that holds the instructions for the living thing. Unlike a computer code, DNA can repair itself, copy itself, and change itself to respond to the environment.

This unit explores many different ideas about how all of the many millions of kinds of living things came to be and how they continue to adapt and change over time. We may never have all the answers about something as complicated as life, but there are many things we can learn though observation and experimentation. We know that living things pass their traits on to their children. We also know that new species can emerge from the recombination of previous species. We can observe many of these changes and adaptations in the natural world all around us. Even though there are still many mysteries around the science of living things, one of the most important discoveries we've made is that all of this variety and change comes from the code written into DNA.

Step I: Library List

Choose books from your library that go with this topic. Here's a list of some favorites and also a list of search terms so you can utilize what your library offers. Read the books with your kids and/or assign them some to read independently. It is from these books your kids will learn most of the facts they need from this unit.

WARNING: This entire topic will include materials about evolution. Some support the Modern Synthesis and some refute it. If you have concerns about either viewpoint, please preview.

Search for: genetics, DNA, Gregor Mendel, evolution, natural selection, Charles Darwin

☺ ☺ ☻*Encyclopedia of Science* from DK. Read "Genetics" on page 364-365 and "Evolution" and "How Evolution Works" on page 308-309.

☺ ☺ ☻*The Usborne Science Encyclopedia*. Read page 338 on "Evolution," page 380 about "Genetics," and pages 382-385 on "Gene Technology."

☺ ☺ ☻*The Kingfisher Science Encyclopedia*. Read page 50 about "Life: Origins and Development" and page 135 about "Genes and Chromosomes."

☺*Gregor Mendel: The Friar Who Grew Peas* by Cheryl Bardoe. The story of the first scientist to systematically investigate how traits are passed from one generation to the next.

☺*The Story of Life: A First Book About Evolution* by Catherine Barr. Simple language and charming illustrations make this book perfect for introducing young children to ideas about how life spread, expanded, and changed over time to fill the earth with the great variety we see today.

☺ ☺*Charles Darwin's On the Origin of Species* by Sabina Radeva. This is a retelling of Darwin's famous book in picture book format.

☺ ☺*Amazing Evolution* by Anna Claybourne. Lovely illustrations and clear, detailed text explain the variety of life on Earth by means of life evolving to meet the needs of new environments.

☺*Tree of Life: the Incredible Biodiversity of Life on Earth* by Rochelle Strauss. Explains how there are millions of different kinds of living things and how they are sorted into groups.

☺*The Wondrous Workings of Planet Earth: Understanding*

Family School Levels

The colored smilies in this unit help you choose the correct levels of books and activities for your child.

☺ = Ages 6-9
☺ = Ages 10-13
☻ = Ages 14-18

On the Web

For videos, web pages, games, and more to add to this unit, visit the Biology Resources at Layers-of-Learning.com.

You will find a link to video playlists, web links, and more.

Bookworms

If you're looking for a family read-aloud, we'd like to suggest this one.

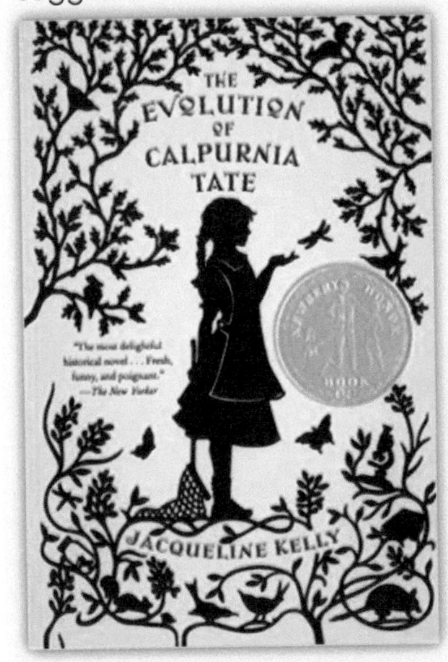

The Evolution of Calpurnia Tate by Jacqueline Kelly is about a young Texas girl who learns a love of nature and science from her grandfather. She evolves through the story as she watches the adaptations of the creatures around her.

Teaching Tip

Whatever your beliefs about evolution, it is important to understand these theories and how scientists explore these topics. You can also teach the difference between macroevolution, changes from one species to another, and microevolution, subtle changes within genes in a population or species. We can observe microevolution; there is no debate about it.

As far as your own beliefs about evolutionary processes, talk about these topics with your kids.

Bookworms

Leviathan by Scott Westerfeld is the first in a series about people who genetically engineer and create animals that function like machines. It's got a definite steam punk vibe, villains who end up as friends, and lots of cliffhanging danger.

Also check out the *Uglies* series by the same author, which is also about genetic manipulation by foolish humans. Ages 12 and up.

Our World and its Ecosystems by Rochel Ignotofsky. Explains biodiversity and so much more.

☺*Evolution: The Story of Life on Earth* by Jay Hosler. This is a graphic novel-style book that teaches the theory of evolution and then discusses the history of life on Earth.

☺ ☻*Genetics: Breaking the Code of Your DNA* by Carla Mooney. Easy to understand and thorough. Discusses cells, cell reproduction, recombining of DNA, and traits.

☺ ☻*On the Origin of Species: Young Reader's Edition* by Rebecca Stefoff. This is Darwin's book, edited and adapted for a modern audience. If you want to read the original, we recommend the illustrated edition with David Quammen.

☻*The Gene: An Intimate History* by Siddhartha Mukherjee. A conversational history of the field of genetics with tons of real world repercussions from eugenics and the Nazis to the future of cancer research. Written for the layman.

☻*Your Inner Fish* by Neil Shubin. Written by the scientist who discovered a fossil halfway between a fish and land animal. A 200-page look at recent discoveries.

☻*The Scientific Approach to Evolution* by Rob Stadler. The author explains the difference between "high confidence science" and "low confidence science." He sorts through the science of evolution, pointing out where the evidence is high confidence and where it is low confidence.

☻*Darwin's Black Box* by Michael J. Behe. Explores the things that Darwin did not know, especially the complexity of a single cell, the complexity of life processes, and many of the problems with the historical theory of evolution. This is a challenge to the Modern Synthesis. Popular level but with some difficult concepts. You can skim parts.

☻*Evolution 2.0* by Perry Marshall. A computer nerd did a ton of research and then took on evolution and genetics from the point of view of "Information Theory," poking holes in the Modern Synthesis. He proposes alternative mechanisms for evolution that are fascinating and only just being explored. It does contain Christian content, mostly to convince Christians to embrace his version of evolution, but the theories are valuable for anyone.

☻*Beginner's Guide to Punnett Squares, Probability in Genetics: Multiplication and Addition Rules, Molecular Biology, Mutations, Microevolution, Evidence of Evolution, Stickleback Evolution*, and *Linked Genes* from Bozeman Science on YouTube. Use these videos as lectures for your high schooler.

Step 2: Explore

Choose a few hands-on explorations from this section to work on as a family. They should be appealing activities that will create mental hooks so your kids remember the information in the unit. Save the rest of the explorations for the next time you do this unit in four years when your kids are older. You can also read the sidebars together and explore some little rabbit trails.

This unit includes printables. See the introduction for instructions on retrieving your Printable Pack.

Genetics

Genetics is the study of how every living thing, from the most basic bacterial cell to complex mammals, uses a code to direct the cells, tissues, and organs to function so the life form thrives. The code, much like a computer code that tells programs what to do, is like a set of instructions for the living thing. The code is known as **DNA** and it is stored in the nucleus of every living cell.

EXPLORATION: Gregor Mendel's Peas

For this activity, you will need:

- Book or video about Gregor Mendel
- "Gregor Mendel" from the Printable Pack (first 2 pages for ages 6 to 13, also add page 3 for ages 14 and up)
- Scissors
- Glue stick
- Colored pencils or crayons

Gregor Mendel was a monk from what is now the Czech Republic. He had a scientific education and spent his time experimenting with pea plants. He wanted to know how living things passed their **traits** to their offspring. For thousands of years, farmers had been purposely breeding animals and plants to develop desirable traits like docility, higher production, or faster maturing rates. Mendel already knew that traits could be selected for. What no one knew was how this happened. Mendel set out to use the scientific method to carefully discover what was happening inside the organism's cells. What he discovered was that traits were passed on in discrete units called **genes**. They were preserved in the offspring in tact, though not always expressed. However, they could be expressed in future generations. So two yellow pea plants could have a green pea plant offspring. Mendel called genes that are always expressed if present **dominant** and genes that were sometimes hidden but still passed on, he called **recessive**.

Memorization Station

Some genetics words to know:

DNA: short for deoxyribonucleic acid, this is the code inside every living cell that directs the cell's activities

Trait: an observable quality in a person, means the same thing as phenotype

Genes: a section of DNA that holds the code for a specific trait

Dominant: the trait that is expressed even if another genetic possibility is present. It is written with a capital letter, T

Recessive: a trait that is only expressed if it is the only one present, it is written with a lowercase letter, t

Famous Folks

Gregor Mendel is the father of genetics.

Like others who rock the scientific boat, his work was ignored until the early 1900s, about 30 years after his discoveries. These discoveries about genetics were made before anyone knew about the insides of cells or DNA.

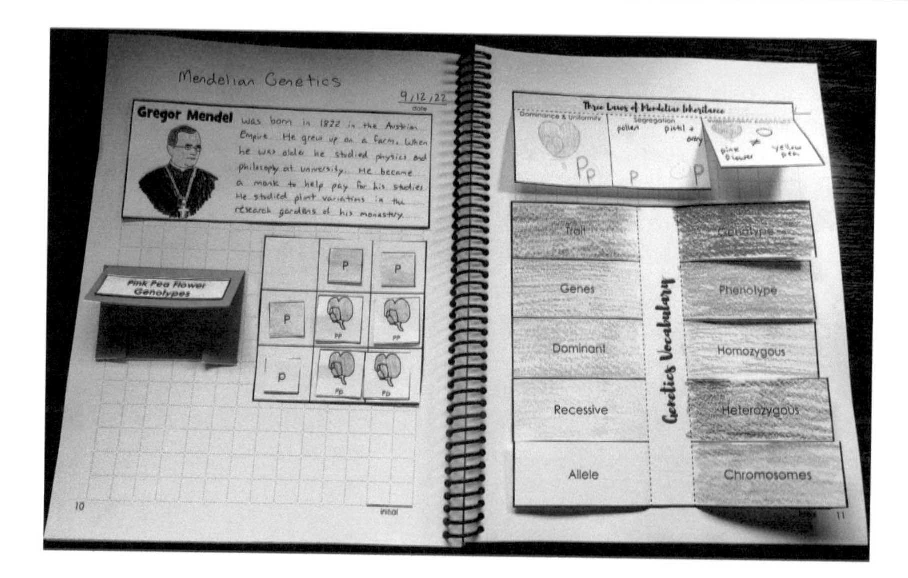

1. Read a book or watch a video about Gregor Mendel.
2. Color the picture of "Gregor Mendel" from the printable. Then, write a short biography of who he was on the lines next to the picture. Cut out and glue this box to the top of a page in your Science Notebook.
3. Next, color all of the small squares and flowers that have an upper case P in them light purple or pink. Color all the flower stems green. Leave the lower case p squares and flowers white. Cut apart all of the small squares. Gregor Mendel tested for several traits of the pea plants he worked with. One trait was the color of the blossoms.
4. Cut out the Punnett square with the nine spaces and the male and female symbols. Glue it to the page in your Science Notebook.
5. Cut out the pocket from the second page of the printable. Fold the pocket on the dashed lines and glue it together. Then, glue it into your Science Notebook. All of the little squares with upper and lowercase P's and flowers go into this pocket.
6. Pick four of the squares that have single letters on them and place them onto the male and female squares in the Punnett square. The male and female genes combine to make the offspring and the dominant trait is expressed. The dominant trait is shown by the upper case letter. P is dominant and expresses as pink blossoms. Put the correct offspring into the Punnett Square spaces.
7. You can repeat this several times with different random parent traits to see how the offspring with turn out.
8. ☻Older teens can also color, cut out, and define the

Fabulous Fact

The laws of Mendelian inheritance are those discovered by Gregor Mendel when he experimented on pea plants.

1. Law of dominance and uniformity: some alleles are dominant and others are recessive. If a dominant allele is present, it will be expressed.
2. Law of segregation: gametes (sex cells) each carry only one allele for each gene.
3. Law of independent assortment: genes are passed on independently of any other genes. One set of genes for one trait has no affect on the genes of a different trait.

Fabulous Fact

DNA can change during the life of an organism. This field is called epigenetics.

Genes do not appear to change, but the parts of the DNA that turn off or turn on the gene expression can change. Specifically, methyl groups can be added to the epigenetic markers (the parts of the DNA that direct the genes whether to express).

For example, infants and children who are exposed to microbes have better ability to fight off disease as adults because markers have been added to their DNA that turn on immune responses.

This is a developing and rapidly changing area of scientific inquiry. Keep en eye on this field!

Three Laws of Mendelian Inheritance and the Genetics Vocabulary from the third sheet. The answers are in the sidebars on the two previous pages.

EXPLORATION: Family Traits

For this activity, you will need:

- "Family Traits" from the Printable Pack
- Card stock
- Scissors
- A group of people, the more the better

Traits are physical characteristics that can be passed down through families from parents to children. As living things reproduce, some of their genes, and therefore, some of the parents' traits, are passed on to their children. Even though you probably have traits from your parents, you also are unique because the combination of traits in you is different from everyone else in the world (unless you are an identical twin). Even though you may have a lot in common with your siblings or parents, there are also traits that are unique to you.

1. Print the "Family Traits" printable on to card stock. Cut apart the cards. Shuffle them into a deck, facing down.
2. Have everyone stand in a circle around the room. Draw a card from the top of the deck and show it. Everyone who has the trait remains standing, while everyone who does not have the trait sits down.
3. Repeat, drawing cards and having people without the trait sit down, until only one person is left standing. How many cards did it take to find a unique person?
4. Reshuffle the cards and play again as many times as you would like.
5. Examine each trait and see how many you have in common with grandparents, parents, or siblings. As you play, discuss how traits run in families and point out some of the traits that run in your family.

EXPLORATION: Gingerbread Genes

For this activity, you will need:

- Gingerbread or sugar cookie dough
- "Gingerbread Cutout" from the Printable Pack (or a person-shaped cookie cutter)
- Frosting in a piping bag
- Colored candies (like M&Ms)
- Two small bowls or cups

Writer's Workshop

Write a journal entry about some similarities and differences you notice between some of the members of your family. What do you have in common? How are you different? How much of your similarities and differences do you think are because of genetics?

Fabulous Fact

The planet's areas of highest biodiversity are in the tropics. The tropical rain forests near the Equator cover less than 10% of the earth's surface but have about 90% of the world's species.

Deep Thoughts

After learning about Mendel's pea experiments, can you think of any ways that understanding traits has changed our food supply on Earth? Our population keeps rising and rising, and if our food-growing technology never improves, it will become more and more difficult to support the population. What things can you see that are being done to produce more food? Do you think we should do this?

Fabulous Fact

Not all traits are Mendelian, that is to say, passed on by one pair of genes. Many traits are actually the result of interactions between several genes, resulting in complex traits.

Eye color is determined by many genes that interact.

When many genes are involved it is called a polygenetic trait. Eye color is a polygenetic trait, but whether your ears are attached or unattached is a Mendelian trait. Mendelian traits are also known as monogenetic traits.

Additional Layer

In 1990, scientists launched the Human Genome Project, a quest to map the entire genetic code of human beings. They finished sequencing the human genome in 2003.

Learn more about the Human Genome Project and how it advanced medicine.

- Masking tape
- Permanent marker
- Oven for baking cookies

Living things have traits that are inherited from their parents. The color of your eyes, the texture of your hair, the quirk of your smile, the way you walk, and your height are all traits that you have inherited. These traits are coded on bits of your DNA called genes.

Genes get mixed up because half of the genetic code comes from a mother and half comes from a father. This is how sexual reproduction creates variety in populations.

1. Roll out your cookie dough and cut out 10 people using the "Gingerbread Cutout" or a person-shaped cookie cutter. Bake the cookies and let them cool completely.
2. Place four of the cookies in a row along a table or counter top. Use the masking tape and a pen to label them "Grandma A," "Grandpa A," "Grandma B," and "Grandpa B." Pipe six dots of icing onto the bodies of each grandparent. Place six red candies on Grandma A, six brown candies on Grandpa A, six yellow candies on Grandma B, and six green candies on Grandpa B. The candies represent the genes that can be passed on to the next generations.
3. Put a cookie beneath each set of grandparents and label them "Mother" and "Father" with the masking tape and pen.
4. Place six of each color of candy from grandparents A into a cup. Randomly pick out six candies from grandparents A to go on the mother. Repeat with

grandparents B to go on the father. Add them to the parents' bodies with the icing.

5. Now, place four cookies beneath the parents. The parents have four children. Put genes that match Mother's and genes that match Father's into a cup. Close your eyes and pick out six. Place these on the first child. Refill the cup with all of Mother's and all of Father's genes. Pick out six candies for the second child. Repeat the process again for the third child and then for the fourth child.
6. Are the children genetically identical? Are they different from each other and from their parents? Sexual reproduction mixes up and scrambles genetic traits, causing variation in populations. Just think! In an actual baby, there are up to 100,000 or so genes, so there are a lot more options and a lot more variety than our simple gingerbread people.

EXPLORATION: Dragon Chromosome

For this activity, you will need:

- "Dragon Chromosome" from the Printable Pack
- "Dragon Genetics Key" from the Printable Pack
- Blank sheet of paper
- Colored pencils or crayons
- Dice - one for each person doing the activity
- Microscope and a "mitosis" or "root tip" prepared slide (optional)
- Science Notebook

A **chromosome** is one long strand of DNA. In human beings, most people have 46 total chromosomes - 23 from the father and 23 from the mother. Most of the time, chromosomes are unraveled so they can be used by the cell. But during cell division, the chromosome is bunched up and organized into an X shape. Each half of the X is called a **chromatid**. The place where the two chromosomes meet is called the **centromere**. Each chromosome contains many genes that determine a **phenotype**, or how the organism looks and behaves. The position on the chromosome where a particular gene is located is called the **locus** (loci is plural).

1. Label the chromatids, centromere, and loci of the "Dragon Chromosome" on the printable. You can see the answers in the image.
2. Follow the directions on the "Dragon Genetics Key" and, using a die, determine which genes your dragon has.

Fabulous Fact

When a soil bacterium called *Sphingobium* was exposed to a pesticide in a lab, the bacteria developed a new way of processing the chemicals so the bacteria could break them down and use them. The *Sphingobium* adapted its normal cell function to take advantage of its environment.

In bacteria, which reproduce very quickly and have relatively simple genetic codes, these adaptations can occur rapidly and be observed in real time. The DNA of the bacteria did not have to mutate to make this adaptation possible.

Memorization Station

Chromosome: a linear strand of DNA and proteins that is inside the nucleus of a cell and contains hereditary information

Chromatid: one of the two identical strands of a chromosome that separate during mitosis

Centromere: the central region of a chromosome that appears during mitosis where the chromatids are held together in an X-shape

Genotype: the genes present in an individual

Phenotype: an observable trait in an individual, like being short or having stripes

Locus: the specific physical location of a gene or DNA sequence on a chromosome

Memorization Station

Allele: alternative forms of a gene found at the same location on a chromosome, represented with letters like T or t

Homozygous: having two identical alleles of a particular gene

Heterozygous: having two different alleles of a particular gene

The Greek root *homo* means the same. The Greek root *hetero* means different.

Writer's Workshop

What is your favorite thing about yourself? Is it one of your personality traits or a physical trait? Do you think you get this trait because of your genetics? Does anyone else in your family have a similar trait? It could be your parents, siblings, grandparents, or other relatives.

Write about your trait and whether you think it's genetic or not.

Deep Thoughts

People often wonder if we will ever enter the realm of engineering babies once we truly understand genetics. For example, if you could choose the traits your children could have, would you? Would you protect them from certain genetic diseases? Would you choose to have a child who is smart, athletic, or attractive? Do you think it would be ethical to choose your child's traits? Why or why not?

3. Draw a picture of your dragon. Remember, only dominant traits are expressed.
4. If you have a microscope, look at a slide of mitosis to see how the chromosomes look in a real cell. Draw a sketch of what you see in your Science Notebook.

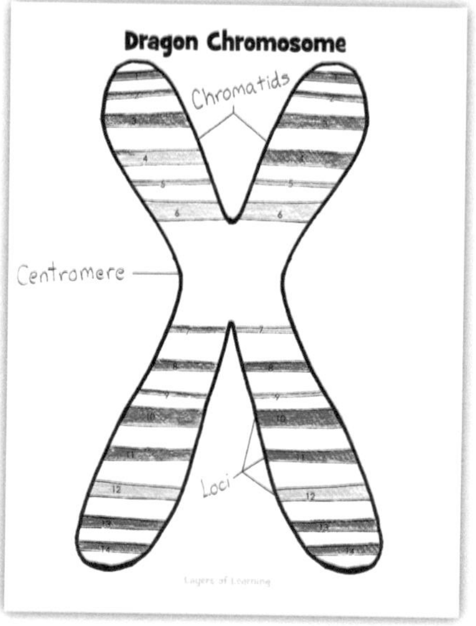

☺ ☻ EXPERIMENT: Punnett Squares and Gene Expression

For this activity, you will need:

- Blood typing kit (one for each person to be tested) or knowledge of each person's blood type
- Knowledge of biological parent's blood types (If you need to, you can invent hypothetical blood types)
- "Blood Typing Punnett Squares" from the Printable Pack
- "Punnett Square Practice" from the Printable Pack
- Answer sheet for "Punnett Square Practice" from the Printable Pack

Alleles are alternative expressions of a gene. For example, there is a gene that determines what your blood type will be. It will either be O, A, or B. The O, A, and B are alleles. Since you have pairs of genes, you have two of these alleles at a time, one possibility that comes from your dad and one that comes from your mom. Here are the possible combinations:

AA AB AO BB BO OO

Alleles can be recessive, dominant, or neutral. A and B are both dominant while O is recessive. A recessive gene is not expressed in the individual. So a person who has AA or AO alleles will have A type blood. A person who has BB or BO alleles will have B type blood. A person who has AB alleles will have AB type blood; both dominant alleles are expressed. The only way to get O type blood is to have OO alleles, because the recessive genes will only be expressed when there are no dominant options present.

A Punnett square is a diagram that is used to predict the genotypes of offspring. In the example below, we are predicting the color of the offspring. One pea is **homozygous** for the green trait and the other is **heterozygous** for the yellow trait, which is dominant.

Parent A has Yg alleles. Since Y is dominant, the parent is yellow.

Parent B has gg alleles. This means parent B is green.

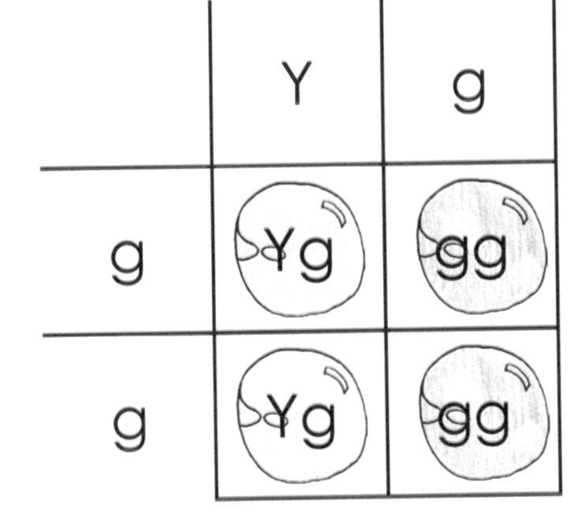

Statistically, half of the offspring will be yellow and half will be green.

1. Find out what your blood type is by using a blood typing kit, consulting records from a hospital or doctor visit where your blood type was determined, or by giving blood and asking the technician what type you have.
2. You also need to know your biological parent's blood types. If you don't know their blood types, see if you can figure out the possible combinations, then use one set of the possibilities for filling out "Blood Typing Punnett Squares."
3. Fill out "Blood Type Punnett Squares" with your information. There are multiple Punnett Squares, but you don't need to use them all. Just use as many as are necessary. If a parent has AB or O type blood, you will need fewer squares than if you have parents with A or B type blood. For example, a person with A type blood could have AA alleles or AO alleles, so you will need more Punnett Squares.
4. Complete the "Punnett Square Practice" sheet. Use the answer key to check your answers. Add this to the Science section of your Layers of Learning Notebook.

EXPLORATION: Population Genetics

For this activity, you will need:

Famous Folks

Thomas Hunt Morgan wanted to prove that random mutation drove evolution.

He bred *Drosophila melanogaster*, fruit flies, hitting them with radiation and chemicals to force mutations more quickly than would happen in nature. He did gets lots of mutated flies, but he never found a benefit.

Along the way, however, he mapped parts of the fly genes and hypothesized linked genes, genes that are inherited together, and the concept of crossing over in meiosis.

Morgan kick started the discipline of genetics.

Morgan's experiments were the first laboratory experiments into searching for evolution. In other words, they were the first high certainty science done in evolution studies.

Fabulous Fact

Even though humans are all very different from each other, all human beings share approximately 99.9% of the same DNA.

Bookworms

The Secret Code Inside You by Rajani LaRocca is a rhyming picture book about how our genes make us who we are instead of poodles or bears or sheep.

For ages 4 to 8.

Additional Layer

One of the benefits of having a lot of biodiversity is that it helps control the spread of diseases. Imagine that you have all of the very same kind of tree in your town. If a disease strikes the variety, it could take out all of the trees because they would all be susceptible to it. However, the more kinds of trees you have, the greater chance that they won't all have the same susceptibilities. The viruses would have to adapt to each new species and that slows down the spread.

The Irish Potato Famine and the United States Corn Leaf Blight epidemic were both largely due to farmers just having one kind of crop. When that crop got diseased, most of them were taken out due to the lack of biodiversity in these areas.

Learn more about monoculture, or the lack of biodiversity.

- "Moth Population" from the Printable Pack
- Scissors
- Bag or bowl to draw moths from randomly
- Science Notebook
- Envelope
- Glue stick

An allele is a variant in a gene. Let's say we have a population of moths and they have two alleles for wing color: an allele for gray wings and an allele for white wings. White wings are dominant. So the alleles look like this: g for gray and W for white.

You can find the frequency of an allele by counting up the alleles in the whole population and dividing by the total number of alleles. In this population, above, each moth has two alleles for wing color.

There are six g alleles out of a total of 8 alleles, so our frequency is 6 ÷ 8 = .75 or 75%. What is the frequency of the W alleles?

Several things can change the allele frequency in a population. One of these is natural selection. If gray moths get eaten more often because they are easy for predators to spot, then the gray allele will probably decrease in the population as a whole.

The second thing that could change the allele frequency is genetic drift. This is based on chance. Perhaps more white moths happen to not reproduce and the frequency of W alleles declines.

The third thing is a bottleneck, a type of genetic drift. A bottleneck is caused by a disaster of some kind. Perhaps there is a forest fire where the moths reside and only a few individuals survive. It's possible that one allele could be wiped out entirely or the frequency could be severely changed so that one allele is rare.

The fourth way allele frequency can change is the founder effect, another type of genetic drift. This is when a few individuals, maybe as few as two or three, migrate and start a new population. Perhaps a few of the moths accidentally take a ride inside someone's car and end up hundreds of miles from the original population. Often, the few individuals will have severely restricted alleles.

1. Cut apart the moth cards from the "Moth Population"

sheet. All of these moths together are your total population of moths. Calculate the allele frequencies in your total population. Write this down in your Science Notebook.

Answers: g =37/68=54% and W = 31/68 = 46%

2. Think of a scenario where gray moths survive but white moths struggle. How would that change your population alleles? Manipulate the moths on the table to show which ones die out. What does your new allele frequency look like? Does this demonstrate natural selection or genetic drift?

   Answer: natural selection

3. Now, put your whole population of moths in a bag or bowl and draw out two. These are the two that survive after a bottleneck or a founder effect. Calculate your allele frequency. What would the second generation look like in frequency of alleles?

   You can draw a Punnett square in your Science Notebook to help you see the proportion of alleles in the second generation.

   Repeat this as often as you like. Are there any scenarios where an allele is completely eliminated? What if an isolated population is reintroduced to the parent population? What will happen to the allele frequency?

4. Glue an envelope into your Science Notebook and put the moth cards into it.

EXPLORATION: DNA Structure Model

For this activity, you will need:

- "DNA Structure" from the Printable Pack for middle grades and up (answer sheet also included)
- Colorful gummy candies or miniature marshmallows (at least four different colors)
- Licorice
- Toothpicks
- Colored pencils or crayons

DNA is made of two backbones made of sugars linked together with amino acids. The amino acids are like rungs on a ladder. The amino acids link the two chains together with hydrogen bonds. Hydrogen bonds aren't very strong, but with billions of them in a row, they make the overall structure of DNA very strong. It's good they're weak though, because they have to be split apart on a regular basis for copying.

Deep Thoughts

Discuss this quote by Donella Meadows, American Environmental Scientist and professor:

Biodiversity can't be maintained by protecting a few species in a zoo, or by preserving greenbelts or national parks. To function properly, nature needs more room than that. It can maintain itself, however, without human expense, without zookeepers, park rangers, foresters or gene banks. All it needs is to be left alone.

Famous Folks

The double-helix structure of the DNA molecule was discovered by James Watson and Francis Crick.

In truth, there were lots of scientists working on it at the same time. Rosalind's Franklin's work was actually pivotal to their discovery, but she is rarely given the credit.

Bookworms

CRISPR is a way to easily edit DNA. It is already being used in plants and animals and has the potential to repair and prevent human genetic diseases.

CRISPR: A Powerful Way to Change DNA by Tolanda Ridge is a children's book about the newest technology in genetics. Ages 8 to 12.

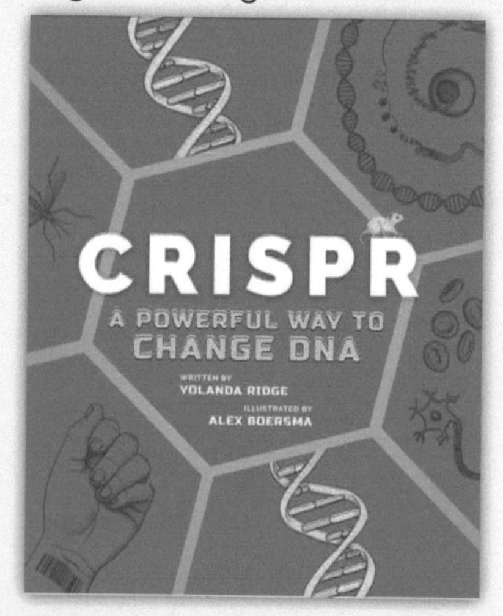

The Code Breaker by Walter Isaacson is a biography of Jennifer Doudna, the scientist who discovered how to easily edit DNA with CRISPR technology.

For 14 and up.

1. Complete the "DNA Structure" sheet from the Printable Pack to learn about the parts of DNA and especially the base pairs: A, T, C, and G. Use the answer key from the Printable Pack to check your work. Younger kids, under 10 years, can skip this.
2. Build a model of DNA. First, separate the gummy candies or marshmallows by color, each in its own container. Label the containers A, T, C, and G respectively. Each color stands for a different amino acid.
3. Thread marshmallows onto the toothpicks, pairing A with T and G with C. Push each end of the toothpick into the licorice so the two sides are connected all up the strand of licorice. Give the licorice a partial twist to make the double helix shape of DNA.

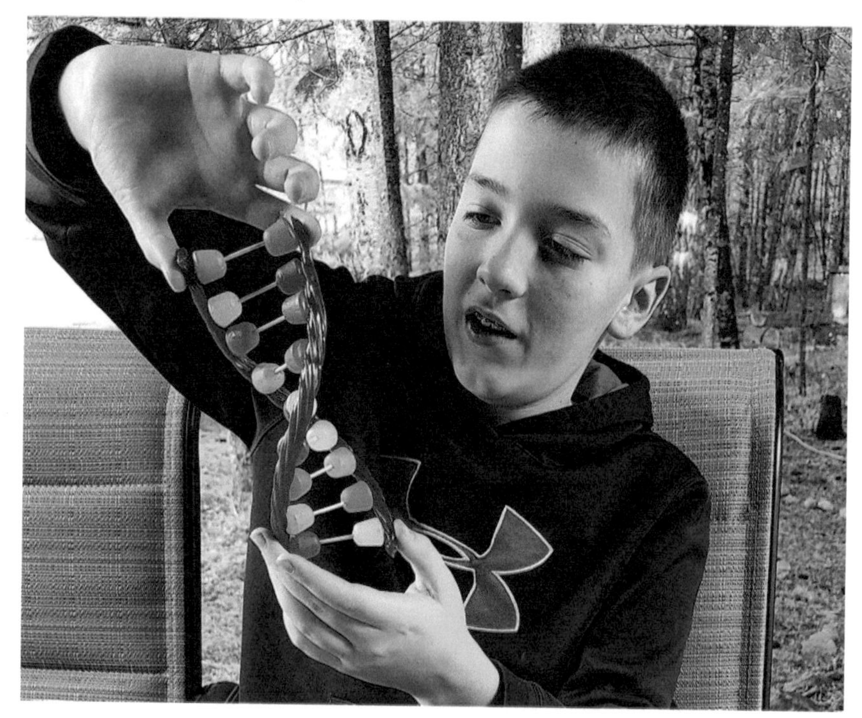

EXPLORATION: Genetic Code

For this activity, you will need:

- Videos about DNA code and how they work from the YouTube playlist for this unit
- "DNA Decoder" and "Decode to Build a Protein" from the Printable Pack
- Pen or pencil
- Scissors

DNA is a code. In the same way a computer is coded, living things are coded. DNA is made up of four molecules called nucleotides: adenine, thymine, guanine, and cytosine, represented by the letters A, T, C, and G. The code is a language. It is information that tells the cell, and ultimately the organism, what to do and how to do it.

Each set of three nucleotides is called a triplet. Each triplet corresponds to an amino acid. Strings of amino acids make proteins.

The code is read by messenger RNA, which decodes it and assigns amino acids, pre-manufactured by the cell, to their spot in the lineup. The amino acids are then assembled inside the ribosomes into proteins. Each protein is made by one section of DNA, called a gene. Typically, proteins have hundreds, or even thousands, of amino acids. That means a protein of 300 amino acids would be coded by 900 DNA base pairs.

1. Watch a video about how the DNA code works. This seems really complicated at first, but some good visuals with explanations makes it so much easier. You may want to watch several videos.
2. Print out "DNA Decoder" and "Decode to Build a Protein" printables. Cut off the bottom portion of the "Decode to Build A Protein" to hide the answer until after you have finished.
3. Use the decoder to find the abbreviations for the amino acids. All the amino acids line up, bonding to one another in order, to make a protein.
4. Invent codes for each other and use the decoder to see if you can reveal the code.

EXPLORATION: Genetic Mutations

For this activity, you will need:

- Video about DNA mutations from the YouTube playlist for this unit
- "Mutant Rock Bugs" from the Printable Pack
- Clean rocks, two per student
- Green paint and paintbrush
- Paper
- Tape
- Black chenille stems (pipe cleaners)
- Googly eyes (or you can just use paper)
- School glue
- Scissors
- Red, green, and black markers

A **genetic mutation** is a mistake in the copying of the genetic code. Sometimes mutations are described as "any change in the genetic code," but this is misleading since the cell actually intentionally changes its DNA code in some situations. A mutation is not desired or intended by the cell.

Writer's Workshop

If someone in your family had a genetic disease, would you want to have genetic testing to find out if you are likely to get the same disease? Would you rather know or not know? Would you have your future children tested to see if they had it too?

Write about what you would do in a journal entry and explain your reasoning.

Bookworms

47 Strings: Tessa's Special Code by Becky Carey is about a little girl born with Down's Syndrome, a genetic disease where people have an extra chromosome.

This is an excellent book both for understanding genetic diseases and for becoming an understanding human being.

For ages 6 to 12.

Memorization Station

Genetic mutation: a mistake in the copying of the genetic code or DNA that persists and is not corrected by the cell

Mistakes are made when DNA is copied, like during mitosis or meiosis. They can also happen if high energy (like from sunlight) or certain chemicals (like from tobacco smoke) come into contact with the DNA. These types of mistakes happen fairly often, but the cell has multiple ways of proofreading and repairing mistakes that have been made or patching up damage that has been caused. The immune system also identifies and destroys mutated cells. This means that actual mutations that remain in the code are, fortunately, rare.

1. Watch a video about DNA mutations. You can find video recommendations in the YouTube playlist for this unit.

2. Cut a piece of paper into long narrow strips and glue the strips together to make one long piece six strips long. This will become a piece of DNA for your mutant bug.

3. Translate your bug's DNA code from the RNA codes given on the "Mutant Rock Bugs" printable. Write the DNA code on your long strip of paper. You have to translate backward from your amino acid RNA code. Remember that A pairs with U/T and G pairs with C.

4. Each of the boxes of the "Mutant Rock Bug" printable tells you what protein chain your bug has in order to meet each of its functions. Follow the directions to craft a healthy bug.

   Note that the proteins, the RNA, the DNA, and the bug are hugely simplified. Real proteins are hundreds of amino acids long.

   Also, pay attention to the part of the DNA that is non-coding, but still does important jobs for the organism. We are only just beginning to understand how the non-coding DNA works.

5. Trade strips with someone else and proofread their code to see if there are mistakes. If there are, they have a mutant bug. There are three types of mutations.

Additional Layer

For a long time it was thought that mutations were the driving force behind evolution, but so far laboratory tests, beginning with the famous fruit fly experiments of Thomas Hunt Morgan in the early 1900s, have failed to show benefits of mutations. Often mutations make no difference, once in a while there is a weak benefit, but overwhelmingly, mutations are bad for the organism. Mutations have not been proven as a mechanism for positive changes in genomes. They have been assumed as the cause of historical changes in genomes, but never directly observed in laboratory tests.

Bookworms

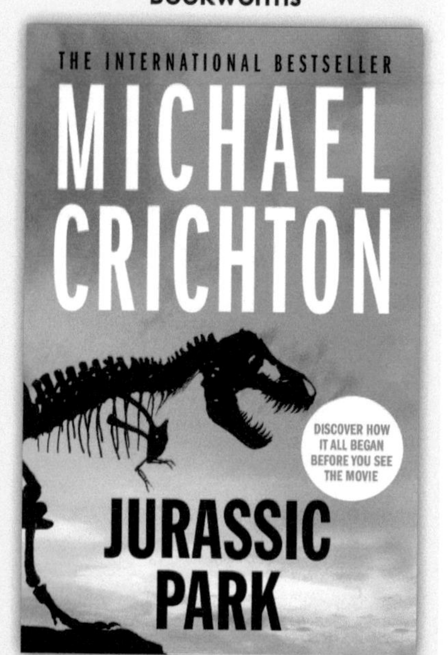

Jurassic Park by Michael Crichton is a novel that uses DNA and basic genetic principles to propose a world in which dinosaurs could be brought back from extinction.

- A substitution means one letter (nucleotide) has been traded out for another. This will change one of the amino acids, which will make the entire protein function differently or not at all.
- An insertion is when one or more extra letters are added to the code. This makes the entire triplet sequence shift and can affect more than one amino acid, changing the protein even more drastically.
- A deletion is when one or more letters are left out. This can also change more than one amino acid and affect the protein drastically.

6. If you have a mutant bug, then think about how the mutation might affect your bug. Craft a new bug to show the mutation(s).

Adaptation

The ability to react successfully to changing conditions is essential for living things. Animals and plants that can't adapt to change go extinct because the Earth is in a constant state of change. In this section, you will learn about the evidence for adaptation that has been discovered through scientifically rigorous experimentation.

EXPERIMENT: Variation

For this activity, you will need:

- 10 or more specimens of a single species: leaves, flower blossoms, snails, worms, moths, or shells of sea creatures, and so on (Choose one easy-to-find species and collect 10 or more specimens. Make sure these are not sensitive or endangered species)
- Hand lenses or magnifying glasses
- Ruler for measuring
- Science Notebook
- Pencil for sketching
- Colored pencils

Variation means there are differences in the DNA of populations. For example, *E. coli* bacteria has the genetic ability to resist antibiotics, but this allele is not usually expressed because normally other traits are more important. However, if most of the population is being killed off by antibiotics, then only the antibiotic resistant bacteria will survive. This leads to an entire population where antibiotic resistance is expressed. But the other alleles are still in the genetic code of the bacteria and when the pressure is removed, the population will go back to being susceptible to antibiotics, though probably not as susceptible as they were before.

Fabulous Fact

There are three major movements in the world of evolution.

Darwinian evolution said that the variety of species on Earth arose from natural selection acting on tiny variations in individuals over a long period of time. Darwin, who knew nothing of the inner workings of a cell or genetics, had no mechanism for this.

The Modern Synthesis, developed in the 1930s, took Darwin's ideas about natural selection and incremental changes and added Mendelian genetics to the mix to explain how tiny changes could occur through random mutations in the genetic code.

In the 1950s, a new movement called the Extended Evolutionary Synthesis added concepts of epigenetics, where the organism is changing its genetic code because of outside stimuli and not just randomly.

Most evolutionary biologists still believe in the Modern Synthesis, though more are coming around to accepting additional mechanisms besides random mutations.

Memorization Station

Genetic variation: differences in genetic codes of individuals in a population

Species fitness: the ability of an organism to thrive and pass its genetic code on to offspring

Memorization Station

Adaptation: a trait that makes a species fit to survive in its environment

Deep Thoughts

Some people say that the theory of evolution proves that there is no God since God isn't needed to create or direct life.

Do you agree with this idea? If evolution is true, does that automatically mean that God is not true?

Explain your reasoning.

Bookworms

Read *How to Clone a Mammoth: The Science of De-Extinction* by Beth Shapiro. High schoolers can explore the idea that extinct species could potentially be brought back to life. Might it be possible as we learn more about genetics and traits?

The higher the genetic variation, the more fit a species is and the more likely it is to survive long term. We call that **species fitness.**

1. Lay the specimens out on a table. Carefully observe the specimens you have collected, looking for variation between the specimens. Choose one specimen each to study, sketch, and describe in your Science Notebook. Take measurements and be precise.
2. Return the specimens to the table, mix them, up and then try to find your exact specimen again. What helped you identify your exact specimen? Take more notes or further sketches as needed.
3. Return the specimens to the table again and mix them up. Trade notes with another student. Find each other's specimen based on the notes and sketches.
4. Think about and discuss the variations you observed.
   a. Some of the variation is from genes and some is probably from environment. Can you tell which are which?
   b. What observable variations might affect the survival of the individual?
   c. Can you see all of the genetic variation?
   d. Why is variation important?

EXPLORATION: Adaptation and Foxes

For this activity, you will need:

- Videos or books about tundra and temperate forest biomes from the Library List or YouTube playlist for this unit
- Videos or books about arctic fox and red fox species, also from the Library List or YouTube playlist
- "Fox Craft Template" from the Printable Pack
- Card stock - white and green
- Orange crayon or paint
- White paper cups
- Black pompom (optional)
- Scissors
- School glue
- Cotton balls

Adaptation means a living thing becomes better suited to its environment by gaining or losing a feature. For example, Arctic foxes live in the tundra and red foxes live in temperate forests. Both species are similar but there are also very important differences. For example, Arctic foxes have thicker fur, an adaptation necessary for surviving the cold of the Arctic. The genes for these adaptations may have been already in the original fox population that spread across the world, but thick fur is only an advantage in the Arctic, so it wasn't expressed much until foxes moved there. It is also possible that the DNA of Arctic foxes changed in response to moving to a colder environment. We can't go back in time and observe the changes happening, so we may never know how it came about.

1. Read books or watch videos about the tundra and temperate biomes and the Arctic and red fox that live in these places. The Arctic and red fox are similar, but there are also some important differences. Which differences make each species better able to survive in their biomes?
2. Cut out two copies of the template pieces from "Fox Craft Templates."
3. Color or paint one of the copies with orange for the red fox. Leave the other one white for the Arctic fox. Also, color or paint one of the paper cups orange and leave the other white.
4. Glue the Arctic fox template pieces on to the paper cups. Use the image, above, as a guide.
5. Trace a circle using one of the paper cups as a guide onto the green card stock. Draw long narrow triangles radiating out from the circle. Cut out the big star shape. Clue it to the bottom of the inverted paper cup with the red fox on it. Bend the pieces up to be grass.
6. Tease cotton balls apart and glue them to the bottom

Famous Folks

Escherichia coli (*E. coli*) are found in the intestines of people and other mammals. They produce vitamin K and help break down food. They also prevent harmful bacteria from colonizing in your gut. Some strains of *E. coli* are harmful and can cause food poisoning, urinary tract infections, meningitis, Crohn's disease, and other infections.

Richard Lenski, an American evolutionary biologist, began growing cultures of *E. coli* in a lab in 1988.

This is Richard Lenski in his lab. Photo by Zachary Blount, CC by SA 4.0 Wikimedia.

Lenski observes the colonies for signs of adaptation. For example, in one experiment, a colony of *E. coli* were exposed to a citrate medium where they normally can't grow, but the bacteria developed the ability to use the citrate as a food source, a sign of an emerging species.

Teaching Tip

Older students can compare all twelve fox species across the world and look for ways they have adapted to their environments.

of the Arctic fox to be snow.

7. Take turns explaining the adaptations each species needed to survive better.

EXPERIMENT: Natural Selection

For this activity, you will need:

- Yarn - yellow, green, and brown
- Scissors
- Timer
- An outside natural area of grass or dirt

All living things have a DNA code that they pass on to their offspring. The offspring mostly look and act like their parents, with tiny variations in each generation. Meanwhile, nature can be harsh. It gets cold, hot, and dry. There are predators and disease. Food is often scarce. This means that only fit living things can survive to pass on traits to the next generation. Generation after generation, nature selects for the traits that allow survival.

Darwin's Galapagos tortoises are an excellent example of this. The larger islands in the Galapagos have soil and lush grass for the tortoises to eat. These tortoises are large and have lovely domed shells.

Tortoise from a grassy island. Photo by David Adam Kess, CC by SA 4.0, Wikimedia

Tortoise from a rocky island. Photo by Nicolas Volker, CC by SA 4.0, Wikimedia

But the smaller islands are rocky and the most plentiful food is a cactus with the green juicy bits on high stalks. A domed shell would prevent a tortoise from reaching its head up high to get the cactus. The tortoises on these islands have a saddle-shaped shell with a high bow over the neck so they can reach up.

Perhaps the genetic variation already in the genome was selected to allow tortoises to survive on the small rocky islands by adjusting the shape of the shell. Or it is possible that the genetic code directing the turtle's shell shape changed in response to the new environment? In any case, the rocky island tortoises are obviously different from the grassy island tortoises. The environment of the tortoises,

Bookworms

Who Was Charles Darwin? by Deborah Hopkinson is a biography about the scientist who first proposed that life could adapt and change to its environment.

For ages 6-10 years.

Famous Folks

Alfred Russel Wallace was a contemporary of Charles Darwin and came up with the theory of evolution and natural selection at the same time. The two scientists wrote to one another and became friends. Wallace encouraged Darwin to publish his ideas about evolution.

Learn more about Wallace and his adventures in the Malay Archipelago.

selected for the type of shell they have. This is **natural selection**.

1. Cut up 40 short (about 5 cm) pieces of yarn in each of three colors: yellow, brown, and green. These yarn pieces represent worms. Without the kids seeing, spread the "worms" around outside in a large, natural area of grass or dirt.
2. Set a timer for 2 minutes and have the kids collect as many worms as they can in that time. The kids represent predators that eat the worms.
3. Bring the worms back inside and tally up how many of each color you collected. Was there a color (or two) that you collected much fewer of? Why? Over time what do you think will happen to your population of worms?

EXPLORATION: Charles Darwin

For this activity, you will need:

- Book or video about Charles Darwin from the Library List or YouTube playlist
- "Charles Darwin" from the Printable Pack
- Scissors
- Colored pencils or crayons

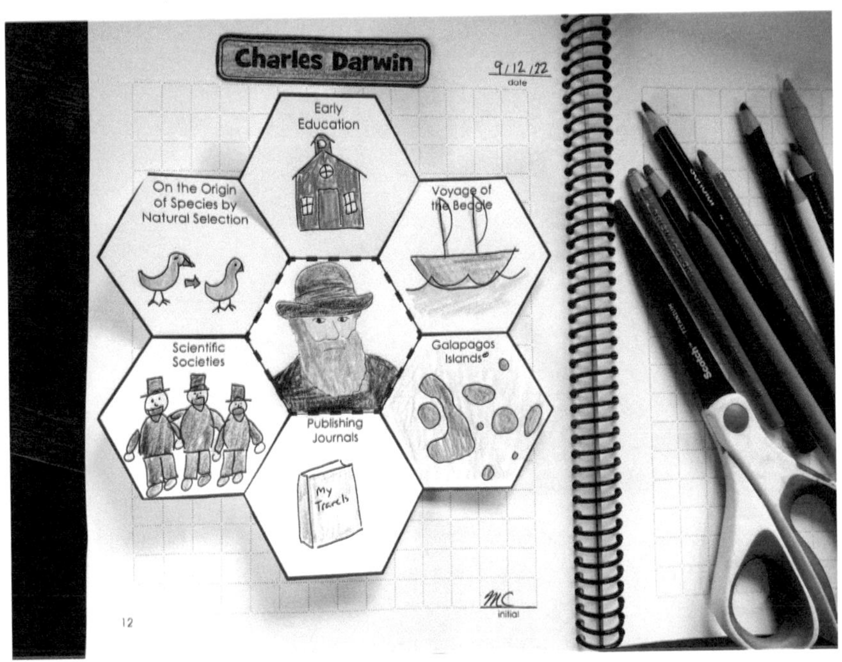

Charles Darwin was a scientist from England. He was born in 1809. At the age of 22 he joined an expedition on the HMS Beagle to sail around the world as the naturalist. His job was to observe nature, make notes and sketches, and collect specimens. During the voyage, he observed

Memorization Station

Natural selection: the process where living things that are better suited to the environment reproduce and pass on their traits to the next generation and those less suited die out or fail to pass on their genes

It is important to note that natural selection is an observed process that is easy to point out in the natural world. Often, it is obvious that certain members are failing to survive because, for example, they are being eaten due to their coloring.

This is the peppered moth. A second variety of the same moth with black coloring became prevalent during the Industrial Revolution in England because of the soot coating buildings and trees. The lighter colored ones got eaten and so the darker colored moths passed on their genes, leading to overall darker coloring in the population of moths.

This is an example of natural selection changing the genetic prevalence of one trait within a species. This sort of natural selection has been observed many times.

We have never observed natural selection developing entirely new traits, however. It can only work with the genetic code already present in a population.

Memorization Station

Evolution: organisms change and diversify through successive generations due to environmental pressures

Additional Layer

The theory of evolution by natural selection was first proposed by Charles Darwin in 1858. People are still debating how the changes happen. Here are some proposed ideas:

1. Random mutations: parts of the genetic code randomly mutate, giving rise to new features. The problem with this mechanism is that beneficial mutations have never been observed.
2. Natural selection: successful organisms reproduce, passing on their features, while the unsuccessful die, leading to changes in the genetics of a population. This has been observed in nature.
3. Transposition: the DNA deliberately changes itself due to information from outside the cell. This has been observed in nature.
4. Horizontal gene transfer: genes are passed from one organism to another completely unrelated organism. This has been observed in nature.
5. Methylation: the DNA intentionally turns sections on or off in response to the environment. This has been observed in nature.

Evolution within a species is common. So far we have not observed drastic body changes in nature or the lab.

an immense variety of life and he began to wonder how so many different plants and animals came to be. Many decades later, he published *On the Origin of Species* where he proposed that living things change in response to their environment, with the strongest surviving and passing on their traits while the weaker died out. Darwin called this process natural selection.

There was a lot Darwin didn't know because he knew nothing about genes, DNA, or the complexity of a cell. But in spite of his limitations, his ideas were a paradigm shift in thinking about living things. Previously, it was assumed that living things were static, that a species stayed exactly as it was. Darwin realized that change was the constant, not stagnation. The idea that living things can and do change over time as a result of environmental pressures is the Theory of **Evolution**.

1. Read a book or watch a video about Charles Darwin.
2. Color the portrait on the "Charles Darwin" printable.
3. Draw illustrations in each hexagon. Then cut out the hexagon and the title.
4. Glue the title and the center of the hexagon to a page in your Science Notebook.
5. Under each flap, write about the different periods in Darwin's life as you learned about him from this exploration.
6. Write down the definition of the theory of evolution somewhere on the page.

EXPERIMENT: Jumping Genes

For this activity, you will need:

- "Jumping Genes" from the Printable Pack
- Crayons or colored pencils
- Scissors
- Glue stick
- Science Notebook

Not only does DNA send information out into the cell, it also receives information. We know that parts of the cell, nearby cells, and even the environment the organism is subject to can cause enzymes to be sent to the DNA and alter it.

One of the ways cells alter their DNA is through transposition, also known as jumping genes. Certain segments of DNA are transposable, they have the ability to move (or jump) from one region to another. Sometimes the transposable section is copied and inserted in another place, so then there are two sections of identical DNA. Other times, the transposable piece is cut out of the DNA and moved to another location so only one copy remains, but in a new location.

Sometimes, transposition results in the deactivation of a gene. For example, peas have an r allele that encodes a certain type of starch, but in some peas, a transposable element has turned off this gene, resulting in the wrinkled peas Mendel observed in his experiments.

Other times, a transposable element will activate a gene. This has been observed in cancerous cells whose out of control growth has been activated when a transposable gene turned on a function that the cell would normally not have. Transposable genes do not usually cause disease, however. They are a normal and frequently used strategy of the cell to cope with environmental changes.

In yet another situation, a transposable element can incorporate new DNA into the genome, such as DNA from a virus or DNA carried by a virus from another organism being permanently acquired by the cell.

It has taken more than 50 years for the concept of jumping genes to be accepted by the scientific community, but today they are thought to be a major mechanism of evolution. In other words, the cell contains the tools to respond to its environment. The changes can be preserved in future generations in sex cells or in asexual organisms.

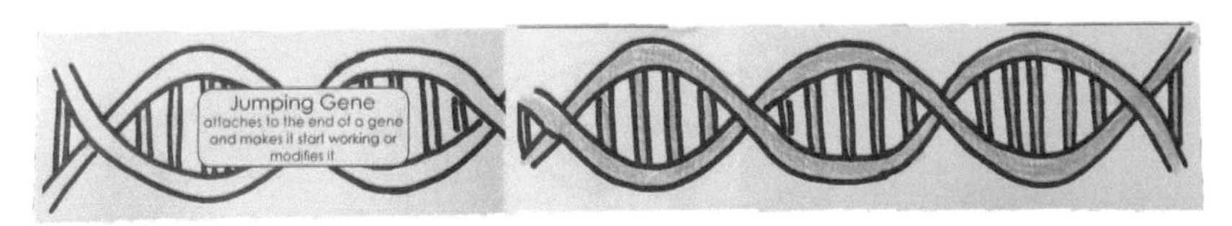

1. Cut the "Jumping Genes" printable apart on the solid rectangle lines.
2. Color the DNA sections blue and the Jumping Genes sections yellow.
3. Fold the jumping genes on the dashed lines. Glue one jumping gene section in the middle of the small DNA sections. Glue the other jumping gene section on the end of the larger DNA section.
4. Glue the jumping genes down into your Science

Famous Folks

Barbara McClintock first discovered transposable genes (jumping genes) during her work with maize in the 1930s through 1950s.

Her research was dismissed for decades because it was so revolutionary. Finally, in 1983, she received the Nobel Prize for her work. Scientists are still learning about transposition and absorbing the implications for evolution and adaptability.

Deep Thoughts

The theory of evolution does not answer the ultimate question - Where did the first life come from?

The leap from non-living elements and molecules to a living, functioning cell is immense. It cannot be accomplished by chance because dozens of beneficial and intricate processes would have to happen simultaneously in order for even the simplest cell to survive.

Scientifically speaking, no one knows how the first life came to be.

Of course, many people have strong beliefs about the origin of life.

What do you think?

Additional Layer

The fossil record shows that life on Earth has not remained static. Millions of species have gone extinct and millions more have emerged.

Usually, species suddenly appear in the fossil record with no evidence of gradual transition. This may be because the fossil record is so imperfect, with most species never becoming fossilized. It also may be because the species really did suddenly appear. We have no way of going back in time and finding out for sure.

Sometimes we do find species that have traits of two different animal or plant groups. The archaeopteryx is a classic example of this.

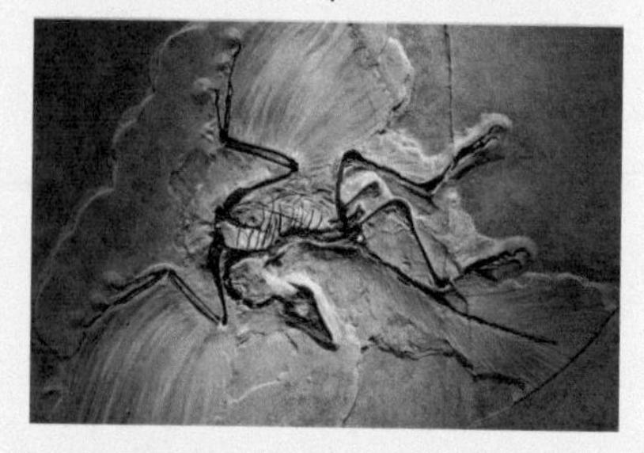

It is a reptile with feathers and wings, thought to be a transition between dinosaurs and birds.

Learn more about the fossil record and evidence of changing species.

Writer's Workshop

Imagine that you could make any wild animal into a friendly variety that was the perfect size. Which animal would you choose to have as a pet? What would life with your puppy-sized elephant or goldfish-sized whale be like?

Notebook and add the page to your table of contents. You can unfold and pull out the DNA to where the jumping genes have been inserted.

EXPLORATION: Horizontal Gene Transfer

For this activity, you will need:

- String or embroidery floss in 2 different colors
- Drinking straw
- School glue
- Science Notebook
- Video about horizontal gene transfer from the YouTube playlist for this unit

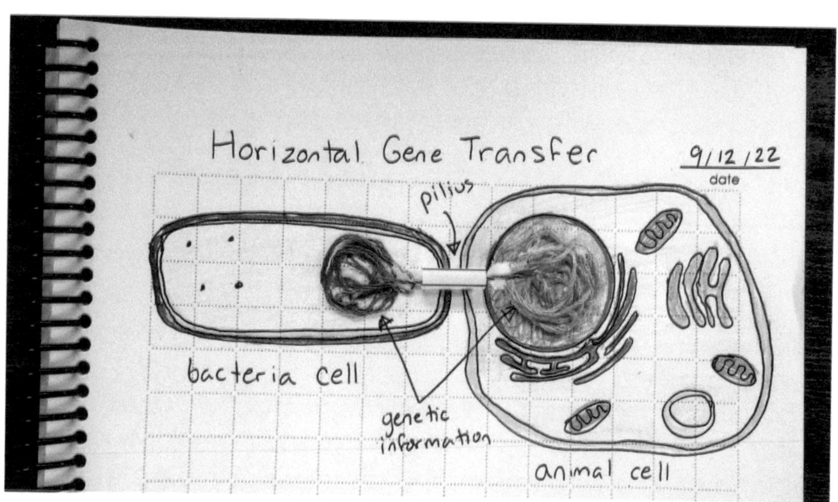

Horizontal gene transfer is when a cell passes genetic information to another, unrelated cell, not a daughter cell. This can happen within a species or between species. It is only partially understood.

Genes can be transferred directly between cells using a pilus, a structure that connects the two cells temporarily. Viruses can transfer genes from one cell to another by picking up genetic material from one organism and then injecting it into another. Or transposable genes, jumping genes, can move between the cells of different species.

Horizontal gene transfer has been detected from bacteria to bacteria, from bacteria to archaea, from bacteria to plant, from plant to plant, from animal to virus, from mammal to mammal, and bacteria to animal. It is not known how often this happens or how much it has influenced the genetics of all of life. It probably occurs more often in organisms that have close relationships—the *Wolbachia*, a parasitic bacteria, and the nematodes it infects, for example.

The significance is that a large section of DNA can be

transferred from one species to another in a single event, leading to big changes in the genome.

1. Draw two cells in your Science Notebook, one a prokaryote, and one a eukaryote. The eukaryote cell should have organelles and a nucleus.
2. Glue some string down in the middle of the prokaryotic cell to represent the DNA. Then glue a second color of string down in the nucleus of the eukaryotic cell to represent DNA.
3. Glue a length of straw down between the prokaryotic and eukaryotic cells. Thread a piece of the prokaryotic DNA string through the straw.
4. Label the two cells. Also label the genetic information.
5. Label the straw piece "pilus." A pilus is a structure that connects one cell to another for the purpose of transferring genetic material. Many species of bacteria have the ability to transfer genetic material this way, not only between cells of the same species but between species and even from bacteria to animal or bacteria to plant.
6. Watch a video about horizontal gene transfer. Take notes on the bottom of your page as you watch.

EXPLORATION: Epigenetics

For this activity, you will need:

- "Epigenetics" from the Printable Pack
- A video about epigenetics from the YouTube playlist for this unit
- Science Notebook
- Colored pencils or crayons
- Scissors
- Glue

Epigenetics is the study of how genes can be turned off and on by adding methyl groups to the DNA. A methyl group is a carbon atom surrounded by three hydrogen atoms. The group can be bonded to sections of the DNA without changing the actual DNA code.

Every cell in your body has all of the DNA, but certain parts of the DNA are turned on to make blood cells and other parts are turned on to make skin cells and so on. These methyl groups are the mechanism that turns on and off sections of DNA.

Besides methylation being pre-programmed for particular cells, there is emerging evidence that environmental

Memorization Station

Epigenetics: the study of changes in organisms caused by modification of gene expression rather than alteration of the genetic code itself

Famous Folks

Recent studies by Michael Wilson of Canada have found that baby mice who were licked right after birth by their mothers had portions of the DNA methylated, which made the mice better able to deal with stress through their lives. The DNA was changed by environment after birth and the behavior was passed down over generations.

The mice below are clones, but one of them was methylated by researchers in a lab. This resulted in a different phenotype as seen by the kinked tail.

These mice are from a lab in Australia run by Emma Whitelaw. Whitelaw has found that the epigenetic markers are heritable from generation to generation. Photo by Emma Whitelaw, University of Sydney, Australia, CC by SA 2.5, Wikimedia.

Writer's Workshop

If you could bring back any extinct species, which one would you bring back and why?

Deep Thoughts

Imagine that a river changed its course and split a population of mice in half, with each half living on its own side of the river. Under these conditions, discuss how speciation might occur over time. Hundreds of years down the road, will the mice populations on each side of the river still be the same? Why or why not? What environmental factors affect how animals form?

Memorization Station

Speciation: the emergence of a new species

Species: a specific kind of living thing that can exchange genes or reproduce with others of its kind

Hybridization: two species sexually reproduce and create a third viable offspring

Allopatric speciation: a population becomes geographically isolated and diverges genetically or behaviorally so that it won't reproduce with the original species

Sympatric speciation: something other than geography divides a population and the two diverge into unique groups

These definitions of species and speciation only apply to sexually-reproducing organisms. Since most life forms reproduce asexually, this is not always useful. What makes a species a species is somewhat subjective.

factors can methylate genes as well. Even more interesting, the methylation can be passed on genetically to future generations even if the methylation happened after the birth of the organism.

This is being studied as a possible mechanism for evolution. It also has bearing on cancer research and other genetic diseases.

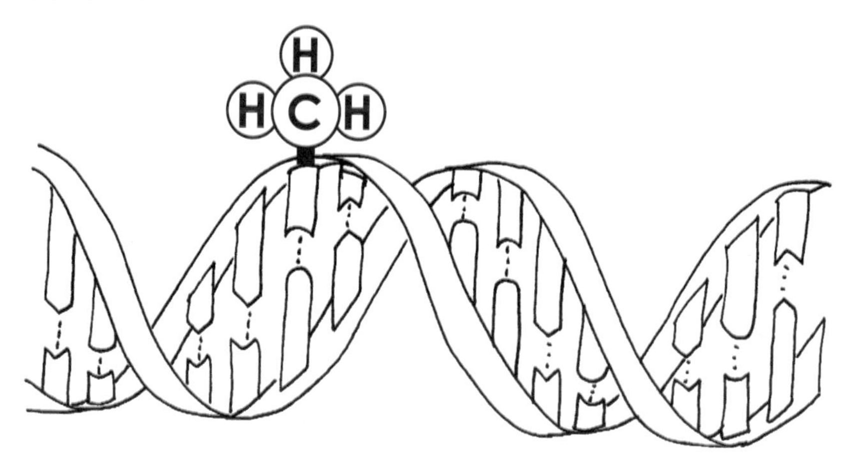

1. Color the methyl group and the DNA strand that is being added to on the "Epigenetics" printable. Put it in your Science Notebook.
2. Watch a video about epigenetics.
3. Take notes about epigenetics during the video.

☻ ☻EXPLORATION: Speciation

For this activity, you will need:

- Internet connection
- "How To Get A New Species" from the Printable Pack
- Scissors
- Glue stick
- Video about speciation from the YouTube playlist for this unit
- Science Notebook

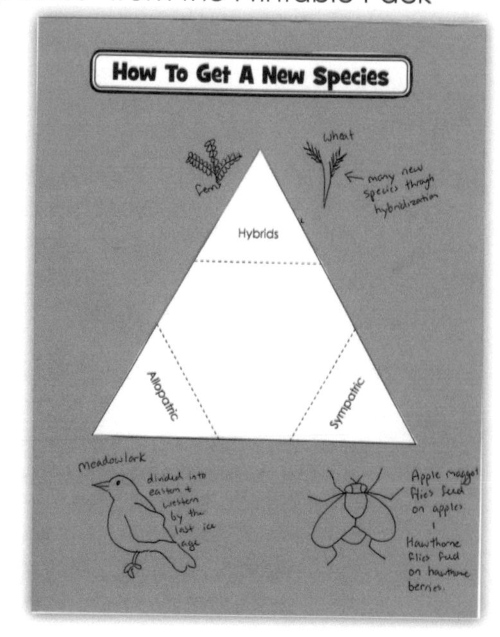

Speciation is when a new species emerges. A **species** is a population that does not normally interbreed with others in the wild. Speciation can happen through hybridization, allopatric speciation, and sympatric speciation.

Hybridization is when two different species sexually reproduce and form a third species. This mostly happens with plants but can happen with animals too. A hybrid population can form very quickly, in one generation.

Allopatric speciation happens when a population becomes geographically isolated from each other. The two parts of the population stop reproducing together and, over time, they drift so that they no longer can mate together.

Sympatric speciation is when a population is divided by something other than geography. The isolation can be because some individuals prefer a certain food source or certain mating rituals over others. Over time, each population changes so that they become distinct species.

1. Cut apart the "How To Get A New Species" worksheet on the heavy lines. Fold the points of the triangles up along the dashed lines to form flaps. Glue the center of the triangle to your Science Notebook, leaving the flaps loose. Glue the title to the top of the page.
2. Under each flap, write the definition of the type of speciation in your own words, using the information above.
3. Go online and watch a video on speciation. While you watch, keep an eye out for examples of each type of speciation. On your paper, draw a picture of organisms that have gone through each type of speciation.

Step 3: Show What You Know

During this unit, choose one of the assignments below to show what you have learned during the unit. Add this work to your Layers of Learning Notebook. You can also use this assignment to show your supervising teacher or your charter school as a sample of what you've been working on in your homeschool, if needed.

There are more ideas for writing assignments in the "Writer's Workshop" sidebars.

Coloring or Narration Page

For this activity, you will need:

- "Biodiversity" from the Printable Pack
- Writing or drawing utensils

1. Depending on the age and ability of the child, choose either the "Biodiversity" coloring sheet or the

Additional Layer

Laboratory experiments on speciation have been conducted mostly using fruit flies, the famous *Drosophila melanogaster* species.

Speciation has also been observed in real time in nature. The apple maggot fly of North America is a great example that you might want to read up on.

This is Rhagoletis pomonella, the apple maggot fly.

You can observe fruit flies. Start by placing an overripe banana in an open jar outside for a few hours in warm weather.

After a few hours have gone by, cover the jar with a paper towel and put a ring around the jar to hold the paper towel in place. Watch your jar every day for a week or two. You can use a magnifying glass if you'd like. Look for evidence of the fruit flies. You usually start by seeing little white specks on the banana (the eggs). Later, you'll notice little wormlike creatures (maggots). They will use the banana as food and grow bigger and bigger. The worms will become pupa, which look like small grains of rice, and then eventually a fly will emerge.

Your flies will keep laying eggs and having more babies if you want to continue your observations.

Unit Trivia Questions

1. Survival of the fittest and evolution were presented by what scientist?

   Charles Darwin

2. Behavioral changes of individuals of a species that allow it to better thrive or survive are called _________.

   a) Environment

   b) Variation

   c) Adaptation

   d) Behavior

3. What plants did Mendel observe to explain how traits are passed on?

   Peas

4. What is the name for one long strand of linear DNA?

   a) Genotype

   b) Phenotype

   c) DNA

   d) Chromosome

5. Which trait is more likely to be expressed - a dominant trait or a recessive trait?

   Dominant, but that doesn't make it impossible for recessive traits to be expressed.

6. True or false - Golden retrievers and Labradors are two different species.

   False - different breeds of dogs can mate and are still the same species even though they are different breeds.

7. A _________ describes the genes present within an individual while a _________ describes the observable traits of the individual.

   genotype, phenotype

"Biodiversity" narration page from the Printable Pack.

2. Younger kids can color the coloring sheet as you review some of the things you learned about during this unit. On the bottom of the coloring page, kids can write a sentence about what they learned. Very young children can explain their ideas orally while a parent writes for them.

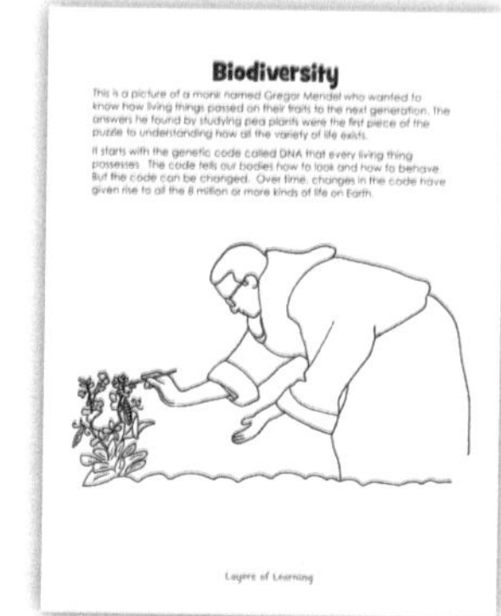

3. Older kids can write about some of the concepts you learned on the narration page and color the picture as well.
4. Add this to the Science section of your Layers of Learning Notebook.

Science Experiment Write-Up

For this activity, you will need:

- The "Experiment" write-up or "Experiment Report Template" from the Printable Pack

1. Choose one of the experiments you completed during this unit and create a careful and complete experiment write-up for it. Make sure you have included every specific detail of each step so your experiment could be repeated by anyone who wanted to try it.
2. Do a careful revision and edit of your write-up, taking it through the writing process, before you turn it in for grading.

Writer's Workshop

For this activity, you will need:

- A computer or a piece of paper and a writing utensil

Choose from one of the ideas below or write about something else you learned during this unit. Each of these prompts corresponds with one of the units from the Layers of Learning Writer's Workshop curriculum, so you may choose to coordinate the assignment you choose with the monthly unit you are learning about in Writer's Workshop.

- **Sentences, Paragraphs, & Narrations:** Choose some of the Memorization Station vocabulary words to use in complete sentences that demonstrate your understanding of what each word means.
- **Descriptions & Instructions:** Describe in a paragraph what Mendel's pea experiments teach.
- **Fanciful Stories**: Choose a species of plant or animal

and write a story about an adventure you must go on to avoid extinction.

- **Poetry:** Write a free verse poem that describes who you are and how you fit into your family.
- **True Stories:** Research a person who has a genetic disease. Write about his or her life and disease.
- **Reports and Essays:** Find three online scholarly journals about biodiversity and write an annotated bibliography that includes all three.
- **Letters:** Write a letter to your future self describing things that you have inherited due to genetics and things that you have learned from your environment (nature versus nurture). Add a few things you'd like to learn, develop, or change about yourself in the next year. Put a date at least a year away on the outside, and wait until that date, then open in and read your letter to yourself.
- **Persuasive Writing:** Create an advertisement either promoting genetically-modified foods or seeking to ban them.

Big Book of Knowledge

For this activity, you will need:

- "Big Book of Knowledge: Biodiversity" from the Printable Pack, printed on card stock
- Writing or drawing utensils
- Big Book of Knowledge

1. Color, draw on, write on, or add to each of the Big Book of Knowledge pages you are using. Only add the printables if you learned these concepts during this unit. If there are other topics you focused on, feel free to add your own pages to your Big Book of Knowledge.

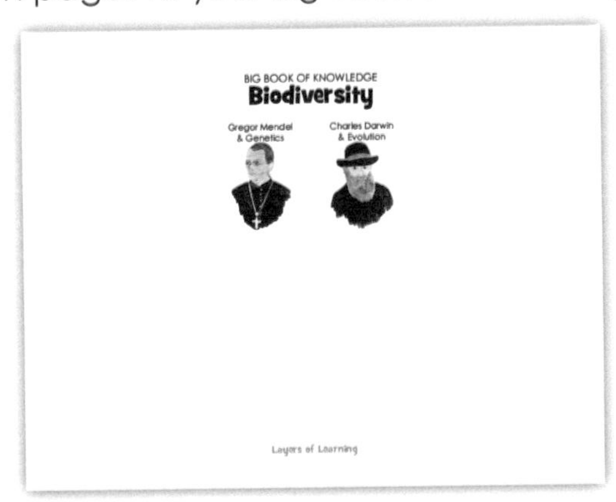

2. Use your Big Book of Knowledge regularly to help you review, quiz, or create games that will help you commit the things you've learned to memory.

Big Book of Knowledge

The Big Book of Knowledge is a book for you, the mentor, to use as a constant review of all of the things you're learning about. You can use it to quiz your kids or prepare tests or review games. Whenever you learn something in Layers of Learning that you want your kids to remember, add it to your Big Book of Knowledge.

Assemble your Big Book of Knowledge in a binder or with binder rings. Divide it into sections for each subject.

In the Printable Pack for this unit you will find a "Big Book of Knowledge" sheet. You can add this sheet to others you collect or create yourself as you progress through the Layers of Learning curriculum. Customize the Big Book of Knowledge to your family by adding facts and topics that you enjoyed exploring as you were learning.

Visit Layers of Learning online to find more information on how to assemble and use your own Big Book of Knowledge.

You will also find cover and section pages to print along with creative games to play with your Big Book of Knowledge to keep school, even the tests, fun!

Unit Overview

Key Concepts:

- There are many kinds of plants that fit into three basic categories: simple plants, conifers, and flowering plants.
- Plants can photosynthesize, turning sunlight absorbed through their leaves into a sugar called glucose, which feeds them.
- Flowering plants reproduce from seeds while most simple plants reproduce from spores.
- Fungi are not plants; they are decomposers that do not photosynthesize.

Vocabulary:

- Leaf
- Stem
- Roots
- Bud
- Flower
- Seed
- Fruit
- Multicellular
- Eukaryotic
- Photosynthesis
- Epidermis
- Mesophyll
- Chloroplasts
- Vascular bundle
- Stomata
- Simple plant
- Fern
- Fiddlehead
- Moss
- Conifer
- Dicots
- Monocots
- Fungi

PLANTS

Plants are living things that grow both on land and in the water. They can grow almost anywhere on Earth, from snow-covered mountains to dry, hot deserts. All plants are multicellular, eukaryotic organisms that can manufacture their own energy from sunlight. Most plants are green because of the chlorophyll contained in every cell of their bodies. It is the chlorophyll that lets them produce energy from sunlight. As part of this process, called photosynthesis, they also release oxygen, the source of the air that animals and people need to breathe.

There are over 320,000 species of plants in every environment from high mountains to deserts to lakes to jungles. They have filled every niche of life and dominate the planet in terms of mass and importance. Plants covered the earth before animals showed up and, it turns out, animals and people are dependent on plants for food, shelter, medicine, and a thousand little things you probably barely think about, from paper and clothing to the wood that made your home and much of the food you eat.

Plants can be divided into categories—simple plants,

conifers, and flowering plants. Simple plants do not produce seeds. They include things like ferns and mosses. Conifers are cone-bearing plants that do not have flowers, like pine trees and ginkgo trees. Flowering plants produce both flowers and fruit. Most of the plants on the planet are in this category. They include grasses, peonies, roses, apple trees, coconut palms, and tomatoes.

Fungi such as yeast, mushrooms, and mold, are not plants or animals. They are in their own category because they have chitin walls instead of cellulose, they are heterotrophs (eat other things), but they can't move on their own. This puts them in neither the animal nor the plant category. We include them in this unit to compare them with plants and contrast them with the upcoming animal units.

Step I: Library List

Choose books from your library that go with this topic. Here's a list of some favorites and also a list of search terms so you can utilize what your library offers. Read the books with your kids and/or assign them some to read independently. It is from these books your kids will learn most of the facts they need from this unit.

Search for: plants, photosynthesis, trees, gardening, flowers, fungus, mushrooms

☺ ☺ ☻*The Usborne Science Encyclopedia*. Read as many of the pages on plants as you would like. The plants sections are found on pages 250-294.

☺ ☺ ☻*Encyclopedia of Science* from DK. Read ""Fungi," "Plants Without Flowers," "Conifers," and "Flowering Plants," on pages 315-319 and "Photosynthesis" and "Transport in Plants" on pages 340-341.

☺ ☺ ☻*The Kingfisher Science Encyclopedia*. Read "Fungi and Lichens," "Plants Anatomy," "Nonflowering Plants," "Flowering Plants," "Fruit and Seeds," "Trees," and "Plants and People" on pages 55-66.

☺ ☺ ☻*Trees, Leaves, Flowers, and Seeds: A Visual Encyclopedia of the Plant Kingdom* by DK. Full of pictures and packed with information, we recommend this book for browsing.

☺*From Seed to Plant* by Gail Gibbons. Simple explanation of where trees, grass, flowers, and our gardens come from.

☺*Hungry Plants* by Mary Batten. All about carnivorous plants, this is sure to interest kids!

Family School Levels

The colored smilies in this unit help you choose the correct levels of books and activities for your child.

☺ = Ages 6-9
☺ = Ages 10-13
☻ = Ages 14-18

On the Web

For videos, web pages, games, and more to add to this unit, visit the Biology Resources at Layers-of-Learning.com.

You will find a link to video playlists, web links, and more.

Bookworms

If you're looking for a family read-aloud, we'd like to suggest this one.

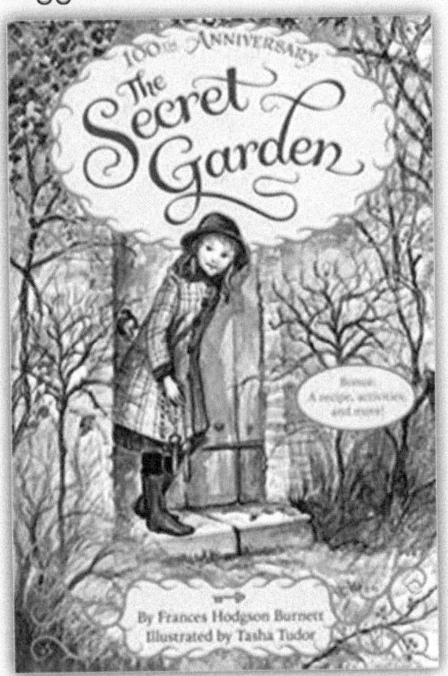

The Secret Garden by Francis Hodgson Burnett is a children's classic about a spoiled little girl who moves to England with distant relatives. Can she grow with the help of a garden? And what other secrets does the manor contain?

Teaching Tip

Throughout this unit, set aside time for nature observation. Have everyone take a chair, their Science Notebooks, and some colored pencils outside. Sit and quietly observe for at least 15 minutes at a time. Sketch plants or flowers and write down observations.

Teaching Tip

Set up a plant exploration station for your kids during this unit with various plants.

Additional Layer

Grow a garden. It could be on your windowsill, back deck, or in your yard. Decide what plants you want to grow and then find out what they need to be healthy.

☺ *Flowers* by Gail Gibbons. Talks about where flowers grow, how botanists group them, and more.

☺ *The Magic and Mystery of Trees* by Jen Green. Adorable illustrations and text about the community of trees.

☺ ☺ *Botany: Plants, Cells, and Photosynthesis* by April Chloe Terrazas. This digs into the deepest level of plants, their cells, how they produce energy, and how they transport materials. Great resource!

☺ ☺ *Nature All Around: Plants* by Pamela Hickman. Charming illustrations and detailed science about the parts of plants, how plants work, and what plants need to thrive.

☺ ☺ *Strange Trees: And the Stories Behind Them* by Bernadette Porquie and Cecile Gambini. Information about some of the oddest trees on the planet with simple drawings on the facing page. Adorable and fascinating.

☺ ☺ *Gardening Lab For Kids* by Renata Brown. This is for those who want to grow a garden. The book takes kids through a series of projects from learning to read a growing zone map to composting and garden art.

☺ ☺ *Fungus is Among Us!* by Joy Keller. A fun rhyming book, with lots of science about mushrooms packed into the pages.

☺ *Trees: A Rooted History* by Piotr Socha. Gorgeous illustrations and tons of info from the science of trees to identification to how trees are used in daily life.

☺ *Understanding Photosynthesis With Max Axiom* by Liam O'Donnell. A graphic-style book with real science.

☺ ☻ *Trees of North America* by C. Frank Brockman. This is an identification guidebook. If you are from outside North America, find a guidebook for your region. Also look for field guides to wildflowers, mushrooms, cacti, or other types of plants that can be found where you live.

☺ ☻ *Botanicum* by Kathy Willis. A stunning book full of illustrations and lots of information about plants of all kinds from horsetails to orchids to crops we eat.

☺ ☻ *Katya's Book of Mushrooms* by Katya Arnold. A lovely little book that will turn you into a mushroom lover. Filled with facts, tips, and tons of enthusiasm.

☻ *Plant* by Janet Marinelli. This is a coffee table book of plants the world over. Great for browsing or as a reference.

☻ *How Plants Work: the Science Behind the Amazing Things Plants Do* by Linda Chalker Scott. Written with a conversational tone, this book delves into the cell and photosynthesis, sharing fascinating tidbits before it diverges into watering, fertilizer, and plant behaviors.

☻*The Hidden Life of Trees* by Peter Wohllenben. This book examines a forest up close, and brings all the interconnectedness of the trees and the things that live among them to your attention.

☻*Eating the Sun* by Oliver Morton. About photosynthesis from the chemistry to the implications for our modern life. For good readers.

☻*The Allure of Fungi* by Allison Pouliot. Looks at fungi from the perspective of ecologists, mycologists, naturalists, farmers, artists, and philosophers. Talks about the history of fungi and humans, especially in Australia.

☻*Plants, Photosynthesis, Plant Nutrition and Transport*, and *Fungi* from Bozeman Science on YouTube. Use these videos as lectures for your high schooler. Have your student take notes from the videos.

Step 2: Explore

Choose hands-on explorations from this section to work on as a family. They should be appealing activities that will create mental hooks so your kids remember the information in the unit. Save the rest of the explorations for the next time you do this unit in four years when your kids are older. You can also read the sidebars together and explore some little rabbit trails.

This unit includes printables. See the introduction for instructions on retrieving your Printable Pack.

How Plants Work

In this section you'll learn about plant cells, how plants move water and nutrients around, and how they make food from sunlight.

☺ ☺ ☻**EXPEDITION: Parts of a Plant**
For this activity, you will need:

- "Parts of a Plant" from the Printable Pack
- Colored pencils or crayons
- Science Notebook
- Plants to observe outdoors

There are over 300,000 kinds of plants in the world and all of

Additional Layer

Plants have a permeable membrane surrounding each cell. This allows some things to go through the membrane while keeping other things out.

Try soaking a few raisins in water overnight and observe what happens. Did any of the water permeate the skin of the raisin?

Expedition

Plan to visit some gardens, nurseries, or forests during this unit. Make arrangements ahead of time so you can have an expert tour guide,

Writer's Workshop

If you could plant something non-living and make it grow and reproduce, what would you plant?

Write about your amazing plant in your Writer's Journal.

Fabulous Fact

Many flowering plants have big, showy flowers, but others have tiny flowers that you might miss altogether if you're not looking closely.

The heads on grass stalks bear itty bitty flowers every spring. Later in the summer the heads will ripen into seeds.

Memorization Station

Leaf: a flattened green outgrowth from the stem of a plant

Stem: the main stalk of a plant

Roots: underground section of a plant body

Bud: a compact growth on a plant that develops into a leaf, flower, or shoot

Flower: seed-bearing part of a plant

Seed: an undeveloped plant embryo

Fruit: seed-bearing structure in flowering plants that is formed from the ovary after flowering

them belong to the kingdom Plantae. Trees, mosses, grass, ferns, flowers, and algae are all in the kingdom Plantae. Within that kingdom, we will explore three groups—simple plants, conifers, and flowering plants. Scientists use the characteristics of plants to classify them into various groups.

1. Look at the "Parts of a Plant" printable and see if you can label each of the parts of the basic tomato plant shown. The answers are shown below. A tomato plant is a flowering plant, the most complex of the plant groups.

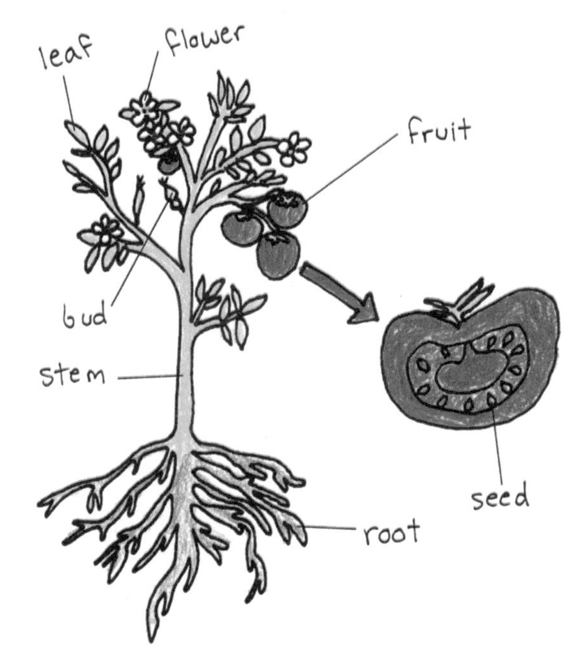

2. Talk about what each of these parts is responsible for.

- **Leaves** make food by taking in sunlight and changing it into energy for the plant. The leaves also release waste oxygen into the air.
- The **stem** supports the plants and carries food, water, and nutrients to the parts of the plant.
- The **roots** soak up water and nutrients from the ground and store them.
- **Buds** turn into flowers.
- The **flowers** create **seeds**, which can then make new plants.
- The **fruit** is where the seed develops and, since animals eat the fruit, they help disperse the seeds.

3. Go outside with your Science Notebook. Look around you for various plants. Do you see any with flowers? Do you see any non-flowering plants? Do you see any trees? If you are in a cool, shaded area you may even see some mosses or ferns. Do you see any aquatic plants that grow under the water? Choose at least one of the plants you see and make a sketch of it in your

Science Notebook. Be as detailed as you can.

4. Label any parts of your sketch that you can, in much the same way you labeled the tomato plant on the printable. If you aren't sure what some of the parts are, do a bit of online research and find out. For example, if you sketched a fir tree, you wouldn't find flowers but rather cones.

EXPLORATION: Plant Cells

For this activity, you will need:

- "Plant Cell" and "Plant Cell Answers" from the Printable Pack
- Book or video on plant cells
- Colored pencils, markers, or crayons
- "3D Plant Cell Model" from the Printable Pack (2 pages)
- Glue stick
- Scissors

All plants are **multicellular**, meaning they are made up of more than one cell. They are also **eukaryotic**, which means that inside the plant's cell you will find a membrane-bound nucleus that contains DNA.

Plant cells are different than animal cells in two big ways. First, plant cells have a rigid cell wall surrounding them. The cell wall helps the plant keep its structure and keeps the water-filled cells from bursting. Second, plant cells have chloroplasts, which are **chlorophyll**-containing organelles where sunlight is transformed into energy.

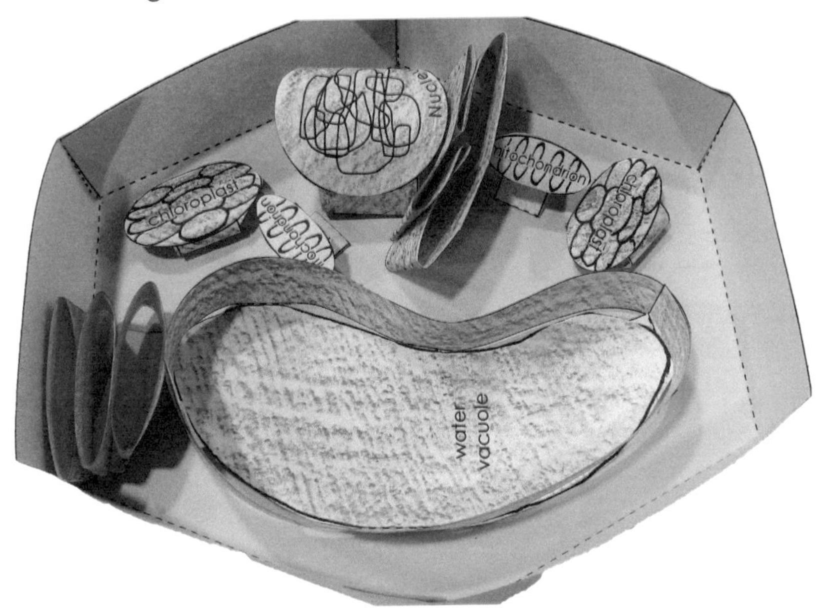

1. Color and label the "Plant Cell" diagram. Use the answers from the Printable Pack as a guide.

2. Watch a video or read a book about plant cells. Make

Additional Layer

Look at some plant cells under a microscope.

You need very thin pieces of plants to see the cells. If you can't get thin enough pieces, try blending some plant parts in a blender for a few seconds with some water and then place a smear of the liquid on a slide.

Basic microscope slide kits for biology also have plant cells. If you purchased a kit, look for "ranunculus," "maize," or "onion" slides.

Whenever you do microscope work, always sketch what you see, as big as possible, in your Science Notebook. Label anything you can. Make sure you write the type of plant and the date on the page.

Memorization Station

Multicellular: a living thing that is made up of more than one cell

Eukaryotic: large, complex cells with a nucleus, membrane bound organelles, and two strands of DNA

Chlorophyll: a green pigment present in all green plants that helps plant absorb light and photosynthesize

Additional Layer

Most states, provinces, and countries have official flowers and trees. Find out which are yours. Can you go out in your community and find live examples of these species?

The Western White Pine is the state tree of Idaho.

Additional Layer

Use some of your pressed flowers and leaves to make a sun catcher.

Paint jumbo craft sticks in bright colors. Place your pressed plants between two pieces of clear contact paper. Then, glue the craft sticks around the contact paper to make a frame.

Hang your sun catcher in a window throughout this unit.

sure you learn about chloroplasts and cell walls.

3. Print the first page of the "3D Plant Model" on green paper and the second page on white paper. Cut out the pieces on the solid lines. Fold on the dashed lines. Glue the green cell wall into a box using the glue tabs to connect everything together.
4. Color the organelles on the white page. Cut them out on the solid lines. Assemble and glue the water vacuole first. Plants have large water vacuoles to keep the plant's shape rigid and provide the water needed for photosynthesis.

   The nucleus and chloroplasts are assembled by gluing the long rectangle into a ring behind the organelle.

   The mitochondria are assembled by folding the long rectangle into an accordion.

   The Golgi body and the endoplasmic reticulum should be glued into rings and then glued to themselves to keep them tightly packed.

   Glue each of the organelles to the inside of the cell.
5. Add the "Plant Cell" worksheet to the Science section of your Layers of Learning Notebook and display your plant cell model throughout this unit.

EXPLORATION: Plant Kingdom Collection

For this activity, you will need:

- "Plant Divisions" from the Printable Pack
- Plant identification guide for your region of the world
- 6 mm (1/4") plywood or board, approximately 20 cm^2
- Sandpaper
- Paints, paintbrushes, and a water cup
- Newspaper
- Thin cardboard (like from a cereal box)
- Heavy duty rubber bands
- Science Notebook

The plant kingdom is typically divided into 14 divisions (sometimes more or less depending on who you ask). In botany, division is used instead of phylum, but it means the same thing. We will focus on just four basic divisions:

- Mosses (Bryophyta)
- Ferns (Pteridophyta)
- Conifers (Pinophyta)
- Flowering plants (Magnoliophyta)

1. Cut the title and leaves out of the "Plant Divisions"

printable. Glue the petiole (stem) of each leaf down to a page in your Science Notebook and glue the title at the top. Under each leaf, draw or write some examples of plant species in each division. You may need to research.

2. Make a plant press. First, cut a thin board (6cm or 1/4 inch) into two pieces that are about 20x20 cm. It should be small enough to fit in a backpack so you can take it out in the field.

   Sand your board so both sides are smooth. Clean off the sawdust, then paint your boards any way you like.

   While the paint dries, cut a thick stack of newspaper so the size is the same as your boards. Cut 5-6 pieces of thin cardboard (like from a cereal box) to size as well.

   Once the paint is dry, assemble your plant press so newspaper sheets are alternated with cardboard pieces. Stack it between the two boards and wrap it with two heavy duty rubber bands.

3. Collect some plants from outdoors. Use a plant identification book to discover what kind each plant is. Draw a sketch and write down the classification. Record the place and date of your collection.

4. Place each plant in the plant press, one plant between each layer of cardboard. Let the plants sit and press for a week, until they are flattened and dry. Carefully open your plant press and pull apart the layers of newspaper. Lay your plants out to dry the rest of the way. Then you can glue them into your Science Notebook or display

Writer's Workshop

Write a paragraph about the life cycle of a plant.

Little ones can keep it simple and talk about flower to fruit to seed to plant.

Older students can include pollination, the development of sex cells through meiosis, and the way seeds form in the ovaries.

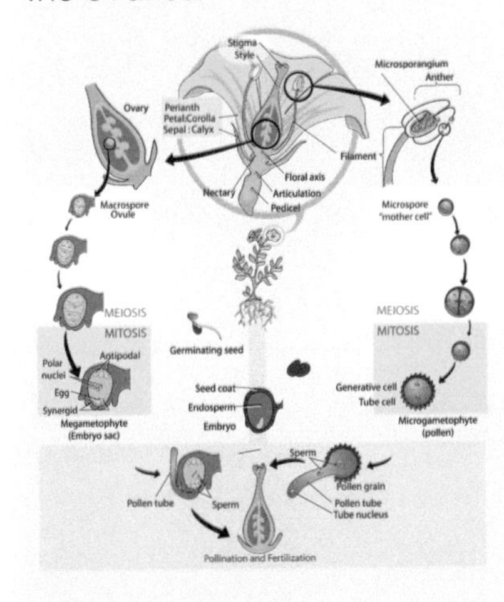

Additional Layer

Use dark colored paper and tape to make a sleeve to fit over one or more leaves of a plant, covering each leaf completely. Leave the leaves covered for three days, then remove the paper covers to see what has happened to the leaves without sunlight.

What does this tell you about how much plants need light?

your favorites in a frame on a wall. Make sure to write down the place and date of collection along with the genus and species name.

EXPLORATION: How Plants Grow

For this activity, you will need:

- Book or video about how plants grow
- Science Notebook
- Pencil and colored pencils
- Clear plastic cup
- Potting soil
- Seed, a large seed, like a bean, works well for this

Most plants grow from seeds or bulbs. Seeds contain embryos from which seedlings begin to develop. Seeds also have food inside of them. When you plant a seed in the ground and provide it with a damp environment, the seed germinates. Water from the ground is absorbed and the seed case bursts open. Roots grow downward while shoots grow upward toward the light above the ground. The shoots eventually emerge out of the ground and a stem and leaves grow from them. Plants then develop flowers or cones or sporangia which produce more seeds or spores so more plants can grow.

1. Watch a video or read a book about how plants grow.
2. Place a seed inside a clear plastic cup with a wet soil. Make sure the seed is against the plastic so you can see it. Observe the seed growth for a week or more as roots and shoots poke out of the seed coat.

Additional Layer

Do you know what gives plants their green color? It's chlorophyll. It is present in nearly all plants, algae, and cyanobacteria. Without it, plants couldn't turn sunlight into sugar.

Learn more about chlorophyll and what happens to green plants in the fall as the days get shorter and plants get less sunlight.

Additional Layer

Make leaf people with toilet paper rolls, colorful paper, glue, leaves collected from outside, markers, and googly eyes.

Wrap the toilet paper roll with colorful paper. Then, glue leaves to the paper rolls to make heads and arms. Glue googly eyes to the leaves. Use a marker to add other details you want.

3. In your Science Notebook, draw and label a diagram that shows the seed growth every few days. Use what you learned from the book or video as well as your own observations.

EXPLORATION: Photosynthesis

For this activity, you will need:

- Book or video about photosynthesis
- Science Notebook
- Pencil and colored pencils
- Plant with large leaves
- Dark colored paper
- Tape
- Container of water

Photosynthesis is the process where plants turn sunlight into sugars. This is how plants make food. Plants absorb energy from the sun through their leaves, take in water through their roots, and obtain carbon dioxide from the air. The sun's energy converts the water and carbon dioxide into a sugar called glucose. In the process, oxygen is produced. From the plant's point of view, the oxygen produced is waste product, but for animals and people, the oxygen allows us to live. Nearly all the oxygen in our atmosphere, in the water, and even in the rocks, comes from living plants.

1. Read a book or watch a video about photosynthesis.
2. In your Science Notebook, draw a diagram of photosynthesis. Younger kids can draw a simplified diagrams and older ones can be more detailed. Color the diagram and add labels to explain what is happening. Older kids should also memorize the photosynthesis equation.

$$6CO_2 + 6H_2O \rightarrow C_6H_{12}O_6 + 6O_2$$

Memorization Station

Photosynthesis: the process plants use to turn sunlight, water, and carbon dioxide into oxygen and energy (in the form of sugar)

$$6CO_2 + 6H_2O \rightarrow C_6H_{12}O_6 + 6O_2$$

Additional Layer

People eat stems, flowers, seeds, leaves, fruits, and roots of plants.

Look through your pantry and fridge or your local grocery store and identify which part of a plant each food comes from.

Flour, rice, corn, oats, and peas are all seeds.

Rhubarb, celery, asparagus, and cinnamon all come from stems.

Lettuce, spinach, and cabbage are leaves.

Tomatoes, apples, grapes, bananas, and avocados are all fruits.

Broccoli is an edible flower.

Carrots, onions,and potatoes are underground roots, bulbs, and rhizomes.

Talk about where your food comes from and then grow a bit at home.

Famous Folks

George Washington Carver was a famous American botanist.

He earned his masters degree in Iowa in 1896 at a time when almost no schools were available for black people. He spent his career teaching at Iowa State University and working on better methods of farming and food production for poor sharecropping farmers in the South (USA). He is most famous for his promotion of the humble peanut.

Bookworms

Read *The Story of George Washington Carver* by Eva Moore. It is a short biography about a man who rose from slavery. Ages 7 to 12.

3. Place one or several of the leaves of a living plant completely submerged in a container of water. Let it sit for 20 minutes in the sunshine, then come back and check on the leaf. You should see bubbles on the leaf surface. This is oxygen that the plant is producing through photosynthesis.

☻EXPLORATION: Leaf Structure

For this activity, you will need:

- "Leaf Structure" from the Printable Pack
- Pencil
- Colored pencils
- Microscope (optional)
- Cross section of a leaf prepared slide (optional)

Leaves are made of layers of cells that each do a different job to help the plant perform the photosynthesis that feeds it and keeps it alive.

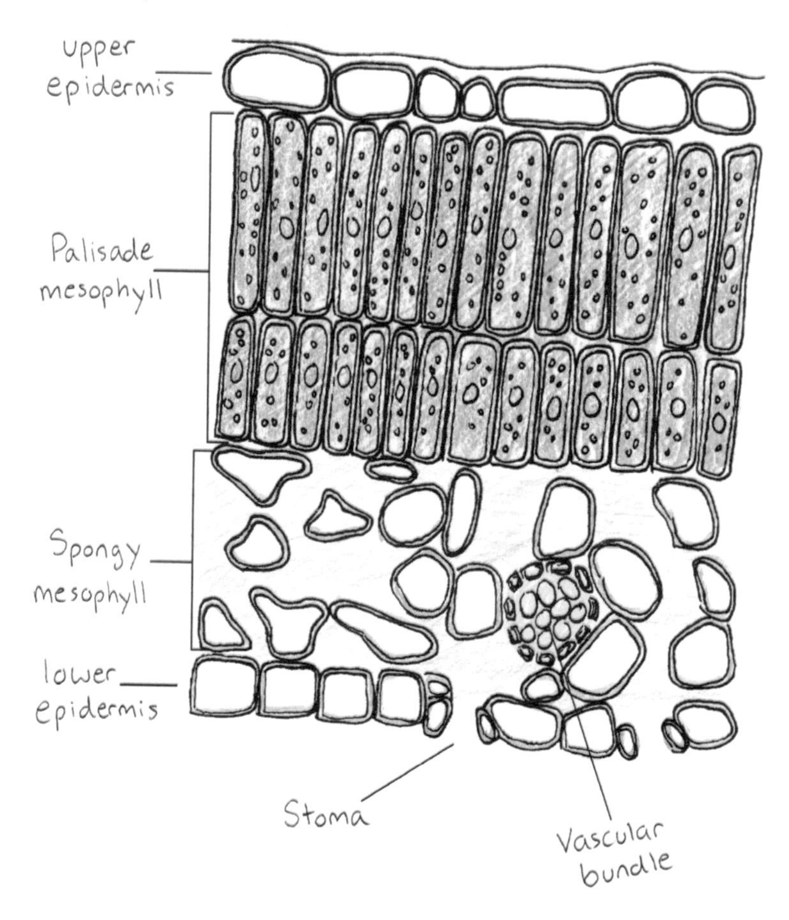

1. The upper layer of cells are called the upper **epidermis.** They protect the leaf and produce wax to coat the leaf. The wax further protects the leaf and keeps water from evaporating too much. On the "Leaf Structure" printable, color the cell walls of the epidermis light

green; leave the rest white because these cells are mostly transparent to let light through. Label this the "upper epidermis."

2. The next layer is the palisade **mesophyll** layer which is full of **chloroplasts**. This is where the photosynthesis mainly takes place. This can be one or several cells thick. The cells in this area are tightly packed and columnar (they look like columns). Color this whole section in a darker green. Label it "palisade mesophyll."
3. The next layer is the spongy mesophyll. This layer's main job is to contain extra water and lots of space for water and gases to move around between cells. The cells are loosely packed. Color the cell walls light green and the spaces between the cells light blue. Label it "spongy mesophyll."
4. The **vascular bundle** in the middle of a leaf is for transporting water and nutrients over larger distances. You can see the larger vascular bundles of a leaf as veins on the leaf. Label the "vascular bundle" and color it yellow.
5. The lowest layer of the leaf is another layer of epidermis. Piercing the lower epidermis are **stomata**. These are like pores on the bottom of the leaf that allow gas exchange so carbon dioxide can enter and oxygen can leave the leaf during photosynthesis. Around each stoma are guard cells that can open and close to control water evaporation. Color the guard cells dark green. Label the "stoma."
6. Put your diagram in your Layers of Learning Notebook in the Science section.
7. Look at the cross section of a leaf through your microscope. Find each of the types of cells you colored in your diagram. Sketch and label what you see in your Science Notebook.

EXPERIMENT: Vascular System

For this activity, you will need:

- 6 white carnation flowers
- 6 glasses or vases
- Food coloring
- Water
- "Xylem and Phloem" from the Printable Pack
- Colored pencils or crayons
- Video about plant vascular systems from the YouTube playlist for this unit

Bookworms

Plant fossils and the history of plants are fascinating but overlooked.

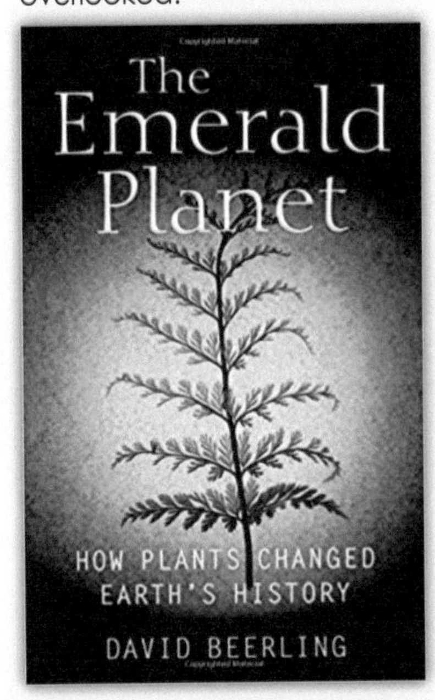

The Emerald Planet by David Beerling talks about the fossil history of plants and what it reveals about climate change in the history of Earth. What could it mean for us and our climate today?

Memorization Station

Epidermis: the outer layer of cells that covers a plant (think of it as the "skin" of a plant)

Mesophyll: the inner tissue of a leaf

Chloroplasts: organelles in a plant cell that convert light into energy through photosynthesis

Vascular bundle: part of the transport system in plants

Stomata: microscopic pores that allow gas and water exchange in plants

Bookworms

Linnea's Windowsill Garden by Christina Björk is an adorable story about a little girl's indoor garden and how she cares for it.

Writer's Workshop

Write a story from the point of view of a honeybee in search of pollen.

Additional Layer

Watch a video or read a book about how a farmer's crops get to the grocery store and your home.

If you can, visit a farm as well.

Plants can be either non-vascular or vascular. Vascular means the plants have bundles of cells that act like straws to transport water and nutrients around the plant. Mosses and algae are non-vascular. Most other kinds of plants—like trees, flowers, grasses, and ferns—are vascular.

A plant's vascular system consists of two types of tubes that run through plants. Xylem moves water and minerals from the roots through the stem and to the leaves. Xylem is made of tough lignin and dead cells. Phloem moves sucrose (plant sugars) from the leaves through the stem and to the roots. Phloem cells are living and use proteins to help move the plant sugars along.

1. Prepare six vases or glasses by filling them 3/4 full of water and adding a different color of food coloring to each glass.
2. Put one cut carnation into each glass of colored water and leave it overnight. In the morning the flowers will have drawn the food coloring up into the flower petals, coloring them.
3. Color the "Xylem and Phloem" diagram from the Printable Pack. Label the upward arrow "xylem" and the downward arrow "phloem." Put the completed printable in your Layers of Learning Notebook in the Science section.
4. Watch a video about plant vascular systems.

EXPERIMENT: Everything A Plant Needs

For this activity, you will need:

- Large clear glass jar, bowl, or box with a wide opening
- Small plant or plants that will stay small - try gnome ivy, moss, golden club moss, miniature fern, spiderwort, friendship plant, starfish flower cactus, nerve plant, African violet, or baby tears
- "Plant Needs" from the Printable Pack
- Potting soil
- Small pebbles or marbles
- 4-6 inexpensive houseplants that are identical (optional)

Plants need light, carbon dioxide, water, and nutrients from the soil to survive.

1. Place a handful of pebbles or marbles into the bottom of a large glass jar. This helps the soil drain and keeps the terrarium from becoming too soggy.
2. Place about 7 cm or so of moist soil on top of the pebbles. Nestle your plant into the soil, pressing firmly so the roots have good contact with the soil. Talk about how soil is a place for the plant to anchor and it also contains water and nutrients.
3. Water your plant as the soil becomes dry to the touch (follow the directions for watering that came with your plant). Keep your plant in a location with indirect light.
4. Make a poster of the things plants need. Start by drawing a plant with roots in the center of the poster, so you fill up most of the space.

   Then, draw and label the other things plants need. Plants need carbon dioxide from the air for photosynthesis, light for photosynthesis, water from soil

Additional Layer

Find out where some of your fruits and vegetables are grown. Did your bananas come from Panama? Are your apples from Washington State in the US? Is your spinach from a local grower? Find some of those places on a map of the world.

Writer's Workshop

Draw a picture of a brand new plant species that you invented. Then write a botanical description of it. What is its life cycle like? Does it have flowers? Is it pollinated by an animal? Does it have interesting features? Where is its habitat?

Additional Layer

Put rhubarb stems in water overnight. The stems burst apart because of turgor pressure. The cells have burst open from over watering.

for photosynthesis and transport, and nutrients to build the plant cells from the soil as well. The major plant nutrients are phosphorus, nitrogen, and potassium, but plants need other trace minerals as well.

Does the plant in your terrarium have everything it needs? Point out how all the plant's needs are being met.

5. Optionally, design experiments with a set of identical plants to see if a plant really does need sun, air, soil, etc. How could you test for each of these things?

Simple Plants

Simple plants do not reproduce with seeds. Sometimes they reproduce asexually and sometimes they produce spores. Most simple plants also lack a vascular system for transporting water and nutrients. They either live in water environments or live as colonies, with each cell acting as an independent agent but cooperating with others. The exception is ferns which reproduce with spores but do have a vascular system.

EXPLORATION: Life Cycle of a Fern

For this activity, you will need:

- "Fern Life Cycle" from the Printable Pack (2 pages)
- Scissors
- Brad
- Forest with ferns or a greenhouse with ferns
- Peat moss
- Plant press or heavy books and paper

Additional Layer

The Maori of New Zealand use an unfurling fern frond as a symbol of life, growth, strength, and peace. They call it the koru.

Make an oil pastel koru. Draw a spiral shape on a large piece of paper so it fills the whole page. Trace over the lines with a black marker so the line is very thick. Color the entire design with oil pastels so the color is very heavy. Water down some black paint and quickly wash it across the entire paper so it fills in wherever the oil pastel missed.

Memorization Station

Simple plant: a plant that has no roots, stems, or leaves

Fern: flowerless plant with feathery fronds and spores

Fiddlehead: tightly coiled fronds of a young fern

Ferns are "simple" plants, which means they have spores instead of seeds. Spores are formed though meiosis inside a specialized bunch of cells called sporangium. The spores are released and scattered onto the ground near the parent plant. The spore grows into a tiny heart-shaped prothalus, which then produces egg and sperm and goes through sexual reproduction. After the egg is fertilized, a new adult fern begins to grow. First, you get a curled up **fiddlehead** which unrolls into an adult plant.

1. Cut out the circle from page 2 of the "Fern Life Cycle." Attach the cut out circle to page 1 of the printable with a brad. Turn the circle to see each part of the fern life cycle. Be able to explain it.
2. If you have a forest near you where ferns grow, take a trip out to see them in the wild. If you don't live near wild ferns, go see some potted ones at a greenhouse. Look for fiddleheads in the spring and spores in the late summer. If you go in the very early spring you might see the prothalus stage in the ground near some adult plants.
3. Sketch what you see in your Science Notebook.
4. Take a sample of a fern frond and press it in a plant press. Once it is dry and flat, glue it into your Science Notebook with the date and place of collection and the species.

Writer's Workshop

Find a flowering plant in your yard or a nearby park. Identify it using a guidebook. Write a detailed description of the plant using only its Latin name. See if someone else can identify the plant from your excellent description.

This is Alaskan Fireweed. It grows well in areas where forest fires have swept through, and that's how it got its name.

EXPERIMENT: Moss Properties

For this activity, you will need:

- Sphagnum moss or moss you collect
- Beaker or jar
- Bowl
- Electronic balance
- Water
- Microscope and/or magnifying glass
- Science Notebook
- Colored pencils

Mosses are non-vascular plants. They do not have xylem and phloem to move water and nutrients around. They do not have a root system either, but rely on absorbing water directly into their leaves. They are small and grow close to the surface. They also must live in places that have abundant water for at least part of the year.

1. Set a beaker on an electronic balance and hit "tare" to zero out the weight. Put a handful of sphagnum moss in the beaker and record the weight.

Additional Layer

Many cultures use flowers in funeral ceremonies. Flowers, with their annual return, often represent rebirth to people.

Ancient Greeks placed a wreath of flowers on the deceased person's head and covered the tomb with flowers.

Ancient Egyptians placed flower necklaces and crowns on their dead.

Modern-day Mexicans use flowers in their Day of the Dead ceremonies just like their Aztec ancestors.

Does your culture use flowers at funerals? What do the flowers mean to you?

Fabulous Fact

Algae are simple plants. They don't have roots, leaves, or a stem system. There are over 400,000 known varieties found in marine and freshwater environments. They produce oxygen which other aquatic life uses.

Many are totally harmless and even beneficial, but some algae can be dangerous. This bio-luminescent algae in the East China Sea is beautiful, but toxic, and it's growing bigger and bigger.

Photo shared by WanRu Chen under CC 3.0 license

Deep Thoughts

There's an old proverb that says "A rolling stone gathers no moss." The rolling stone is a metaphor for people who do not settle down in one place. What do you think it means?

2. Remove the moss. Put the moss in a bowl and add water. Let the moss sit for ten minutes so it has time to absorb the water.

3. Put the moss back in the beaker on the electronic balance and weigh it again. How much water weight can your moss absorb?

4. Moss leaves are just one cell thick. Examine the moss closely with a microscope or magnifying glass. Draw what you see in your Science Notebook. Color and label your illustration.

EXPERIMENT: Moss Propagation

For this activity, you will need:

- Moss you find locally
- Buttermilk
- Sugar
- Water
- Blender
- Paintbrush
- Spray bottle of water for misting
- Science Notebook & pencil
- Colored pencils

Mosses produce spores which are distributed by the wind to produce new colonies of mosses. There is a second way that mosses can propagate. Each moss cell is totipotent, which means a single moss cell can regrow an entire moss plant.

1. Find some moss growing in your yard or nearby. Observe carefully the surface the moss is growing on and the conditions. Is the moss growing on soil, wood, or stone? Is the moss in the shade, partial shade, or sunlight? How

much water does this place have now and how much would it have at other seasons of the year? Take some notes and add sketches as needed.

2. Then take a small sample of the moss. Don't destroy the whole colony.
3. Mix the moss sample with a half cup of buttermilk (the acid helps the moss get started), a teaspoon of sugar, and a cup or two of water. Blend it up in your blender.
4. Spread a thin layer of the moss mixture on a surface like the one you found it originally growing on and with similar conditions.
5. Mist the moss once or more a day until it is established, several weeks at least. This works well in a garden near your house where you can keep an eye on the moss. Try coating a piece of garden art like a statue or a bird bath with the moss mixture.
6. Make sketches of the moss every few days as it establishes itself. Color and label your drawings.

Conifers

Conifers are bushes and trees that form cones for their seeds and do not have flowers. Many conifers are evergreens, never shedding all of their leaves at one time. They have needles or scales instead of broad leaves.

EXPLORATION: Layers of a Tree

For this activity, you will need:

- Cross section of a tree, preferably a conifer with the bark still on (optional)

Memorization Station

Moss: a small, flowerless, green plant that lacks true roots, grows in damp habitats, and reproduces with spores

Conifer: a tree that is cone-bearing, has evergreen needles, and grows in cool areas

Fabulous Fact

Moss naturally filters the air and helps remove pollutants. Often when we think about the plants that give us oxygen and benefit our planet, we forget the huge role that mosses, algae, and marine plants like seaweed play. The earth has an amazing variety of plant life.

Expedition

Learn about the differences between evergreen and deciduous trees, then go out in a forest and see if you can identify some.

Fabulous Fact

Dating a tree using its rings is called dendrochronology.

Additional Layer

Learn all about the life cycle of a pine tree and make a simple diagram in your Science Notebook.

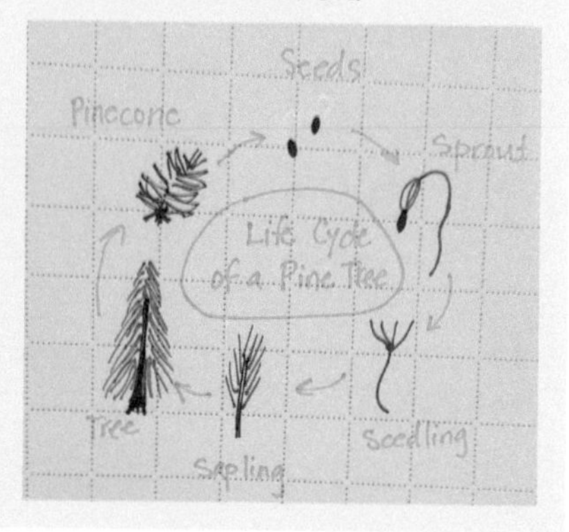

Deep Thoughts

In recent years, severe wildfires have prompted people to consider when forest management should include more controlled fires.

Do you believe we should have controlled fires or is intentionally burning forests just destroying more of our planet?

Read a news article about each side, then discuss what you think.

- Science Notebook
- Toilet paper tube
- Corrugated cardboard
- Colored paper - green, light brown, dark brown
- Crayons or colored pencils
- Scissors
- School glue

A tree's trunk has specialized layers that help the tree to thrive. From the inside out, the layers are heartwood, sapwood, cambium, inner bark, and outer bark.

1. If you have access to a cut piece of a conifer tree, examine a cross section. Carefully peel away the layers of bark and notice how it is composed. Look for the tree rings. Each ring represents one year of growth. How old was your tree when it died?
2. Make a sketch of your tree cross section in your Science Notebook.
3. Make a model of a tree's layers. First, the inner layer is the heartwood. This is the supporting pillar of the tree trunk. It is made of dead cells, but it is very strong, because those cells are tough cellulose fibers, intertwined and bound together with lignin. A toilet paper tube can represent the heartwood of your tree.

4. Next comes the sapwood. These are the xylem cells that move water up a tree's trunk and out to the needles or leaves. These cells are young and still fresh. As they age, they will become heartwood. Represent this on your model with a layer of corrugated cardboard, cut to wrap around your tree trunk.
5. Next is the cambium. This is the layer of cells that is actively growing this year. It produces both bark and sapwood for the tree. Represent this with a sheet of green paper cut to size to fit around your tree.
6. The fourth layer is the inner bark. This is the phloem, that brings nutrients and sap from the leaves and transports it around the tree and down to the roots. It is short-lived and later becomes bark. Represent this with a piece of light brown paper wrapped around your trunk.
7. The final layer is the outer bark. It can be quite thick and grows in distinctive patterns for each species of tree. It controls the moisture level of the trunk and branches and protects the tree from injury and pests. Draw a pattern all over a piece of dark brown paper and glue it to the outside of your trunk to represent this final bark layer.
8. Display the layers of a tree model on a shelf until the end of this unit.

EXPEDITION: Needles & Cones

For this activity, you will need:

- A wooded area with conifers to explore
- "Needles and Cones Genus Identification" from the Printable Pack
- Tree Identification guidebook for your region (optional)
- Science Notebook and pencil
- A fresh cone that you gather from the forest

The needles and cones on every species of tree are unique and so a tree can be identified by examining these parts. The needles of a conifer are specialized leaves. They are thin and spiky to help the tree retain moisture and withstand cold weather while easily shedding snow. You tend to find conifers in dry or cold environments.

The cones of a tree are its "flowers" and the place where seeds are produced. The large woody cones are female. In the spring, you can also see the male cones, which are herbaceous, usually yellow or light brown, and smaller than the female cones. The male cones produce pollen, which is spread by the wind to pollinate female cones.

Memorization Station

Learn tree anatomy when you color "Parts of a Tree" from the Printable Pack.

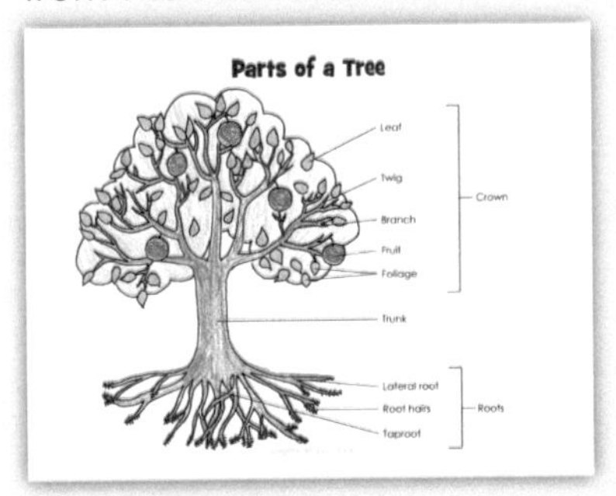

Cover up the labels and try to name the parts without looking.

Bookworms

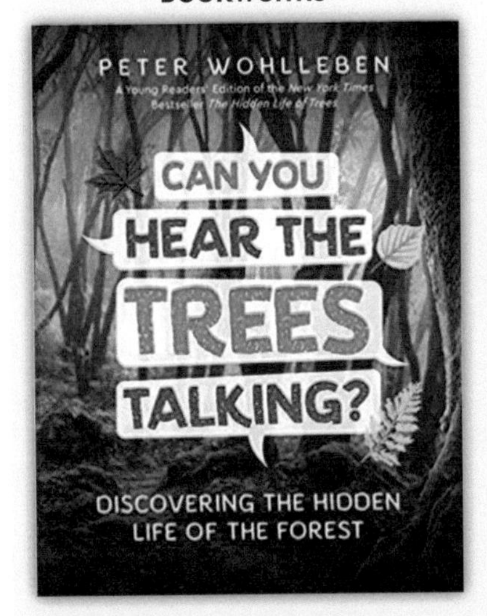

Can You Hear the Trees Talking: Discovering the Hidden Life of the Forest by Peter Wohlleben is an interactive science book for kids. It's not just about trees, it also talks about other life within the forest and how it all interacts. It includes quizzes and has beautiful photographs. You'll also enjoy *Peter and the Tree Children* by the same author. In the story, a forester teaches all about trees while helping a little squirrel.

Fabulous Fact

The main function of a pine cone is to keep the seeds of the tree safe. They close their scales to protect the seeds until it's time to open up and release them so the seeds can germinate. Learn more about pine cones.

Bookworms

The Giving Tree by Shel Silverstein is a classic tale about a boy and a tree. It's also a great book for discussions about happiness, selflessness, and love.

1. Visit a wooded area with conifer trees. Gather and compare different types of conifer needles and cones: cedar scales, fir, pine, cypress.
2. Use the "Needles and Cones Genus Identification" from the Printable Pack to identify what genus of tree you have. Not all the conifer genera are on the identification guide, just the four most common. Also, this will be most useful if you live in the Northern Hemisphere. In other areas, just use a local guidebook to identify the trees.

   By the end of the activity, your kids should recognize the difference between fir, pine, spruce, and cedar, or the common genera in your area.
3. If you have a tree identification guide for your area, you can discover what species you have as well. In your Science Notebook, sketch one or more of the tree species you identified. Be sure to make a careful close-up sketch of the needles and how they are arranged on the branch as well as an overall sketch of the whole tree.
4. Choose one fresh cone to bring home. Trees produce cones in the late summer, so if you are exploring any other time of year, you will have to settle for an old cone. Let the cone dry out and examine the seeds (old cones will have dropped their seeds, but you can see where they once were). The seeds have thin, light "wings" that help them disperse in the wind.
5. Draw a detailed sketch of the cone you brought home.

EXPEDITION: Tree Ecosystem

For this activity, you will need:

- A tree with low branches that you can reach from the ground
- Light colored bed sheet
- Science Notebook and pencil

Trees are habitats for many different kinds of creatures. Birds and squirrels nest in trees, of course, but there's much, much more.

1. Choose a tree where you can reach the lower branches. Hold a light colored sheet taut just under a branch. Shake the branch, catching things on the sheet.
2. Carefully lower the sheet to the ground, examine the sheet to see what was living in the tree. What other animals have you observed living in or around the tree?
3. Draw some sketches of the things you find living in the tree in your Science Notebook.

Flowering Plants

Most plants you are familiar with reproduce with flowers. Sometimes the flowers may be tiny, like those of grass, and sometimes they are large and showy, like a rose. The flower produces the seeds that allow sexual reproduction of the plant.

EXPERIMENT: Anatomy of a Flower

For this activity, you will need:

- "Parts of a Flower" and "Parts of a Flower" answers from the Printable Pack
- Pencil
- Colored pencils
- Tissue paper - green plus a bright color of your choice
- Chenille stems (pipe cleaners) - yellow, orange, green
- Scissors
- Flower with a showy bloom—lily or iris (optional)
- Magnifying glass (optional)
- Scalpel (optional)
- Cutting mat or cutting board (optional)
- Science Notebook

Famous Folks

Agnes Arber was a British botanist who spent her life studying plants. She focused on monocots and aquatic plants and released nearly a hundred publications.

Additional Layer

Make a lapbook about plants with a green file folder.

Add in foldable papers that show plant parts, plant life cycles, or the things a plant needs.

Draw and write notes right on the folder as well.

Tape or glue pressed plant specimens to the folder.

Present the lapbook to an audience.

Many plants have flowers that they use for reproduction. A flower has sepals on the outside that held the blossom in place when it was a bud. The petals attract pollinators. In the center is a pistil where the pollen is deposited. Surrounding the pistil are several tall stamens, where the pollen is produced. Below the pistil there is a swelling, which is the flower's ovary. This is where the seeds will develop.

1. Label the "Parts of a Flower" printable. Use the "Parts of a Flower Answers" as a guide.
2. Cut a yellow chenille stem in half. Cut an orange chenille stem in thirds. Wrap both yellow pieces of chenille stem around a green chenille stem, about 4 cm from the end of the green, so that all 4 ends are poking up from the green stem. Then, wrap one cut orange chenille stem around the green and yellow so that one end of the orange sticks up in the middle of the yellow pieces.

   The yellow chenille stems represent the stamens, where pollen is produced. The orange represents the pistil where pollen is deposited to fertilize the flower. The green pipe cleaner is the stem of the flower.

3. Lay seven layers of bright colored tissue paper and one layer of green tissue paper on top of one another. Cut all eight layers into rectangles approximately 15 x 30 cm.
4. Arrange the sheets so the green is on the bottom. Fold all eight sheets accordion-style, beginning with the long edge.
5. Wrap the shorter end of the green chenille stem around the center of the accordion-folded tissue paper, making sure the green is on the bottom, closer to the longer end of the green stem.
6. Carefully pull up and spread each layer of tissue paper so it surrounds the center.

Additional Layer

Pollinators are animals that move pollen from one plant to another, helping flowering plants through their life cycle. Bees, bats, ants, butterflies, and birds can all be pollinators.

Having a healthy group of pollinators is essential for ecosystems and food production.

Use three identical small dishes. Cut paper card stock flowers out of three colors. Place a dish on each flower. Fill the containers half full of sugar water. Is there a specific color that the pollinators prefer?

Variation: use a hummingbird feeder with each port surrounded by a different color paper flower. Is there a preference?

Writer's Workshop

Make a Venn diagram that compares annual plants with perennials. What do they have in common and what differences exist? You may have to do some research to find out.

The sepals are the green tissue paper that surrounds the bottom of the flower.

7. ☻With older kids, dissect a real flower and find the parts from the printable. Use a scalpel and cutting board to carefully cut the flower into pieces. Use a magnifying glass to closely examine the flower. Diagram and record your findings in your Science Notebook.

Teaching Tip

Kids learn to take notes by taking notes. The first step is to have them copy from the teacher as she or he writes information on a board. If you don't have a wall chalk board or white board, you can use a smaller hand-held one and get the same benefits.

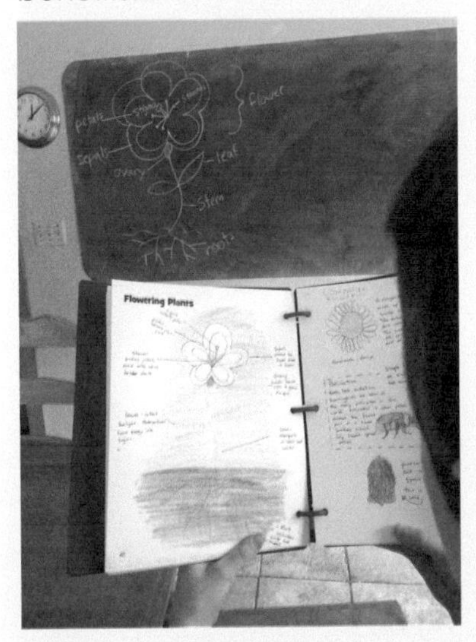

The second step is to take notes from a lecture or video without the teacher specifically telling the student what to write. Students aren't usually ready to independently take notes without direction until about age 14 or so.

☻**EXPLORATION: Life Cycle of a Flowering Plant**

For this activity, you will need:

- *The Tiny Seed* by Eric Carle - check your library or look for a YouTube video of the book being read aloud
- Paper
- Paints
- Scissors
- Glue stick

1. Paint several sheets of plain white paper in these colors:
   - Greens
   - A bright color - orange and yellow or pink and purple
   - Blues
   - Yellows
   - Browns

   Let all the paint dry completely.
2. Read *The Tiny Seed* by Eric Carle. Talk about how a flower makes a seed, a seed sprouts, the sprout grows into a new plant, then produces flowers to make more seeds.
3. Cut and tear your colored papers apart to make seeds,

Famous Folks

John Muir was a naturalist who saw the value of plants and animals and pushed for wilderness preservation.

Writer's Workshop

Use the "Plant Report Template" from the Printable Pack to write your own plant report.

Memorization Station

Dicot: flowering plant with embryonic leaf pairs

Monocots: flowering plant with a single embryonic leaf

On the left is a monocot; on the right is a dicot. Photo shared courtesy of Peter Halasz under CC 3.0 license.

soil, water, seedlings, the sun, and flowers. Glue the pieces to one or more sheets of paper to tell the story of a seed.

EXPERIMENT: Dicots and Monocots

For this activity, you will need:

- Dry beans - pinto, kidney, Lima, or black
- Corn seeds - the kind to plant in the garden
- 2 small bowls
- Water
- Scalpel
- Cutting board or cutting mat
- Magnifying glass
- Microscope (optional)
- Paper towels
- Zipper plastic bags - sandwich or quart size - or a clear jar or bowl
- Science Notebook

Flowering plants can be divided into two broad groups: dicotyledons (**dicots** for short) and monocotyledons (**monocots** for short). Dicots have two (di) embryonic leaves inside the seed. Monocots have one (mono) embryonic leaf inside the seed. When the plants grow up, dicot leaves have branching veins running through them while monocots have long parallel veins. Dicots have an odd number of petals on their flowers and monocots have an even number of petals on their flowers.

1. Soak a handful of bean and corn seeds in water overnight.
2. Examine a few of the seeds with a magnifying glass or microscope. Split them in half so you can see the inside

of the seeds. You may need a scalpel to split the seeds open. Beans are dicots and corn is a monocot. Can you see the embryonic leaves inside the seed?

Sketch the two types of seeds in your Science Notebook. Label them with the appropriate names.

3. Soak a piece of paper towel and place it inside a plastic bag. Slide 3-5 bean seeds down into the bag between the side of the bag and the wet paper towel. Use seeds you did not dissect. Repeat with corn seeds in another bag. Set the seeds in indirect sunlight.
4. Watch your seeds for a week as they develop into seedlings. Sketch the process every few days in your Science Notebook. Do you notice any differences between the dicot and the monocot plants?

EXPEDITION: Visit a Botanical Garden

For this activity, you will need:

- A botanical garden near you - check with universities, research organizations or cities with botanical gardens
- Science Notebook and pencil
- "Botanical Garden Scavenger Hunt" from the Printable Pack

Botanical gardens are dedicated to growing a wide collection of plants, usually from many parts of the world, though some may be dedicated to a specific region or type of plant.

1. Wander around a botanical garden, reading the informational signs and examining the plants. Stop to sketch a plant here or there that you find interesting. Make sure to record the type of plant.
2. Use the "Botanical Garden Scavenger Hunt" to help you notice some interesting features of the botanical garden. You can print a sheet for each person and have a competition to see who can get the most, or you can do the scavenger hunt together as a group.

Fungus

Fungi are not plants. They are in their own kingdom, the fungi kingdom. Fungal cell walls are made of chitin, while plants are made of cellulose. Fungi don't photosynthesize either. Instead, they get food by absorbing nutrients through their cell walls. Some common fungi you have heard of are mushrooms, yeast, and mold (mould). There are thousands of varieties of fungal life.

Famous Folks

There are many famous botanists, but much of what we know about plants has actually come from indigenous people across the globe who didn't have a lab but observed and experimented with plants.

Additional Layer

Bake a nice loaf of homemade bread and then smell your kitchen. That familiar smell is mostly thanks to yeast. The yeast transforms the sugar in the dough into carbon dioxide gas, which makes the bread rise. Alcohol is produced in the reaction and then evaporates while baking. Every time you bake bread you can thank the fungus, yeast.

Memorization Station

Fungi: plant-like organisms that do not make chlorophyll (singular = fungus)

Mushrooms, yeasts, and molds are all fungi

EXPLORATION: Fungus Features

WARNING: Do not eat mushrooms you find in the wild, even if you think you know what it is, unless you have an experienced guide helping you. Mushrooms are notoriously tricky to identify and mistakes are often deadly.

For this activity, you will need:

- A book or video about fungi
- "Fungus Features" from the Printable Pack
- Mushroom guidebook
- A forest or moist outdoor area to search for mushrooms
- Science Notebook

Fungi are eukaryotic like animals and plants, but the similarities don't go much further. They are made of chitin instead of cellulose, proteins, or lipids. They are the primary decomposers in most environments. They eat by secreting enzymes into the material they are feeding on and then absorbing the nutrients into their cells.

1. Start by reading a book or watching a video about fungi. Pay attention to how fungi get their nutrition and how their bodies are organized, including the mycelium.
2. While you read or watch, take notes on the "Fungus Features" page. There are some labeled diagrams already on the page. You can add more interesting facts in the blank spaces.
3. Go outside and search for mushrooms. Make sketches of the mushrooms you find. Take notes about where the mushroom is growing. Moist soil? Leaf litter? On a dead log? Use a mushroom guidebook to identify the type of mushroom. If you are in an area where collecting is permissible, collect a few to bring home.

EXPERIMENT: Spores

For this activity, you will need:

- A mushroom, from the wild or from the grocery store
- White paper
- Knife and cutting board
- Jar or bowl large enough to fit over the mushroom
- Artist's fixative spray
- Magnifying glass

Additional Layer

People harvest shelf fungi to use as a canvas for paintings. If you have shelf fungi growing in a forest near you. You can try this too. Shelf fungi are also sold at art supply stores for this purpose.

Fabulous Fact

Ringworm is a common fungal infection among pets and people. It is caused by the *Trichophyton mentagrophytes* fungus, not a worm.

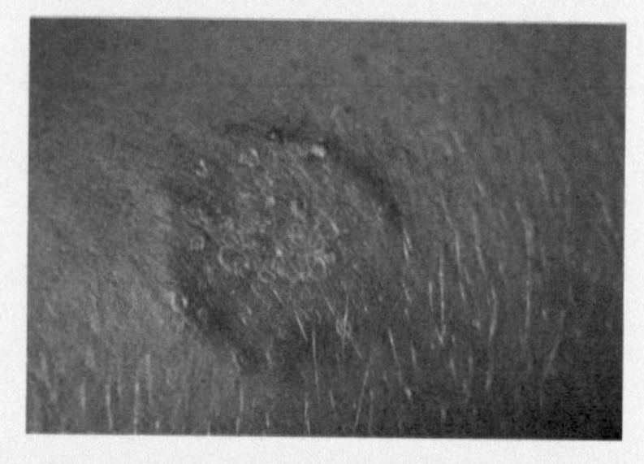

It is usually treated with a topical anti-fungal cream. It spreads easily from person to person and, commonly, from pets to people.

The ringworm fungus hangs out in the soil waiting for contact with an organism. Then you pet your pet and you have ringworm rashes. There's really no way to prevent it, but you can be vigilant about treating it on pets and people the moment it appears.

Learn about the life cycle of this fungus and make an illustrated fact sheet about it.

One method of fungal reproduction is through spores; they can be sexual or asexual. Spores are developed in the gills of a mushroom or on tall stalks of a mold. The spores are distributed by either wind or water.

1. Cut the stalk off a mushroom near the cap. Lay the cap on a sheet of white paper, cover it with an upturned jar or bowl, and let it sit overnight. In the morning you will see the spores have fallen out of the cap onto the paper, making a pattern.

   The exact pattern and color of a spore print is unique to each species and is a key to identifying them.

2. Spray artist's fixative over the spore print to set it in place if you want to keep your spore print.
3. Look at the spores and parts of the mushroom, including the gills, with a magnifying glass.

Additional Layer

Buy a mushroom growing kit.

The mushrooms are edible once you have grown them.

Additional Layer

Mushrooms often form a fairy ring. Find out why.

EXPERIMENT: Grow Mold

WARNING: If anyone in your family has a mold allergy, either skip this or wear a face mask while working on this experiment. Dispose of the mold in the trash, wrapped in plastic or paper to keep the spores from spreading around your house.

For this activity, you will need:

- Piece of bread
- Plastic bag or transparent container with a lid
- Science Notebook
- Colored pencils
- Magnifying glass or microscope

Mold is a multicellular fungus made up of white hyphae threads. It looks fuzzy because of the millions of tiny threads. The spores it produces are different colors, like reds or blues or greens.

1. Put a piece of bread inside a plastic bag and let it sit out on a counter or windowsill until mold begins to grow. Watch the mold growth over several days. Sketch what you see each day. Color and label your drawing.
2. Examine the mold with a magnifying glass or microscope. Draw detailed sketches of your close-up view.
3. Design an experiment to determine the best growing conditions for mold. You can test bread with preservatives vs bread without, moisture or dry, in the light vs in the dark, or different temperatures.

Fabulous Fact

Mold grows best in warm, moist conditions, but it will grow in colder conditions as well. It just takes longer.

Our refrigerator slowed the growth of the mold on these strawberries, but it couldn't stop it entirely.

Drying foods does stop mold entirely, however. Mold requires moisture.

Unit Trivia Questions

1. Which part of a plant provides support?

   Stem

2. Which of these are decomposers?

   a) Oak trees

   b) Shelf fungi

   c) Daisies

   d) Strawberry plants

3. A compact growth on a plant that develops into a leaf, flower, or shoot is called a ________.

   Bud

4. Recite the photosynthesis equation.

   $6CO_2 + 6H_2O \rightarrow C_6H_{12}O_6 + 6O_2$

   (6 Carbon dioxide + 6 Water yields Glucose + 6 Oxygen)

5. How many embryonic leaves do dicots have?

   Two

6. Give an example of a monocot.

   Grass

7. What pigment helps plants absorb light and gives them their green color?

   Chlorophyll

8. What part of a plant produces the seeds?

   a) Roots

   b) Leaves

   c) Stem

   d) Fruit

9. Name the layers of a tree's trunk.

   Heartwood, sapwood, cambium, inner bark, outer bark

Step 3: Show What You Know

During this unit, choose one of the assignments below to show what you have learned during the unit. Add this work to your Layers of Learning Notebook. You can also use this assignment to show your supervising teacher or your charter school as a sample of what you've been working on in your homeschool, if needed.

There are more ideas for writing assignments in the "Writer's Workshop" sidebars.

Coloring or Narration Page

For this activity, you will need:

- "Plants" printable from the Printable Pack
- Writing or drawing utensils

1. Depending on the age and ability of the child, choose either the "Plants" coloring sheet or the "Plants" narration page from the Printable Pack.
2. Younger kids can color the coloring sheet as you review some of the things you learned about during this unit. You might talk about how plants use sunlight to reproduce or about simple, conifer, and flowering plants. On the bottom of the coloring page, kids can write a sentence about what they learned. Very young children can explain their ideas orally while a parent writes for them.
3. Older kids can write about some of the concepts you learned on the narration page and color the picture as well.
4. Add this to the Science section of your Layers of Learning Notebook.

Science Experiment Write-Up

For this activity, you will need:

- The "Experiment" write-up or "Experiment Report Template" from the Printable Pack

1. Choose one of the experiments you completed during this unit and create a careful and complete experiment

write-up for it. Make sure you have included every specific detail of each step so your experiment could be repeated by anyone who wanted to try it.

2. Do a careful revision and edit of your write-up, taking it through the writing process, before you turn it in for grading.

☻ ☻ ☻ Plants Quiz

For this activity, you will need:

- "Plants Quiz" from the Printable Pack (& Answer Key)

Give each student a copy of the quiz to see how much information he or she recalls about plants. You can add or remove questions as needed, depending on what you focused on during the unit.

☻ ☻ ☻ Big Book of Knowledge

For this activity, you will need:

- "Big Book of Knowledge: Plants" printable from the Printable Pack, printed on card stock
- Writing or drawing utensils
- Big Book of Knowledge

1. Color, draw on, write on, or add to each of the Big Book of Knowledge pages you are using. Only add the printables if you learned these concepts during this unit. If there are other topics you focused on, feel free to add you own pages to your Big Book of Knowledge.

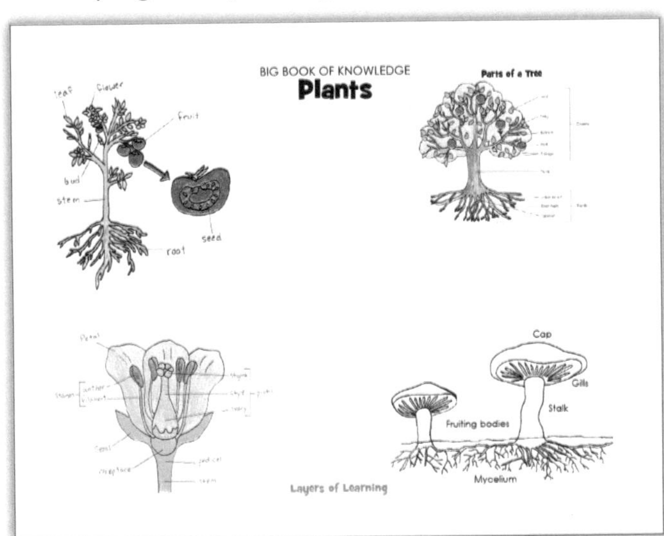

2. Use your Big Book of Knowledge regularly to help you review, quiz, or create games that will help you commit the things you've learned to memory.

Big Book of Knowledge

The Big Book of Knowledge is a book for you, the mentor, to use as a constant review of all of the things you're learning about. You can use it to quiz your kids or prepare tests or review games. Whenever you learn something in Layers of Learning that you want your kids to remember, add it to your Big Book of Knowledge.

Assemble your Big Book of Knowledge in a binder or with binder rings. Divide it into sections for each subject.

In the Printable Pack for this unit you will find a "Big Book of Knowledge" sheet. You can add this sheet to others you collect or create yourself as you progress through the Layers of Learning curriculum. Customize the Big Book of Knowledge to your family by adding facts and topics that you enjoyed exploring as you were learning.

Visit Layers of Learning online to find more information on how to assemble and use your own Big Book of Knowledge.

You will also find cover and section pages to print along with creative games to play with your Big Book of Knowledge to keep school, even the tests, fun!

INVERTEBRATES

The invertebrates group includes all the animals without a backbone. Insects, jellyfish, worms, slugs, lobsters, and squids are all invertebrates. The majority of animals, more than 90%, fit into this group. The insect group alone has more than a million species. There are over 2,000 species of corals and 22,000 different kinds of segmented worms. This group is massively diverse and includes many different body features, life cycles, and habitats.

There are six phyla from this group that we will focus on: Arthropoda, Mollusca, Cnidaria, Porifera, Echinodermata, and Annelida.

Arthropoda is the largest phylum of any animal and includes insects, spiders, crabs, and scorpions. These are animals with an exoskeleton and jointed legs.

Mollusca animals have a mantle with a large body cavity and a shell, either outside for protection or a small inner one that acts as a beak. Mollusks include octopuses, clams, snails, slugs, cuttlefish, and squids.

Cnidaria are jellyfish and coral. They have soft bodies filled with jelly and a life cycle that has a swimming medusa stage and a sessile polyp stage.

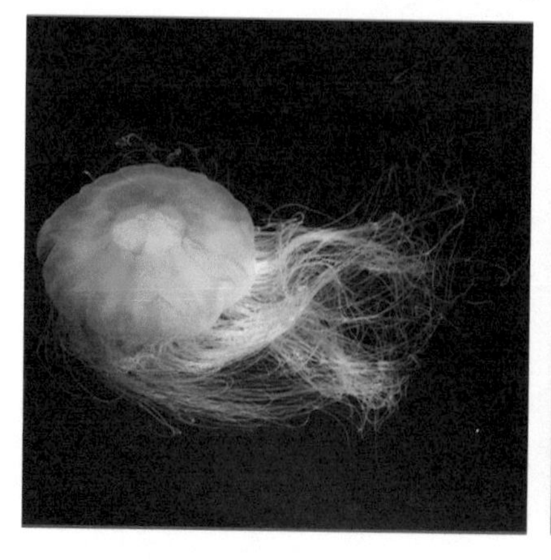

Unit Overview

Key Concepts:
- Invertebrates are all of the animals without a backbone—many sea creatures, insects, worms, and spiders.
- There are millions of invertebrate species and they outnumber vertebrates many times over.
- Many invertebrates have senses, developed brains, complex behaviors, and intelligence.

Vocabulary:
- Invertebrates
- Sessile
- Motile
- Orifice
- Radial symmetry
- Bilateral symmetry
- Endoskeleton
- Exoskeleton
- Segmented
- Head
- Thorax
- Abdomen
- Ventral
- Dorsal
- Anterior
- Posterior
- Inferior
- Superior
- Proximal
- Distal
- Aquatic
- Terrestrial
- Colony
- Symbiotic
- Social insects
- Pheromones
- Metamorphosis
- Migration
- Cephalothorax

Scientific Skills:
- Students should learn to respect the creatures they are experimenting with and observing during this unit.

Porifera includes sponges, a group that has differentiated cells, but no organs, and relies on water flowing through their bodies for food capture and waste removal.

Ecinodermata are animals that have radial symmetry, live in the sea, and can regrow their bodies from small pieces. They include sea stars, sea cucumbers, sea urchins, and sand dollars.

Annelida includes animals with soft, tubular, segmented bodies. They include earthworms, leeches, and tubeworms.

Step I: Library List

Choose books from your library that go with this topic. Here's a list of some favorites and also a list of search terms so you can utilize what your library offers. Read the books with your kids and/or assign them some to read independently. It is from these books your kids will learn most of the facts they need from this unit.

Search for: invertebrates, worms, insects, spiders, crabs, jellyfish, squid, octopus, sea star, tide pools, and other animals you are interested in

☻ ☻ ☻ *Encyclopedia of Science* from DK. Read "Jellyfish, Anemones, and Corals," "Worms," "Arthropods," "Mollusks," and "Seastars and Sea Squirts" on pages 320-325.

☻ ☻ ☻ *The Kingfisher Science Encyclopedia*. Read "Marine Invertebrates," "Mollusks," "Worms," "Crustaceans," "Spiders, Centipedes, and Scorpions," and "Insects" on pages 70-77.

☻ ☻ ☻ *The Usborne Science Encyclopedia*. Read "Body Structure" and "Body Coverings" on pages 300-303, "Animal Senses" on page 320-323, and "Life Cycles" on page 328.

Family School Levels

The colored smilies in this unit help you choose the correct levels of books and activities for your child.

☻ = Ages 6-9
☻ = Ages 10-13
☻ = Ages 14-18

On the Web

For videos, web pages, games, and more to add to this unit, visit the Biology Resources at Layers-of-Learning.com.

You will find a link to video playlists, web links, and more.

Bookworms

If you're looking for a family read-aloud, we'd like to suggest this one.

Project Mulberry by Linda Sue Park is about a young American girl and her science fair project -- growing silkworms like her mother did back in Korea. Is it too strange for an American science fair though?

Fabulous Fact

This butterfly is an invertebrate because it does not have a backbone. Can you think of other invertebrates?

Teaching Tip

Invertebrates are any animals that do not have a backbone, but invertebrates are not actually a taxonomic category. The term is used for convenience in educational courses like this one where we want to group animals.

It does lead to the unfortunate impression that invertebrates are somehow deviants from the "normal" vertebrate animals. However, in the real world, invertebrates are far more normal in terms of numbers of individuals and numbers of species. Because invertebrates tend to live in the ocean or are much smaller than vertebrates, they are not as obvious to people.

Going outside and making a point of seeing many of the invertebrate animals in our world is an invaluable experience for children.

☺ ☺ ☻*Spineless* by Susan Middleton. A coffee table-style book with gorgeous detailed photographs of marine invertebrates. The photos are accompanied with just enough text to spark interest. At the back of the book is a visual index of each species with details.

☺ ☺ ☻*Golden Guide: Insects* by Clarence Cottam. This is a tiny field guide to identify insects of North America. If you live elsewhere, look for a guide in your area.

☺*A Butterfly is Patient* by Dianna Aston. Each 2-page spread is one topic like "A Butterfly Is Creative" and then gives scientific facts about butterflies.

☺*From Caterpillar to Butterfly* by Deborah Heiligman. Explains the life cycle of a butterfly.

☺*An Earthworm's Life* by John Himmelman. Lovely illustrations and sparse text of worms and the things they do.

☺*Spiders* by Gail Gibbons. The perfect amount of facts and text paired with Gibbons' adorable illustrations.

☺*Insect Detective* by Steve Voake. Teaches fun facts about backyard insects and how to find them and the signs they have been there. Gets kids outside and looking.

☺*Big Book of Bugs* by DK. Full-color photos and text with plenty of info for curious kids. Not just bugs, but all kinds of insects and spiders. Mostly focuses on exotic, weird species.

☺*Small Wonders: Jean-Henri Fabre and His World of Insects* by Matthew Clark Smith. A picture book biography of one of the most important naturalists of all time,

☺*Star of the Sea: A Day in the Life of a Starfish* by Janet Halfman. Colorful picture book about what sea stars do, what they eat, how they regrow lost arms, and more.

☺ ☺*Bugs: A Stunning Pop-up Look at Insects, Spiders, and Other Creepy Crawlies* by George McGavin. Full-color insects pop up from the pages full of facts.

☺ ☺*Seashells, Crabs, and Sea Stars: Take Along Guide* by Christina Kump Tibbits. Tells about the animals and where to find them. Includes activities. Look for others in the series.

☺ ☺*Spiders* by Seymour Simon. A favorite science author!

☺*Pocket Genius: Bugs* by DK. A pocket-sized book filled with pictures and facts about insects, from the phylum they belong in to how their bodies are constructed.

☺*The Octopus Scientists* by Sy Montgomery. Gorgeous underwater photos and the perfect amount of text.

☻*How Insects Work: An Illustrated Guide to the Wonders of Form and Function From Antennae to Wings* by Marianne Taylor. Full-color photos. Chapters on evolution, body plans, senses, movement, feeding, respiration, and circulation.

☻*Buzz, Sting, Bite: Why We Need Insects* by Anne Sverdrup-Thygeson. Discusses insect anatomy, mating, and reproduction, and the behaviors of insects.

☻*Octopus: The Ocean's Intelligent Invertebrate* by Jennifer A. Mather, Roland C. Anderson, & James B. Wood. Written by three marine biologists, this book goes into intricate detail about the octopus, its body, and its behavior.

☻*Jellyfish: A Natural History* by Lisa-ann Gershwin. The author has discovered over 200 new species of jellyfish. The book is divided into chapters on anatomy, lifestyle, taxonomy, ecology, and human interactions with jellyfish. All through the book, individual species are highlighted and explained. Thorough and fascinating.

☻*Spiders of the World* by Norman I. Platnick, et. al. Written by six world leading spider experts, this book explains ecology, tons of illustrations, color photos, spider behavior, and more. An excellent reference and browsing book.

☻*Dr. Eleanor's Common Spiders* by Eleanor Spicer Rice. Conversationally takes you on a tour of spiders in the world.

☻*Animals* from Bozeman Science and *Exploring Invertebrates* from Nature League on YouTube. Use these videos as lectures for your high schooler. Have your student take notes from the videos.

Step 2: Explore

Choose a few hands-on explorations from this section to work on as a family. They should be appealing activities that will create mental hooks so your kids remember the information in the unit. Save the rest of the explorations for the next time you do this unit in four years when your kids are older. You can also read the sidebars together and explore some little rabbit trails.

This unit includes printables. See the introduction for instructions on retrieving your Printable Pack.

☻☻☻EXPLORATION: Invertebrate Phyla

For this activity, you will need:

- "Invertebrate Phyla" and the following page of animal cards from the Printable Pack

Memorization Station

Memorize the six invertebrate phyla from this Exploration and what kinds of animals fit into each group.

- Porifera -sponges
- Cnidaria - jellyfish and coral
- Echinodermata - sea stars and sea urchins
- Arthropoda - insects, spiders, and crabs
- Mollusca - snails, clams, and squids
- Annelida - earthworms and leeches

Try creating your own rhyme to help you remember.

Deep Thoughts

Each phyla of animals represents a different body plan. For example, segmentation or having an exoskeleton and six limbs are two different types of bodies. There are currently 36 phyla and therefore 36 recognized body plans.

Think about the six different body plans of invertebrates that are featured in this unit. What makes them different enough to put in separate categories?

Scientists are constantly changing and adjusting the taxonomic categories of life as new knowledge is gained. For example, Linneaus only included two kinds of invertebrates: Vermes (worms) and Arthropods (insects, crabs).

Think about the differences between flat worms and segmented worms. Are those profound enough to be considered different body plans?

Memorization Station

Invertebrates: animals without a backbone

Sessile: fixed in one place, cannot move about

Motile: able to move about

Orifice: an opening in the body from the outside leading inside, like a mouth, ear, nostril, or anus

Radial symmetry: a central axis around which the body is equally arranged

Bilateral symmetry: a body that can be divided into two identical, but mirror image halves

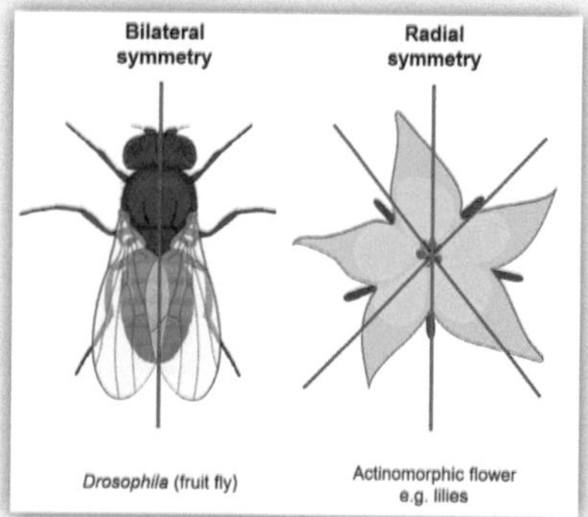

Image by Charl Hutchings, CC by SA 4.0, Wikimedia

Endoskeleton: an internal skeleton

Exoskeleton: a rigid outer covering that protects the bodies of some animals

Segmented: a body formed of distinct parts

Head: The front or upper part of an animal where the brain and sensory organs are concentrated

Thorax: central body cavity of an animal where the organs for circulation and respiration are located

Abdomen: the lower part of the body where the digestive organs are located

- Colored paper
- Scissors
- Colored pencils or crayons
- String
- Glue stick
- Paper plate
- Hole punch

There are over 30 phyla of **invertebrates**, but we are only going to focus on six of them.

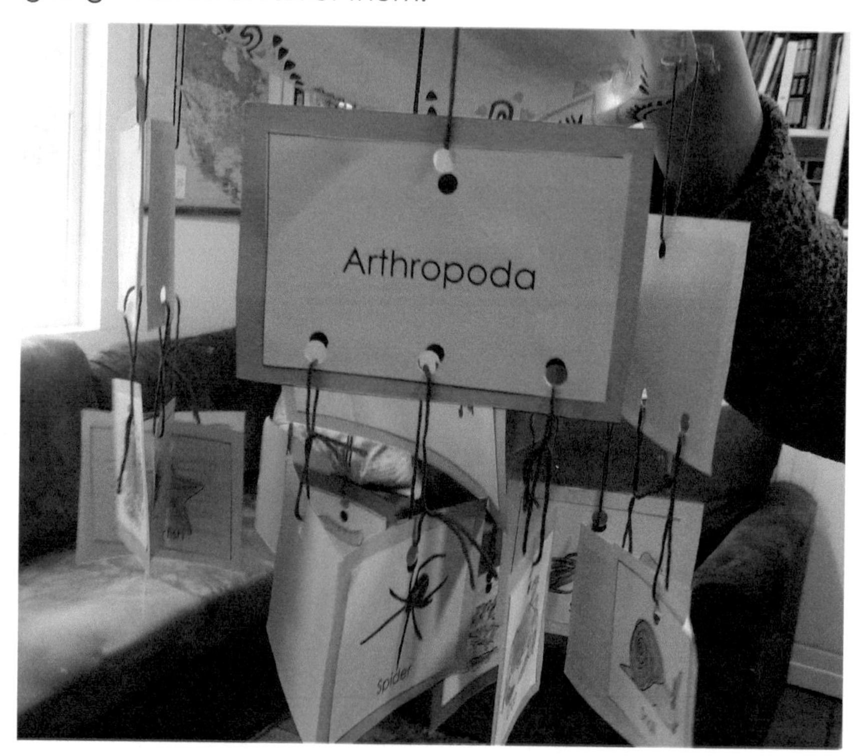

- Porifera are sponges. They have pores and channels through which water circulates, bringing them food and oxygen. They are **sessile** (cannot move themselves about) and have no distinct tissues or organs.
- Cnidaria (the c is silent) include sea anemones, corals, and jellyfish. Their bodies are made of jelly-like mesoglea between two layers of a thin epithelium (skin). There are two forms: swimming **motile** (can move about) medusae and sessile polyps. They have a single **orifice** and cavity for eating and expelling waste.
- Echinodermata are sea stars, sea urchins, and sea cucumbers. They all live in the ocean. They have **radial symmetry** and an **endoskeleton** just beneath the skin.
- Arthrodpoda consists of insects, crabs, and spiders. These animals have a hard **exoskeleton** and a body **segmented** into parts: usually **head**, **thorax**, and **abdomen**. They also have pairs of jointed limbs.
- Mollusca includes snails, clams, and squids. They have a mantle with a large cavity used for breathing and excretion, a shell (either on the inside like with squid or on the outside like a snail), a spiny tongue called a radula, **bilateral symmetry**, and soft unsegmented bodies.

- Annelida are segmented worms like earthworms and leeches. They are bilaterally symmetrical.

1. Use the "Invertebrate Phyla" and the animal cards that follow to make a mobile of the invertebrate groups. Start by coloring the animals. Then cut apart all the rectangles on both sheets along the solid lines.
2. Glue each rectangle to a colored piece of paper. You can use a different color for each phyla or do the whole mobile in one color. Cut out around the colored paper, leaving a margin on each rectangle.
3. Hole punch each piece on the black dots. Hole punch around the rim of a paper plate six times. Tie the phyla around the rim with string, then tie the animals of each phyla to their phylum name. Tie the title to the center of the plate.

Worms & Pill Bugs

Worms are part of phylum annelida and pill bugs are part of phylum arthropoda. We group them here only because we didn't have space for separate sections.

EXPERIMENT: Earthworms and Soil
For this activity, you will need:

- "Experiment" or "Experiment Report Template" from the Printable Pack
- Live earthworms collected from your garden or purchased (bait shops or online)
- 4 different varieties of soil with varying amounts of organic matter (moist loam, dry soil, sand, mud, etc.)
- Medium box - like a shoe box

Earthworms are in the phylum annelida. Annelids have segmented bodies with a digestive tract that runs the length of their bodies from mouth to anus. They also breathe through their skin, which is why they must stay moist at all times.

Remember, these are living creatures and should be treated with respect. Be gentle with them and return them to the outdoors when you are finished experimenting. Also, remember that if worms get dried out, they will die; keep them in a bit of moist soil the entire time.

1. Put four types of soil in four corners of a box. Put earthworms in the box and see which type of soil they go for. Before you release the worms, come up with a hypothesis. Make sure you think through your reasoning.

Bookworms

A House For Hermit Crab by Eric Carle is a charming picture book about a hermit crab who outgrows his house and has to find a new one.

Additional Layer

Worms are any animal with a long cylindrical tube shape and no backbone. Segmented worms (Annelida), like earthworms, are just one phylum. There are also Nemotoda (round worms) and Platyhelminthes (flatworms). Many of these species are parasitic, living in the bodies of other animals, sometimes humans.

This is a giant tubeworm colony under the sea, near the Galapagos Islands. This is Rifitia pachyptila of Annelida.

Look up some round worms or flat worms to find out more.

Memorization Station

When doing dissections, directions may be given about the anatomical location of organs or body parts. Memorize what these mean:

Ventral: the belly

Dorsal: the back or spine

Anterior: the belly or the head (in fish)

Posterior: the back or tail (in fish)

Inferior: lower down or further from the center of the body than another body part

Superior: higher or near the center of the body than another body part

Proximal: where an appendage (leg, arm, tentacle) attaches to the body

Distal: the terminal end of an appendage

Teaching Tip

Dissections can be gross, but they're also important for students because:

- Students get firsthand knowledge of anatomy.
- Concepts of organs and other body parts are best understood in real life.
- The reality of a once-living thing is vastly different from a diagram.
- It gives insight into how human bodies work by comparison or by contrast.
- If approached correctly, dissections should increase awe and respect for life and its intricate balance.

2. Make careful observations and record what you see. Write up your findings on an experiment form.

   What do you know about what earthworms need to survive? Why do you think they chose the type of soil they did?

EXPERIMENT: Earthworm Dissection

WARNING: This activity uses sharp tools and includes powerful chemicals. Children should be taught to use tools correctly and to never eat during this lab. Thoroughly wash hands and all surfaces well after the experiment.

For this activity, you will need:

- "Parts of an Earthworm" from the Printable Pack and the "Parts of an Earthworm Answers"
- Worm dissection kit (Or you can find an online dissection if the gross-out factor is too much. Nothing beats the real thing though!)
- Magnifying glasses
- Gloves
- Goggles

The most common species of earthworm is the *Lumbricus terrestris*, also known as a nightcrawler. They can be found all over the world and are seen on lawns in the night and very early morning. Unlike most worms, they feed mostly on the surface. They also surface during rain if the soil becomes too saturated for them to breathe. They are usually the species in dissection kits.

1. Label and color the "Parts of an Earthworm" sheet. Use the "Parts of an Earthworm Answers" as a guide.
2. First, find all the external parts of the earthworm; use a magnifying glass to look closely.
3. Use the instructions that came with your dissection kit to dissect the worm carefully, finding all the parts.
4. Use a magnifying glass to look closely at the parts of the earthworm.
5. Clean up by throwing the specimen and any disposable items that came with the kit, plus your gloves, into the trash. Carefully clean tools, surfaces, and hands with soapy water.

EXPERIMENT: Worm Senses

For this activity, you will need:

- Live earthworms collected from your garden or purchased (bait shops or online)
- Medium box - like a shoe box
- Soil
- Other materials as desired
- Science Notebook

Invertebrates are often seen as being much simpler than vertebrates. They have less developed brains and bodies and, often, we think of them as having few senses. But is that true?

Remember that these are living creatures and should be treated with respect. Be gentle with them and return them to the outdoors when you are finished with your experiments.

Deep Thoughts

Many people have an ethical problem with animal dissections because they involve the death of animals.

Write up a list of reasons for and against animal dissections for students. Be as honest as you can be.

Discuss whether you think animal dissections are ethical or not.

Teaching Tip

Teach animal dissections along with lessons on the life of an organism and its role in its habitat. This is how respect and awe are fostered instead of callousness.

Bookworms

Wiggling Worms at Work by Wendy Pfeffer teaches how worms benefit the soil.

The Earth Moved: On the Remarkable Achievements of Earthworms by Amy Stewart is perfect for 14 and up.

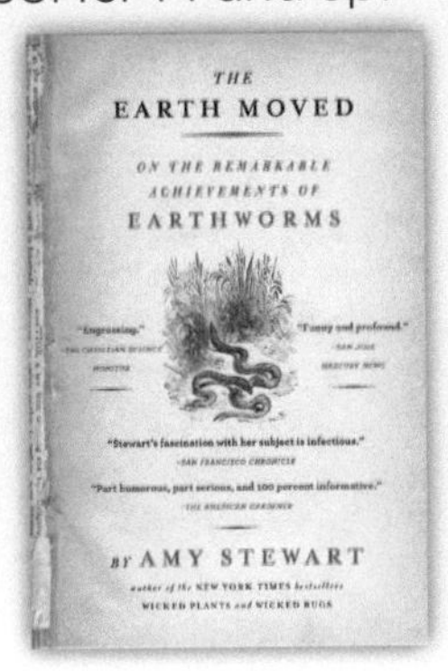

Additional Layer

Animals live in places with the climate that suits them. There are no scorpions in the Rocky Mountains because the habitat is too cold for them.

Think about how climate affects living things.

You can learn more about climate in *People & Planet: North America* and in *Earth & Space: Seasons & Climate*.

Also, remember that if worms get dried out they will die so keep them in a bit of moist soil the entire time.

1. Put a thin layer of moist soil in the bottom of a box and then put the worms you have collected in the box. For each test, draw diagrams of how you set up the test, come up with a hypothesis before you run the test, make careful observations, and run your experiment a few times to be sure the results are accurate.
2. Design an experiment in the box to test for how earthworms respond to light. Hint: use a lamp and a dark space in the box.
3. Design an experiment to test how earthworms respond to touch. Hint: use garden soil in one half of the box and another texture (like sand or mud) in the other half of the box.
4. Design an experiment to test how earthworms respond to sound. Hint: add deeper soil and see if worms come to the surface if sounds are made near the surface.
5. Design an experiment to test if earthworms can taste or smell. Hint: use vinegar in one section of the box to see how earthworms respond to it.
6. Make notes and diagrams in your Science Notebook.

Additional Layer

In Biology: Ecology you will learn more about habitats and how important they are for living things.

EXPERIMENT: Pill Bug Habitat

For this activity, you will need:

- Two small identical boxes - paper milk cartons work well
- Scissors
- School glue
- Rubber bands
- Magnifying glass
- Science Notebook
- Various materials like two different types of soil, small rocks or "logs" or leaf litter (varies)
- Pill bugs

Pill bugs, also known as potato bugs, roly-polies, woodlice, and sow bugs in different parts of the world, are crustaceans, a class of the phylum arthropoda. Other crustaceans include crabs, shrimp, prawns, lobsters, and barnacles. Crustaceans all have a hard exoskeleton, which they have to molt in order to grow. They are different from insects, which are also arthropods, because they have more than six legs. Most crustaceans are **aquatic**, they live in water, but pill bugs are **terrestrial**.

1. Go outside and find several pill bugs. They can be found under logs and rocks pretty easily. Put the pill bugs in a jar. Remember that these are living creatures and should be treated with respect. Be gentle with them and return them to the outdoors when you are finished with your experiments.
2. Examine the pill bugs with a magnifying glass. Draw a sketch of what you see in your Science Notebook.
3. Build a two-sided habitat for pill bugs to see what conditions they like best. Cut the tops off of two boxes, leaving high sides. Cut windows in the two boxes at the same height. Glue the two sides of the boxes with the windows so the windows line up. Wrap a rubber band around the two boxes so the boxes are tightly pressed together as the glue dries.
4. Fill the bottom of the two boxes with soil to the bottom of the windows so the pill bugs can travel between the two sides.
5. Design some experiments using your two boxes, pill bugs, and various materials. You want to test for these types of preferences:
   - Moist versus dry
   - Hiding places versus exposure
   - Light versus dark

   Write down a hypothesis before each trial. Take good notes in your Science Notebook.

Sea Creatures

The sea is full of invertebrates. In the water, there is no particular need for a skeleton to support the body, so creatures without any rigid support structure can thrive.

EXPERIMENT: Sea Star Dissection

WARNING: This activity uses sharp tools and includes powerful chemicals. Children should be taught to use tools

Memorization Station

Aquatic: an animal or plant that lives in the water

Terrestrial: an animal or plant that lives on land

Bookworms

Next Time You See a Pill Bug by Emily Morgan is all about these little arthropods.

Kids learn about the these roly poly bugs and also get some good ideas of ways to observe them in their own back yards.

For ages 5 to 10.

Fabulous Fact

Many sea creatures that are no longer considered fish have the word "fish" in their names in older sources. This is because the definition of fish has changed over the years, but the names stuck. A fish used to be just about any creature that lived in the sea.

Currently, we only call cold-blooded creatures with a backbone that breathe under the water with gills a fish.

Deep Thoughts

Once an old man saw a young man throwing sea stars into the sea. The old man asked why.

"The tide is going out, if I don't throw them into the sea they will die."

"There are miles of beach and millions of sea stars. You can't possibly make a difference," said the old man.

The young man bent down, picked up a sea star, and threw it into the sea, "It made a difference to that one."

Discuss this story and what you think it means.

Memorization Station

Colony: a community of animals or plants of one kind living closely together

This brain coral looks like one organism, but it is actually millions of tiny coral polyps living together in a colony.

Symbiotic: two different organisms living closely together and benefiting one another

Coral and algae have a symbiotic relationship. The algae gets somewhere to live and the coral gets food.

correctly and to never eat during this lab. Thoroughly wash hands and all surfaces after the experiment.

For this activity, you will need:

- "Parts of a Sea Star" from the Printable Pack and "Parts of a Sea Star Answers"
- Sea Star dissection kit (Or you can find an online dissection if the gross-out factor is too much. Nothing beats the real thing though!)
- Gloves
- Goggles (to protect eyes from splashing chemicals)

Sea stars are echinoderms, animals that live in the ocean and have radial symmetry. Most sea stars have five arms, each of them identical. They do not have blood, but instead, circulate oxygen, carbon dioxide, and nutrients around their bodies with a water vascular system.

1. Label and color the "Parts of a Sea Star" sheet. Use the "Parts of a Sea Star Answers" as a guide. Color the digestive system green. Color the water vascular system blue. Color the reproductive organs yellow.
2. First, find all the external parts of the sea star: feet, spines, arms, madreporite, and central disc. Use a magnifying glass to see structures up close.
3. Use the instructions that came with your dissection kit to dissect the sea star carefully, finding all of the parts.
4. Clean up by throwing the specimen and any disposable items that came with the kit, plus your gloves, into the trash. Carefully clean tools, surfaces, and hands with soapy water.

☻ ☻ ☻EXPLORATION: Build A Coral Polyp

For this activity, you will need:

- "How To Draw A Coral Polyp" from the Printable Pack
- Paper
- Colored markers
- Model Magic (from Crayola) or air-dry clay
- Poster paint - brown, tan, orange, and/or green
- Paintbrushes
- World map or globe

Corals look more like plants than animals until you get up close and find that corals are actually made up of a **colony** of millions of individual polyps. Each polyp is an animal that has a mouth, tentacles, and a body.

1. Draw a coral polyp with the "How To Draw A Coral Polyp" from the Printable Pack. The coral you are

drawing is a hexacoral, it has six tentacles. All hexacorals have six (or multiples of six) tentacles in radial symmetry around their bodies. Other kinds of coral are octocorals, with eight tentacles.

2. Make a coral polyp model out of Model Magic. Start by thinning some modeling compound into a sheet and making a tube. Then break the compound into six equal pieces and roll each piece into a tentacle. Attach the tentacles around one end of the polyp. Let the polyp model dry overnight.

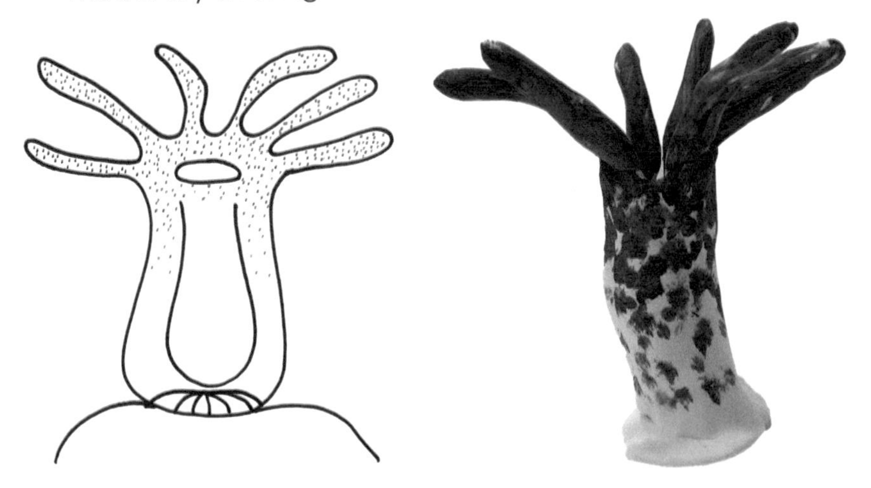

3. Paint your polyp. Most corals are brownish. The color comes from **symbiotic** algae called zooxanthellae. They live in tissues of the polyps and provide up to 90% of nutrients, while the polyp protects the algae and gives it a place to live plus carbon dioxide. When corals lose their algae, they become white and are said to be bleached. They usually die without their algae.

4. Most coral reefs are located between 20° north and 20° south latitudes. The Great Barrier Reef off the east coast of Australia is in this zone as is the Belize Barrier Reef in the western Caribbean Sea. Find them on a map.

   Most species of coral like shallow warm waters with few nutrients. But there are species of coral that live all over the ocean. The ones in the deeper, colder seas have barely been studied and little is known about them.

EXPLORATION: Life Cycle of a Jellyfish

For this activity, you will need:

- "Jellyfish Life Cycle" from the Printable Pack (2 pages)
- Dark-colored paper
- Tissue Paper - various colors
- Clear adhesive contact paper
- Scissors

Additional Layer

Pacific Islanders in countries including the Solomon Islands and Fiji free dive for mollusks such as the *Trochus* sea snail and pearl oysters. The inner shells of these creatures are covered in mother of pearl, which is sold internationally. The mother of pearl is used in making jewelry and buttons.

This is a Trochus shell. Their harvest is unregulated in most Pacific nations and can can bring in more than $1 million in US dollars annually.

Learn more about the Pacific Island people in *Mapping Our World: Pacific Islands*.

Writer's Workshop

Use watercolors to paint your own jellyfish scene.

Over the top of your painting, write a poem about jellyfish that includes these rhyming words and phrases somewhere within your poem:

- Out of sight, light
- Sting, bring
- Wave, save

Fabulous Fact

Octopuses have the ability to edit their RNA, mostly in their central nervous system, in real time. They do this to adapt to temperature changes.

Not much about this is understood as octopuses are only beginning to be studied seriously.

Additional Layer

Cuttlefish are part of the mollusca phylum. They have the most advanced form of camouflage of any animal found so far. They change their skin color, pattern, and texture to match their surroundings and to signal their emotions.

Look up a video of a cuttlefish to learn more about these amazing creatures.

- School glue
- Crepe paper streamers in whatever color you like

Adult jellyfish swim and drift along in the oceans. They produce eggs which are fertilized inside the medusa of the females. The fertilized eggs hatch into free swimming planula larvae which then settle to the ocean floor. The larvae attach to a surface, like a rock or a coral reef, and develop into a polyp. They feed and grow and then, in the spring, they form buds which split off to become ephyra larvae which float in the sea and grow into mature jellyfish.

1. Print the 2 pages of the "Jellyfish Life Cycle" onto dark-colored paper or draw your own outline of each stage onto dark-colored paper.
2. Cut out the center of each shape with scissors. Press the shapes down onto clear contact paper. There will be a window with the sticky contact paper in the center.
3. Cut or tear small pieces of colored tissue paper. Press the pieces down onto the sticky contact paper in the middle of each shape until the contact paper is completely covered.
4. Trim around each shape, cutting through the contact paper and the dark paper so there is a thick border of dark paper around your shapes.
5. Glue long tentacles made of streamers along the bottom edge of the large medusa. Hang all of the jellyfish cycle pieces in a window in order to let the light shine through.

EXPLORATION: Octopus Investigation
For this activity, you will need:

- Video or book about octopuses
- "Octopus Anatomy" from the Printable Pack
- Frozen octopus from a seafood or Asian market
- Large bin filled with water (optional)
- Magnifying glasses
- Science Notebook
- Colored pencils

Octopuses are in the mollusca phylum, along with squids, snails, and clams. They have soft bodies that can squeeze into tiny crevices. All octopuses also have eight arms. They have complex nervous systems and are one of the most intelligent of invertebrates.

1. Read a book or watch a video about octopuses to get some background information.
2. Get an octopus from a seafood market, let it thaw. Put it in a large bin of water to see how it moves. With the help of the "Octopus Anatomy" sheet, examine and identify its body parts: the mantle, the arms, the mouth, the eyes, and the suckers. Use magnifying glasses to get a closer look.
3. Draw a sketch of what you see in your Science Notebook. Make notes around your sketch of things you have learned about octopuses.

EXPLORATION: Crab Habitat

For this activity, you will need:

- Library books, videos, or internet sites on crabs
- Medium box, like a shoe box
- Paint & brushes
- Air-dry clay
- Construction paper
- School glue
- Scissors

Crabs live in every ocean all over the world and there are freshwater species as well. A few types of crabs live most of their lives on land. Most crabs are omnivores, eating algae, mollusks, worms, fungi, other crustaceans, and bacteria.

Additional Layer

Make an octopus hat with construction paper. First, make a band to go around the child's head. Attach a large semi-oval for the mantle and long pieces of curled construction paper for the legs. Glue the front legs up above the child's face.

Read a book or watch a video a second time to find facts about octopuses. Write a fact on each arm of the octopus.

Teaching Tip

You can switch from crabs to a different invertebrate if you can't find information on crabs or if you have an interest elsewhere.

Famous Folks

Crab scientists are called carcinologists. Horton H. Hobbs was a carcinologist who went to school to get a music degree. In a science class, he had to dissect a crayfish and fell in love. He eventually got a PhD and the nickname "Crawdaddy." Learn more about him.

Bookworms

Seashore Life by Lester Ingle and Herbert S. Zim is an excellent and easy-to-use identification guide for shores and tide pools.

Expedition

Natural history museums have displays about animals. They are usually full of preserved creatures propped up in display cases with descriptions and natural backgrounds. See if there is a natural history museum near you to see real specimens.

Additional Layer

Entomologists are scientists who study insects.

These women are capturing and counting butterflies in the Cascade Mountains of Oregon, USA.

1. Research crabs and find out where they live and what they eat. You may want to choose a particular species like fiddler crab, pea crab, or Japanese spider crab.
2. Make a diorama in a shoe box that shows their habitat and food sources. Use paint, paper, air dry clay, or any other materials you like.
3. Explain your diorama and what you learned about crabs in front of an audience.

EXPEDITION: Visit An Aquarium or a Coastal Tide Pool

For this activity, you will need:

- An aquarium to visit near you
- A guide to tide pool animals (optional)

Visit an aquarium or coastal tide pool towards the end of this unit, after you have some knowledge of the animals you will be seeing. Go prepared with one or two good questions to ask staff or to find out from signs or observation.

If you are heading to a tide pool, take along a guidebook to help you identify the species. In an aquarium, the creatures will be identified for you.

Insects

EXPLORATION: Insect Orders Collection

For this activity, you will need:

- Clear jar with a lid or a canning ring and fabric
- Nail and hammer (optional)
- Piece of scrap wood
- Paint and paintbrush (optional)
- Butterfly net (optional)
- Magnifying glass
- Science Notebook
- Pencil
- Colored pencils
- Insect identification handbook for your region

The best way to learn to identify insects correctly is to learn the orders. There are 29 different orders of insects. We are going to learn about eight of them, including the most common insects.

- Coleoptera - beetles
- Diptera - true flies, mosquitoes
- Hymenoptera - bees, wasps, ants
- Lepidoptera - moths and butterflies
- Hemiptera - true bugs like aphids and shield bugs
- Dictyoptera - cockroaches, termites, mantids
- Orthoptera - grasshoppers, crickets

- Odonata - dragonflies and damselflies

In this exploration, you will make an insect collection. Instead of killing the insects and pinning them to a board, you will observe them and draw them while they are alive and then let them go. (If you want to make a permanent collection of insects instead, keep them in a sealed jar in the freezer overnight to kill them).

1. First, prepare an insect viewing jar. You need a transparent jar with a lid. Set the lid on a piece of scrap wood and pound a nail through it in several places to make air holes. Make sure the holes aren't too big. Paint the lid of the jar for decoration if you like. Or, if you have canning jars, use a piece of fabric or screen with the ring to enclose the top.
2. Try to find at least one specimen from each order of insects. Remember to flip over rocks and logs (put them back when you're done) and look under leaves and on the bark of trees. Some insects can be trapped by luring them with bait like sugar, a bit of tuna fish, a dish of water, or light on a dark night.
3. Draw the specimens you find in your Science Notebook and/or photograph them. Use an insect identification book to find out what kind of insect you have. Often you can identify down to the genus, but getting the exact species can be difficult. Do your best. Then let them go.

   Remember that these are living creatures and should be treated with respect. Take good care of them while you study them and return them to their habitat as soon as you are finished.

Memorization Station

Memorize the insect orders. Try writing the orders on one set of cards and drawing a picture of an example insect from the order on another set of cards. Then play matching games.

Famous Folks

Ana Botsford Comstock was an American illustrator who focused on nature and especially insects. She would study an insect under a microscope and then draw it in detail. Her illustrations were used in entomological reports and textbooks.

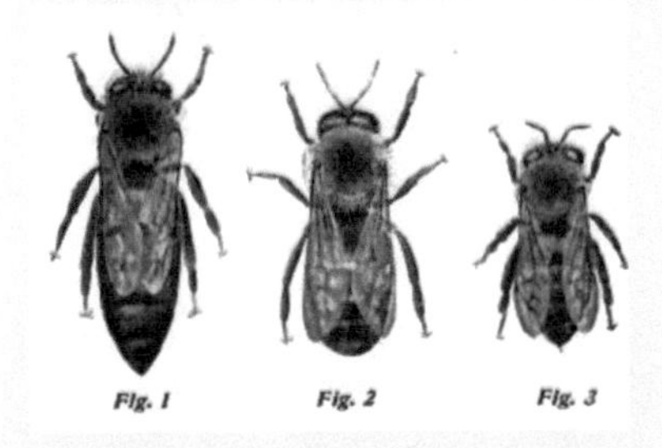

She also believed in students getting outside and doing nature studies and sketches on their own.

On the Web

There are several online insect identification guides including insectidentification.org and bugguide.net for those who live in the US. You can also use Google Lens to identify an insect if you have a good clear picture.

Writer's Workshop

As you observe insects, spend time writing about your observations in your Science Notebook.

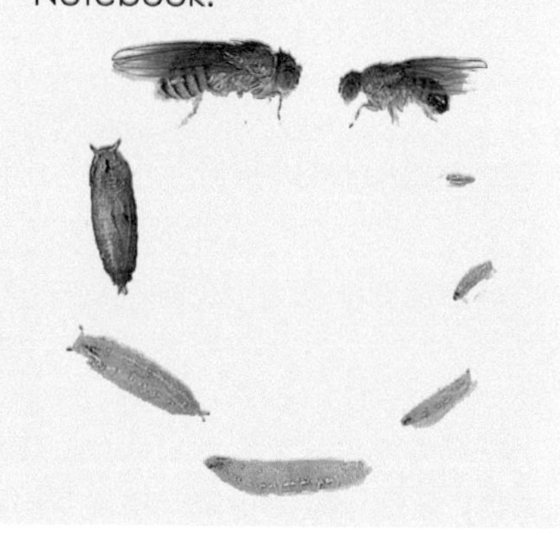

Memorization Station

Throughout this unit, observe insects, draw them carefully in your Science Notebook, then practice labeling the parts—head, thorax, abdomen, antennae, six legs, compound eye, wings—to help you learn them.

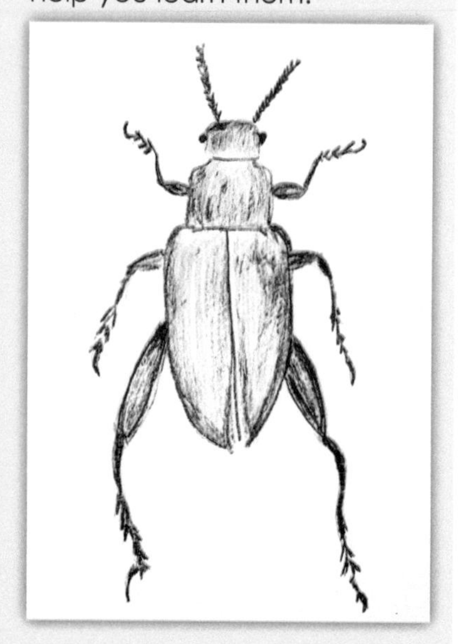

Repeat this again and again using either real insects or photographs of insects as your models to draw. You can even make a quiz game out of labeling the various body parts of insects as you learn them.

EXPLORATION: Fruit Flies

For this activity, you will need:

- Petri dish or jar
- Nylon stocking and a rubber band
- Banana
- Magnifying glass
- Colored pencils or crayons
- Science Notebook

You can observe live insects up close by hatching fruit flies.

1. Read a book or watch a video about insects.
2. Peel a banana and put a piece of it outside in a petri dish or uncapped jar during warm weather. Watch it over several days and when several small fruit flies are seen inside, cover the top with the stocking and secure it with the rubber band.
3. Continue to observe what happens in your jar. You should be able to observe the life cycle - eggs, larvae (maggots), pupae, and then adult flies.
4. Identify the head, thorax, abdomen, exoskeleton, antennae, and compound eyes.
5. Notice needs, behaviors, and appearance.
6. Sketch a fly, label its parts, diagram its life cycle, and write more notes about it in your Science Notebook.
7. Release your flies outside when you're finished.

EXPERIMENT: Grasshopper Anatomy

WARNING: This activity uses sharp tools and includes powerful chemicals. Children should be taught to use tools correctly and to never eat during this lab. Thoroughly wash hands and all surfaces after the experiment.

For this activity, you will need:

- "Grasshopper Anatomy" and "Grasshopper Anatomy Answers" from the Printable Pack
- Pencil
- Colored pencils
- Internet connection
- Grasshopper dissection kit (optional)
- Gloves
- Goggles (to protect eyes from chemical splashes)

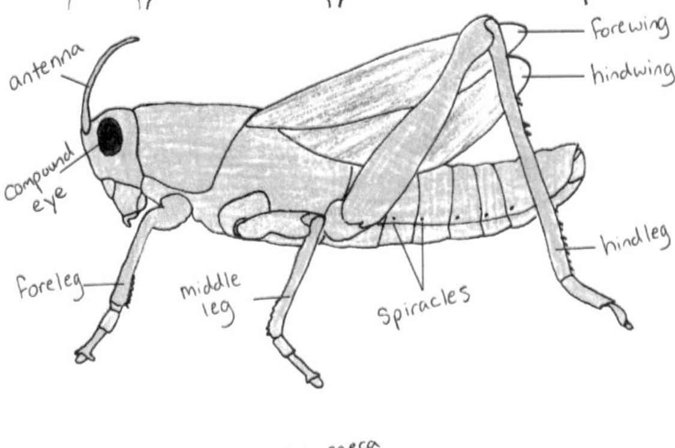

Insects have three basic body parts—a head, thorax, and abdomen—whether it is a grasshopper, ant, or beetle. They also have an exoskeleton, six legs, antennae, and compound eyes.

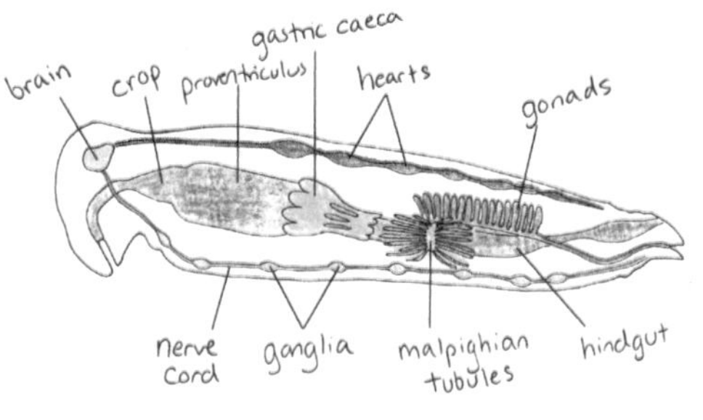

Grasshoppers are in the order orthoptera. They are herbivorous (plant eating). They have powerful hind legs that let them jump far distances to escape being eaten. They hatch from eggs, becoming nymphs, which look like adult grasshoppers except much smaller and lacking wings. They molt five times as they grow into adult grasshoppers.

1. Label and color the "Grasshopper Anatomy" sheet, using the "Grasshopper Anatomy Answers" as a guide. Look up "spiracles" online to find out what they are and why a grasshopper has them.

2. Choose any insect to draw in your Science Notebook and label its main body parts.

3. Optionally, you can use a grasshopper dissection kit to learn more about insect anatomy. Examine the outside of your grasshopper specimen and find all of the parts. Follow the directions in your grasshopper dissection kit to dissect the grasshopper. Find as many of the internal organs as you can. Clean up by throwing the specimen and disposable parts of the kit into the trash. Wash tools, surfaces, and hands with soapy water.

Writer's Workshop

Compare the life cycle of a butterfly with the life cycle of another insect, like a honeybee. What do they have in common? What are the differences?

Deep Thoughts

Many people around the world eat insects as a regular part of their diet. Insects are complete proteins with many vitamins and minerals plus fiber and unsaturated fat.

This is an insect food stall in Bangkok, Thailand. Photo by Takoradee, CC by SA 3.0, Wikimedia.

Have you ever eaten an insect? In the United States, in particular, this is seen as disgusting, but we eat honey, regurgitated bee spit, as a matter of course.

Discuss whether or not you think insects should be a normal part of our diets.

- Are there good reasons for normalizing this food source?
- Are there good reasons for avoiding it?

Famous Folks

Henry Walter Bates was a British entomologist who traveled to the Amazon Basin to collect insects. He collected 14,712 species, 8,000 of which were new to science.

He was especially interested in mimicry, where one animal looks like another to stave off predators.

EXPERIMENT: Moths and Colored Lights

For this activity, you will need:

- 3 lamps
- 3 different colors of light bulb (for example, a normal white bulb, a blue bulb, and a red bulb)
- A stable surface outdoors to set each of your lamps on
- Extension cords to plug in each of your lamps

Moths are in the order Lepidoptera, along with butterflies. You can tell the difference between moths and butterflies by looking closely at their antennae. Butterflies have thin antennae while moths have feathery antennae. Also, most moths are nocturnal (active at night) while most butterflies are diurnal (active during the day).

1. Set up 3 lamps, each with a different color of light bulb outdoors on a porch or in a yard after dark. Make sure the three bulbs are several feet away from each other or run the tests separately.
2. Record how many moths are attracted to each color.
3. Make a graph of the data.

Fabulous Fact

Did you know that no one knows why moths are attracted to porch lights?

There are hypotheses that the moths normally navigate by the moon, but artificial light disrupts and confuses them.

Do you think this sounds like a promising hypothesis? Can you think of any others?

Memorization Station

Social insects: live in colonies, divide their labor, and are all descended from the same queen

Pheromones: substances secreted by a member of a species to communicate with another member of the same species

EXPERIMENT: Insect Groups & Communication

For this activity, you will need:

- A nearby ant colony
- Food source such as syrup or peanut butter
- Small plate
- Ruler
- Magnifying glass
- Insect identification guide for your region
- Watch or timer
- Science Notebook

Bees, ants, and termites, all members of the order Hymenoptera, are **social insects** that live in colonies. Each member of the colony has a specific role—the queen who lays the eggs, those who guard the queen, those who gather food, and others. The members of the colony communicate with one another through smells, or **pheromones**, which they detect with their antennae. For example, an ant who has found food will leave a pheromone trail behind for other members of the colony to follow to the food source.

1. Find an ant colony near your home.
2. Place a small amount of food in the center of the bottom of a small plate approximately 20 cm from the ant colony.
3. Find a place nearby where you can watch the ants but not obstruct their travels to the food source. Draw a map of the ant colony and the food source, including the distance, in your Science Notebook.
4. Observe the ants for one hour. While you observe, take notes in your Science Notebook and sketch the ants or the scene. You may want to observe the ants more closely with a magnifying glass.
   - How long did it take them to establish a trail to the food?
   - Do the ants of the colony all follow the same trail?
   - How long did it take from the time the first ant discovered the food until the second ant followed the trail?
   - What is the weather like?
   - What species of ant are you observing?
   - How long did it take the food to run out?
   - How many total ants visited the food source?
5. Discuss how you could expand this experiment to learn more about ant communication and pheromones.

EXPEDITION: Beehive

For this activity, you will need:

- Local beekeeper or beehive

1. Make arrangements to go visit a local beekeeper and observe the hives. They generally love showing their hives.
2. Go prepared with questions about bees, life cycles, social behaviors and roles, insect anatomy, and more.

Additional Layer

You can build your own temporary ant farm.

Place a smaller jar upside down (with its lid on) inside a larger jar. The purpose is to take up space so the ants will be more likely to build tunnels against the sides of the jar where you can see them. Poke holes with a small nail in the lid of the larger jar. Make sure the holes aren't big enough for ants to crawl through.

Add ants and soil to the large jar with a funnel. Fill it to within an inch of the top of the jar. Place a bit of water on a bottle cap or a leaf, and some food (dead bugs or sugar water) in with the ants. Screw the lid on.

Wrap black paper around the jar and tape it in place. Leave the ants alone for a few days to get settled. Then, take the paper off to observe the ants. Give them fresh water and food regularly.

You can do experiments to see what type of food they like best, how they respond to light, and so on.

Writer's Workshop

Write a descriptive paragraph imagining about what it would be like if you were inside of a chrysalis.

Additional Layer

Insects are an important part of human cultures. Aborigines in Australia eat witchetty grubs, the larvae of the *Endoxyla leucomochla* moth.

Aboriginal boy eating a witchetty grub. Photo by Ed Gold, CC by SA 4.0.

Learn more about the insects used by Aborigines.

You can also learn more about Aboriginal people in *Middle Ages History: Pacific Peoples.*

Fabulous Fact

Honeybees are famous for their highly structured social organization. There is one queen bee, who lays all the eggs and motivates the rest of the hive to do its work.

The workers who gather nectar from flowers, build and clean the hive, and care for the queen are females. The drones, of which there are only a few have one job: to fertilize the queen.

Learn more about hive organization and how bees communicate.

EXPLORATION: Butterfly Life Cycle

For this activity, you will need:

- White printer paper
- Paint in various colors
- Construction paper
- Marker
- Uncooked rice
- School glue
- Small colored pompoms
- Scissors
- Craft stick
- Chenille stems (pipe cleaners)
- Book or video about butterflies
- Butterfly kit (optional)

Butterflies and moths undergo complete **metamorphosis**. This means their body form completely changes through four stages of their lives. First, it's a tiny egg on a leaf. Then, it hatches into a caterpillar that grows and grows, molting its exoskeletons several times. Then, it wraps itself into a cocoon, or pupa, during which its body changes from a caterpillar form to a winged adult butterfly.

1. Watch a video or read a book about butterfly life cycles.
2. Fold a piece of printer paper in half, hot dog-style, the short way. Draw a butterfly shape with the fold at the center of the butterfly, where the body would be. Open the paper again so the crease is in the middle of the paper.
3. Drop blobs of paint across one half of the paper. Don't cross the fold in the middle! Fold the paper in half along the center line again and squish the paint between the two halves of the paper. Open the paper up and let it dry completely for several hours or overnight.

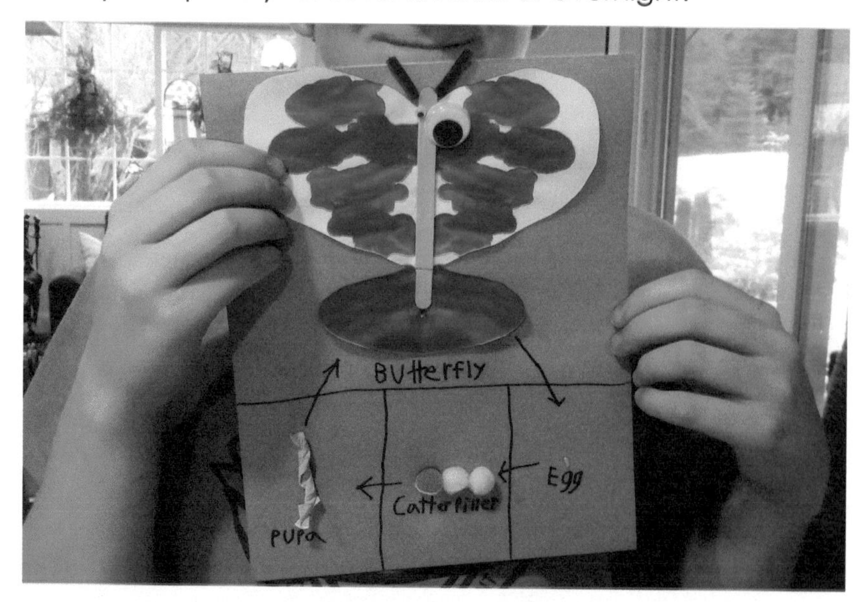

4. Cut out the butterfly while the template is folded to make two identical halves. Glue the butterfly to the

upper part of the paper. Glue a craft stick down the center of the wings to be the butterfly's body. Glue on black or brown chenille stems to be the antennae. Label this section "butterfly."

5. Draw a horizontal line beneath the butterfly on your paper. Then, divide the lower rectangle into three smaller boxes.
6. Glue a grain of rice in the box on the right and label it "egg."
7. Glue colored pompoms in a row to the middle box and label it "caterpillar."
8. Take a small rectangle of green or brown paper and twist the ends opposite directions to make a "pupa" that you can glue on the box on the right.
9. Explain the butterfly life cycle to a friend or family member using your craft.
10. Optionally, you can also observe this entire cycle with a butterfly kit.

EXPLORATION: Monarch Migration

For this activity, you will need:

- Video about the Mariposa Monarch Butterfly Reserve
- "Monarch Butterfly Migration" from the Printable Pack
- Colored pencils or crayons
- Scissors
- Glue stick

Memorization Station

Metamorphosis: a profound body change from the immature stage to the adult

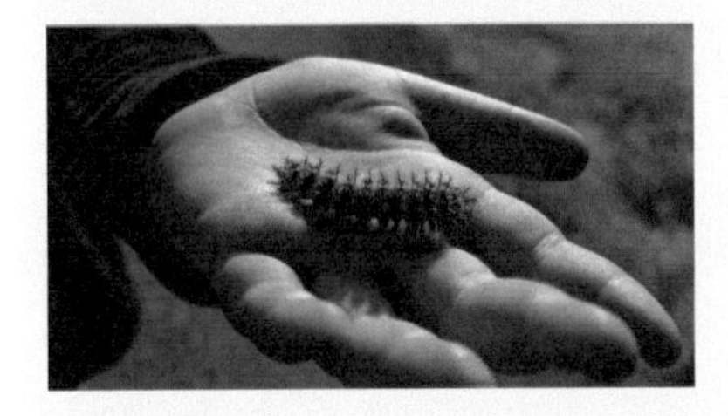

Fabulous Fact

Remember how biological classification is organized?

Here is the organization for the European honeybee:

- Domain - Eukaryota
- Kingdom - Animalia
- Phylum - Arthropoda
- Class - Insecta
- Order - Hymenoptera
- Family - Apidae
- Genus - *Apis*
- Species - *A. mellifera*

Deep Thoughts

Migration is amazing.

- How do the animals know where to go?
- How do they navigate?
- How is this information passed to the next generation?
- How can insects who have never seen the goal, such as monarchs born in the north, return to their homeland in Mexico?

Some insects, like butterflies, locusts, dragonflies, and some beetles **migrate**, or move long distances seasonally in search of food and warm temperatures. Usually, the migrating insects move in swarms or large numbers of individuals. The individuals that migrate south in the winter die before the spring migration which is undertaken by different individuals.

1. Watch a video about the Mariposa Monarch Butterfly Reserve and the migration of monarch butterflies.
2. Color, cut out, assemble, and paste the "Monarch Butterfly Migration" book into your Science Notebook.

Fabulous Fact

Soil mites are tiny arthropods, about the size of a pin head or even smaller. Many are too small to see without a microscope.

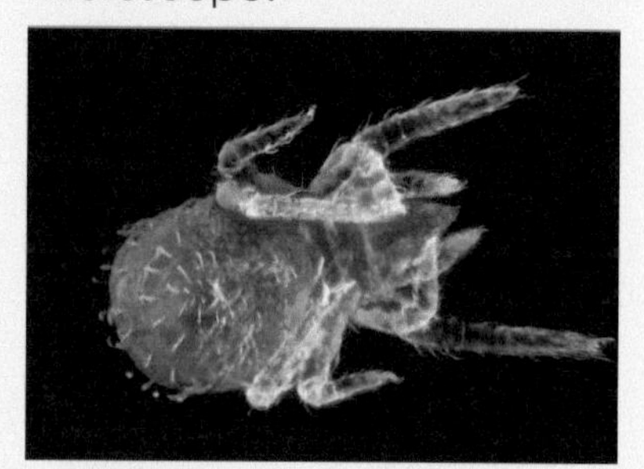

This is a microscope image of a soil mite. Photo by Alexander Klepnev CC by SA 4.0, Wikimedia

They love soils rich in organic matter and are almost always found in compost piles. They have eight legs and are related to spiders and ticks. There are several species of soil mites, most of which are decomposers and are good for plant growth.

Writer's Workshop

Write a what-if journal entry inspired by insect anatomy.

What if you had six legs?

What if you had antennae?

What if you had wings and could fly?

Deep Thoughts

Where is a place on Earth's surface that you think it must be impossible for something to live? Go look up that place and find out if it is indeed barren.

Consider Earth's systems and cycles that you have learned about so far. How does life fit into those systems?

EXPERIMENT: Tiny Life in the Dirt

For this activity, you will need:

- Plastic jug or soda pop bottle
- Scissors
- Screen material or mesh fabric
- Trowel or small shovel
- Soil from outdoors, collected to a depth of 5 cm
- Lamp with a bendy neck (an incandescent light you can direct at the soil - you need both heat and light)
- Magnifying glass
- Microscope (optional)

Tiny invertebrates live in the soil, many of them insects. In the 1880s, an Italian scientist named Antonio Berlese (Ber-lay-zee) developed a way to collect these tiny organisms. Today, we call his contraption a Berlese funnel.

1. Cut a plastic jug or soda pop bottle in half with scissors.
2. Place the jug's narrow end inverted into the bottom end to make a funnel over a basin.
3. Put the screen into the funnel/jug to provide a stable

place for the soil to rest.

4. Place a soil sample into the funnel, resting on top of the screen. Direct a light down on to the soil in the funnel. The heat and light will drive the tiny creatures down into your collection jar. Let the light sit for at least 3 hours, or overnight.

5. Observe and record the creatures you collect. Use a magnifying glass and/or microscope to get a good look. You need to know both the numbers of creatures and the types. Even if you don't know the species, you can give them your own names or identifiers.

   When you are finished, you can let the creatures go back outside.

6. ☻ ☻For older kids: repeat the experiment several times using soil from different sorts of locations—under leaf litter, in a dry area, in a vegetable garden, somewhere that has had pesticide or fertilizer used recently, in the shade, and in the sun, for example. Track how many and which kinds of creatures you find from each location. Do you have a hypothesis you could test further?

Arachnids

☻ ☻ ☻**EXPERIMENT: Life Cycle of a Spider**
For this activity, you will need:

- Book or video about spiders from the Library List or YouTube Playlist
- "Life Cycle of a Black Widow Spider" from the Printable Pack
- Colored pencils or crayons

Spiders are arthropods. They have a hard exoskeleton. Since they have eight legs and no antennae, they aren't insects; they are in their own class: Arachnida. Arachnids

Memorization Station

Migration: the long distance seasonal movement of animals from one location to another

Teaching Tip

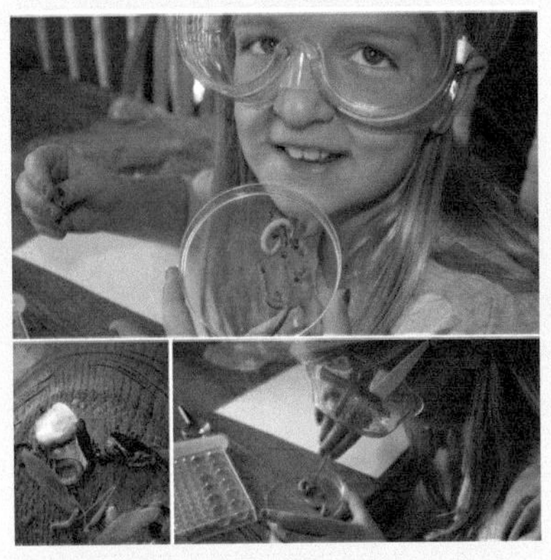

Take a day and just spread out books, science supplies, and plenty of insects to explore.

If possible, take this outside in a natural setting and bring along your Science Notebook. Compare what you find in the books with what you are looking at in front of you and discuss some of the things you have been learning about together.

Some free time coupled with the right supplies and inspiration will often create amazing learning opportunities.

Additional Layer

Make a cute spider with a paper plate, paint, googly eyes, and chenille stems.

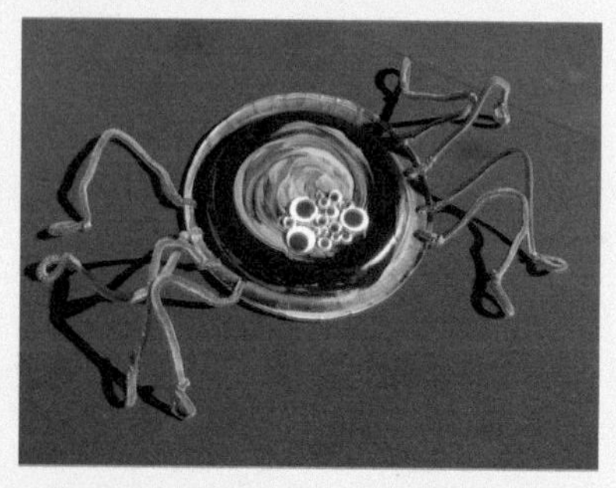

If you stick the eyes in the wet paint, you won't need glue. Punch holes around the rim of the plate to thread the legs through.

Memorization Station

Cephalothorax: the fused head and thorax of spiders and some crustaceans

Deep Thoughts

Have you heard the saying "wouldn't hurt a fly"?

What do you think it means?

Do you think it is wrong or immoral to squish a bug?

Explain your reasoning. Is it the value of the bug's life you are protecting or the innocence of the human's soul? Or both?

are predators and carnivores. They often kill their prey with venom and then spit digestive juices from their stomachs onto the body of their dead prey, dissolving it before sucking up the pre-digested bits.

Black widow spiders mate in spring and early summer. The male is about half the size of the female. Though the black widow does sometimes eat her mate after breeding, this is not always or even usually the case. When the female black widow does feed on her mate it is because she is low in nutrients and needs the energy.

After mating, the female black widow weaves a round or pear-shaped nest of silk into which she deposits her eggs. The web of the black widow is random and messy in appearance, not symmetrical and neat like some spiders' nests. Female black widows can mate and lay three or more egg sacs a season. Hundreds of eggs are laid in each egg sac. The mother black widow guards the egg sac for up to a month until it hatches.

When the baby spiders hatch, they begin to eat one another for nutrients, so only a few from each egg sac survive. The survivors leave the egg sac within a couple of days of hatching. When they leave they often "balloon." The baby spiders spin a long filament of web and it catches a breeze and floats them away from home.

The baby spiders molt several times over the summer and fall. Over winter, they mature and, in the spring, they are ready to find their own mates and start the cycle over again.

1. Watch a video or read a book about spiders. If you'd like, you can choose one specific species to research. We'll used the black widow in this Exploration, but you can select any species you're interested in to learn about as well.
2. Read and color the "Life Cycle of a Black Widow Spider" printable.
3. Add it to the Science section of your Layers of Learning Notebook. Explain in your own words how a black widow grows from an egg to an adult to laying more eggs.

EXPLORATION: Spider Hunt

WARNING: In this activity you are searching for live spiders. Spiders are always venemous, though the vast majority are harmless to humans. If you have dangerous spiders in your area (like widows or recluse), spend time helping your

children identify these spiders before you go out searching so no accidents occur. Gloves and long sleeves/pants are a great idea.

For this activity, you will need:

- "Spider Anatomy" and "Spider Anatomy Answers" from the Printable Pack
- Collection jar (from the "Insect Orders Collection" Exploration)
- Flashlight (optional)
- Magnifying glass
- Spider identification book (spiders are often included in insect identification books)
- Science Notebook

One class of animal inside the order Arthropoda are the Arachnida. Spiders, scorpions, ticks, and mites are all arachnids. Arachnids have eight jointed legs and a body divided into two parts: the **cephalothorax** and the abdomen. The more than 48,000 species of spiders are in the order Araneae.

1. Complete the "Spider Anatomy" sheet. Use the "Spider Anatomy Answers" as a guide. Knowing a little bit about spider anatomy is essential to identifying them. Pay special attention to the number of eyes and the way they are arranged. That is the first step to identification.

2. Go outside and search for spiders. Spiders can be found in bushes or long grass, hiding amongst dead wood, and under porches or up in the eaves of a porch or shed. If you are brave, you can take a flashlight out at night and shine it down at long grass from up by your eyes, looking for shining spider eyes. Catch some of the spiders in collection jars to bring home.

3. Observe the spiders you found closely with a magnifying glass. Use an identification guide to find the species or at least the genus. Sketch the spiders you found in your

Fabulous Fact

Millipedes and centipedes are Myriapoda, a sub-phylum of Arthropoda.

This centipede is the Chinese Red-headed. Centipedes are easiest to find in moist woods, especially in decomposing logs. Many centipedes have a venomous bite that can make you very ill.

Famous Folks

Australian arachnologist, Robert Raven, was just a boy when his father, a mine engineer, told him about spider-infested caves and searing away the webs with a lighted roll of newspaper. "The spiders fall off and down the back of your neck," his father told young Robert.

Robert has been petrified of spiders ever since. He decided to make the study of spiders his life's work in order to face his greatest fears.

Science Notebook. When you are finished observing the spiders, return them to their habitats outside.

EXPLORATION: Spider Webs

For this activity, you will need:

- Video about spider webs from this unit's playlist
- Spray bottle with water
- Camera
- Printer
- Scissors
- Glue stick
- Science Notebook

Not all spiders make orderly webs. This mess is also a typical web.

All spiders spin silk from spinnerets. They use the silk strands to climb, to build nests for their eggs, to trap food, and to store food for later. Among spiders that spin webs, each web design is unique to that species of spider. Webs can be beautiful as well.

1. Watch a video about spider webs.
2. Go outside and search for spider webs. It can be easier to find webs in the early morning when dew has fallen on them, making them sparkle in the sun. If you find a web that does not have dew, spray it gently with a mist of water so it shows up better.
3. Take photos of the webs you find. Did you find more than one style of web? Did you find the spider that made the web? If you are very quiet and watch for a little while, you may get to the see the spider in action, either building a web or hunting for food.
4. Print your favorite photos of the webs and glue them into your Science Notebook. Record the date and location that you found them and identify the genus of spider, if possible.

Writer's Workshop

A lot of people are afraid of spiders. Are you? Write about an animal you are afraid of. Describe in detail what about that animal scares you.

Deep Thoughts

There are more than 45,700 known species of spiders. Scientists have proposed more than 20 different ways to classify these spiders within the Arachnid family since 1900. Taxonomy, or the groups we put living things in, can change depending on which characteristics we think are the most important.

Which characteristics do you think are most important in sorting them? Habitat? Hunting? Mating? Body shape?

Famous Folks

Sir Vincent Wigglesworth was a British entomologist who studied metamorphosis.

Photo CC by SA 4.0, Wikimedia

He discovered obscure things like the hormone that triggers metamorphosis.

Step 3: Show What You Know

During this unit, choose one of the assignments below to show what you have learned during the unit. Add this work to your Layers of Learning Notebook. You can also use this assignment to show your supervising teacher or your charter school as a sample of what you've been working on in your homeschool, if needed.

There are more ideas for writing assignments in the "Writer's Workshop" sidebars.

Coloring or Narration Page

For this activity, you will need:

- "Invertebrates" from the Printable Pack
- Writing or drawing utensils

1. Depending on the age and ability of the child, choose either the "Invertebrates" coloring sheet or the "Invertebrates" narration page from the Printable Pack.
2. Younger kids can color the coloring sheet as you review some of the things you learned about during this unit. On the bottom of the coloring page, kids can write a sentence about what they learned. Very young children can explain their ideas orally while a parent writes for them.
3. Older kids can write about some of the concepts you learned on the narration page and color the picture as well.
4. Add this to the Science section of your Layers of Learning Notebook.

Science Experiment Write-Up

For this activity, you will need:

- The "Experiment" write-up or "Experiment Report Template" from the Printable Pack

1. Choose one of the experiments you completed during this unit and create a careful and complete experiment write-up for it. Make sure you have included every specific detail of each step so your experiment could be repeated by anyone who wanted to try it.
2. Do a careful revision and edit of your write-up, taking

Famous Folks

Thomas R. Odhiambo was a Kenyan entomologist who studied ways to control insect pests without pesticides.

One of the buildings on the ICIPE campus. The organization started out in corrugated buildings that flooded every time it rained. Photo by ICIPE - Nairobi, Kenya, CC by SA 2.0, Wikimedia.

His most important contribution was the establishment of the International Centre of Insect Physiology and Ecology (ICIPE), the Third World Academy of Sciences, the Kenyan National Academy of Sciences, and the African Academy of Sciences. He overcame huge obstacles to achieve a place where young Africans could perform their own advanced research closer to home.

Additional Layer

Learn which dangerous spiders live in your area and how to identify them. Learn what to do if you get bitten. Some spider bites are very dangerous and painful.

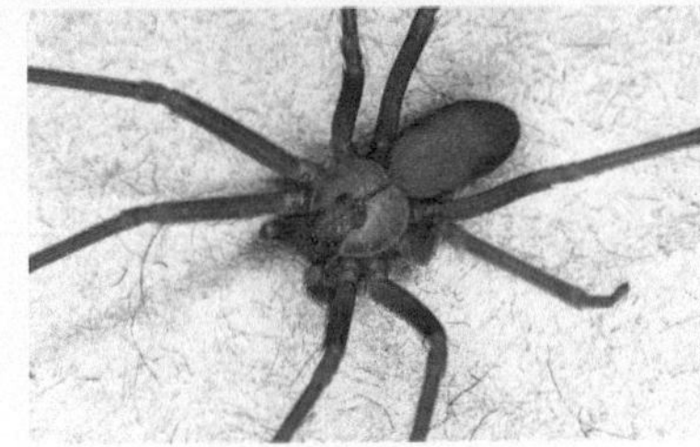

This is a brown recluse.

Unit Trivia Questions

1. Invertebrates do not have:
   a) Eyes
   b) Senses
   c) Brains
   d) A backbone
2. Match the phylum with the animals that fit in it.

Porifera	Snails & squid
Cnidaria	Sponges
Echinodermata	Jellyfish
Arthropoda	Worms
Mollusca	Sea star
Annelida	Insects & crabs

3. Name three invertebrates that live in the sea.
   sea star, octopus, squid, coral, jellyfish, etc.
4. The hard outer covering on an insect is called the _______.
   Exoskeleton
5. Which animal has a radial body plan?
   a) Earthworm
   b) Jellyfish
   c) Grasshopper
   d) Crab
6. Describe the life cycle of one animal you studied in this unit.
   Answers vary
7. True or false - All arthropods live in the sea.
   False. Spiders, insects, and pill bugs are arthropods as well as crabs and lobsters.
8. An insect has ____ legs and three body parts including a ______, _______, and ______.
   6, head, thorax, abdomen

it through the writing process, before you turn it in for grading.

Block Tower Quiz

For this activity, you will need:

- A block tower game
- Permanent marker
- Big Book of Knowledge

1. Write a question mark on one side of about half of the blocks for the block tower game.
2. A student pulls a block from the tower. If it is blank, place it on top. If it has a question mark, the mentor asks a question from your Big Book of Knowledge. If the student can answer the question then he or she keeps the block and play proceeds to the next student. If the student can't answer the question, then the answer is given aloud and the block is placed on top.
3. Keep playing until your block tower tumbles. The student with the most blocks collected wins.
4. You can re-use the game with any subject.

Writer's Workshop

For this activity, you will need:

- A computer or a piece of paper and a writing utensil

Choose from one of the ideas below or write about something else you learned during this unit. Each of these prompts corresponds with one of the units from the Layers of Learning Writer's Workshop curriculum, so you may choose to coordinate the assignment you choose with the monthly unit you are learning about in Writer's Workshop.

- **Sentences, Paragraphs, & Narrations:** Orally describe the most impressive thing you learned in this unit.
- **Descriptions & Instructions:** Write similes about four different animals from this unit. Use the similes in writing four descriptive paragraphs.
- **Fanciful Stories**: Write a fable using some of the animals from this unit as your characters.
- **Poetry:** Pick a phylum, then come up with a list of alliterative words to match the phylum name. Write a poem using your list as inspiration.
- **True Stories:** Write a detailed setting for the true story about the life of an animal from this unit.
- **Reports and Essays:** Draw and label a detailed picture of the external or internal anatomy of one animal from this unit.
- **Letters:** Write a letter to a politician as though you are a specific invertebrate. What law do you most need?
- **Persuasive Writing:** Pick a book you read during this unit. Write a book review about it.

☻☻☻Big Book of Knowledge

For this activity, you will need:

- "Big Book of Knowledge: Invertebrates" printable from the Printable Pack, printed on card stock
- Writing or drawing utensils
- Big Book of Knowledge

1. Color, draw on, write on, or add to each of the Big Book of Knowledge pages you are using. Only add the printables if you learned these concepts during this unit. If there are other topics you focused on, feel free to add your own pages to your Big Book of Knowledge.

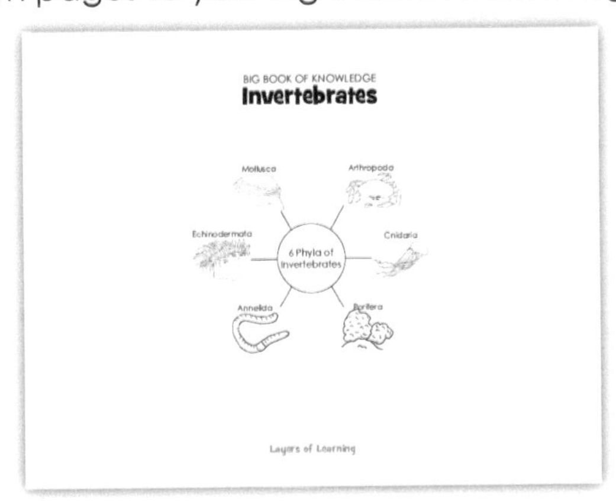

2. Use your Big Book of Knowledge regularly to help you review, quiz, or create games that will help you commit the things you've learned to memory.

Big Book of Knowledge

The Big Book of Knowledge is a book for you, the mentor, to use as a constant review of all of the things you're learning about. You can use it to quiz your kids or prepare tests or review games. Whenever you learn something in Layers of Learning that you want your kids to remember, add it to your Big Book of Knowledge.

Assemble your Big Book of Knowledge in a binder or with binder rings. Divide it into sections for each subject.

In the Printable Pack for this unit you will find a "Big Book of Knowledge" sheet. You can add this sheet to others you collect or create yourself as you progress through the Layers of Learning curriculum. Customize the Big Book of Knowledge to your family by adding facts and topics that you enjoyed exploring as you were learning.

Visit Layers of Learning online to find more information on how to assemble and use your own Big Book of Knowledge.

You will also find cover and section pages to print along with creative games to play with your Big Book of Knowledge to keep school, even the tests, fun!

Unit Overview

Key Concepts:
- Fish, amphibians, and reptiles are all cold-blooded vertebrates that rely on their environments for temperature regulation.
- Fish, frogs, and reptiles all have unique life cycles and body plans. Comparing them is useful in understanding how animals are related and how they differ.

Vocabulary:
- Vertebrate
- Fish
- Amphibian
- Reptile
- Warm-blooded
- Gills
- Spinal chord
- Cold-blooded
- Predator
- Oviparous
- Ovoviparous
- Viviparous
- Life cycle
- Egg
- Fertilization
- Juvenile
- Adult

Scientific Skills:
- Living things need to be treated gently and with respect when studying them or experimenting with them.
- Dissections are important for understanding the way living things work. They should be done alongside the study of the living animals so that respect is increased.
- Experiments on live animals to observe their behavior and life cycle can be done ethically.

COLD-BLOODED VERTEBRATES

Cold-blooded vertebrates are animals with a backbone that use their environment to regulate their body temperature. This includes fish, amphibians, and reptiles.

Fish must live in the water because they breathe with gills, special organs that extract oxygen from water. They can be found in most bodies of water from ponds and streams to nearly every zone of the ocean. There are over 34,000 different species of fish known to science, more than all the other vertebrate groups put together.

Amphibians live part of their lives in water, breathing through gills, and then undergo metamorphosis and develop lungs after which they must breathe air. They return to the water to lay their eggs.

Reptiles live their entire lives as air breathers, but some of them spend a great deal of time in the water as well. They lay their eggs on land and some give birth to live young. They have tough, scaly skin that prevents their bodies from drying out.

Step I: Library List

Choose books from your library that go with this topic. Here's a list of some favorites and also a list of search terms so you can utilize what your library offers. Read the books with your kids and/or assign them some to read independently. It is from these books your kids will learn most of the facts they need from this unit.

Search for: fish, sharks, amphibians, frogs, reptiles, snakes, turtles, lizards, and any specific species that you are interested in.

☻ ☻ ☻ *Encyclopedia of Science* from DK. Read "Fish," "Amphibians," and "Reptiles" on pages 326-331.

☻ ☻ ☻ *The Kingfisher Science Encyclopedia*. Read "Fish," "Amphibians," and "Reptiles" on pages 78-83.

☻ ☻ ☻ *The Usborne Science Encyclopedia*. Read "Moving in Water" on pages 304-305 and "Breathing" and "Internal Balance" on pages 314-317.

☻ ☻ ☻ *Reptiles and Amphibians* by Hobart M. Smith. This is an inexpensive field guide to identifying species of reptiles

Family School Levels

The colored smilies in this unit help you choose the correct levels of books and activities for your child.

☻ = Ages 6-9
☻ = Ages 10-13
☻ = Ages 14-18

On the Web

For videos, web pages, games, and more to add to this unit, visit the Biology Resources at Layers-of-Learning.com.

You will find a link to video playlists, web links, and more.

Bookworms

If you're looking for a family read-aloud, we'd like to suggest this one.

Bartleby of the Mighty Mississippi by Phyllis Shalant is about a pet turtle whose owner accidentally leaves him at a pond. Can Bartleby find food and escape predators as he travels from his pond down the mighty Mississippi?

Bookworms

The Line Tender by Kate Allen is a coming of age novel about a girl whose marine biologist mother died in the course of her work. Lucy has been hanging on ever since, but then the sharks return to Cape Cod and Lucy has to understand her mother's work to save herself and her father. Both funny and moving. For ages 9 and up.

Fabulous Fact

Scientists who specialize in studying fish are called ichthyologists.

Ichthyologists need a bachelor's degree in marine biology, zoology, or a related degree. Most also get a master's degree or a doctorate.

and amphibians in North America. If you live on another continent, find a field guide for your area.

Learn to Draw Reptiles & Amphibians by Diana Fisher. Use this book as practice for when you go out in the field to sketch or to add great images to reports and projects.

Fish Everywhere by Britta Teckentrup. A colorful picture book about fish that live deep, deep in the ocean, on coral reefs, and even in the desert.

Shark Lady: The True Story of How Eugenia Clark Became the Ocean's Most Fearless Scientist by Jess Keating. A picture book about the woman who studied sharks, revealing they were more than vicious man-eating predators.

Manfish: A Story of Jacques Cousteau by Jennifer Berne. A magical picture book about the foremost oceanographer of the 20th century.

Frogs by Gail Gibbons. From habitats to life cycle, this is a perfect beginner science book about frogs.

Sea Turtles by Gail Gibbons. Simple text introduces young children to sea turtle life cycles and the eight living species of sea turtles. Also discusses conservation efforts.

Fish by Steve Parker. This is a DK Eyewitness book, lots of full-color images surrounded by fascinating text.

Awesome Snake Science!: 40 Activities for Learning About Snakes by Cindy Blobaum. Especially for kids who are into snakes, this book has hands-on activities for nearly everything a snake does, from its fangs to the way it moves.

The Magnificent Book of Reptiles and Amphibians by Tom Jackson. An oversized book with a vintage feel. Each double-page spread highlights one reptile or amphibian with a large colorful illustration and facts about the species. Look for others by this author.

Fanatical About Frogs by Owen Davey. Gorgeous illustrated book about frogs from itty bitty ones that can fit on your fingernail to frogs the size of a dinner plate.

Frogs, Toads, and Turtles by Diane Burns. Intended to be a book that you take outside to find wildlife in the wild of North America. Also check out *Snakes, Salamanders & Lizards* by the same author.

Bone Collection: Animals by Rob Colson. Examines the skeletons of dozens of vertebrate animals including fish, reptiles, amphibians, birds, and mammals. Includes two models to build skeletons of an elephant and crocodile.

☻*The Ultimate Book of Sharks* by Brian Skerry. Teaches about dozens of different shark species, shark life cycles, shark anatomy, and shark behaviors. Lots of full-color images and high interest for kids.

☻*Everything You Need to Know About Snakes* by DK. Full of facts and full color images, it covers snake defenses, babies, fangs, swallowing prey, and much more.

☻*Ultimate Reptileopedia: The Most Complete Reptile Reference Ever* by Christina Wilsdon. Lots of full-color images and fascinating facts packed into pages that highlight different species one by one.

☻*Mission: Sea Turtle Rescue* by Karen Romano Young. Explains the life cycle and biology of sea turtles and how they are in danger, then gives kids some things they can do to help.

☻*What A Fish Knows* by Jonathan Balcombe. Explores the way fish, think, learn, experience emotions, have relationships, feel pain, and so much more.

☻*The Field Herping Guide: Finding Amphibians and Reptiles in the Wild* by Mike Pingleton. This is a guide on how to find and observe reptiles and amphibians in the wild. Get this book or one like it if you want to really spend time in nature observing wildlife.

☻*Secrets of Snakes: The Science Beyond the Myths* by David A. Steen. Discusses the biology of snakes and tackles many of the myths and misconceptions around the world's most feared animals.

☻*Lizards of the World: A Natural History* by Mark O'Shea. A book that highlights and photographs lizards from 80 different lizard families. Discusses each lizard and its unique features.

☻*The Field Guide to Drawing and Sketching Animals* by Tim Pond teaches principles of how to draw live animals in the wild, in zoos, aquariums, and other real life situations. If you want to do nature journaling or observe animals closely, you need this book.

☻ *Response to External Environments, Plant and Animal Defense Mechanisms* and *Animal Behavior* from Bozeman Science on YouTube. Use these videos as lectures for your high schooler. Have your student take notes from the videos.

Additional Layer

Fish was a staple food of Europeans in the Middle Ages. Consumption grew so high that kings and lords had to start regulating fishing to prevent the extinction of the fish stocks. Learn more about this time period in *Middle Ages History: High Middle Ages Europe*.

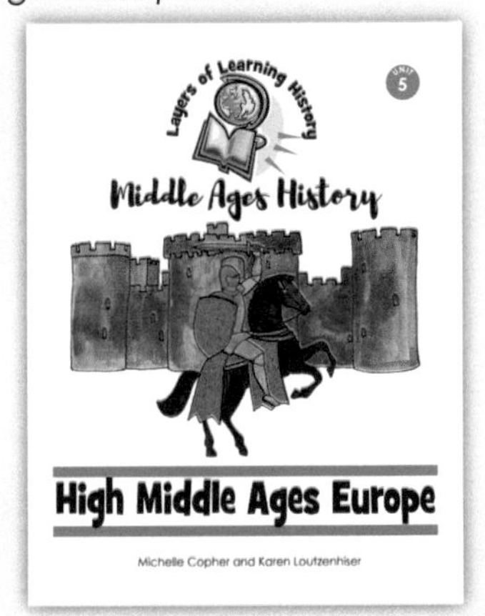

Bookworms

The Shark Caller by Zillah Bethell is about Blue Wing who is grieving for her parents' deaths.

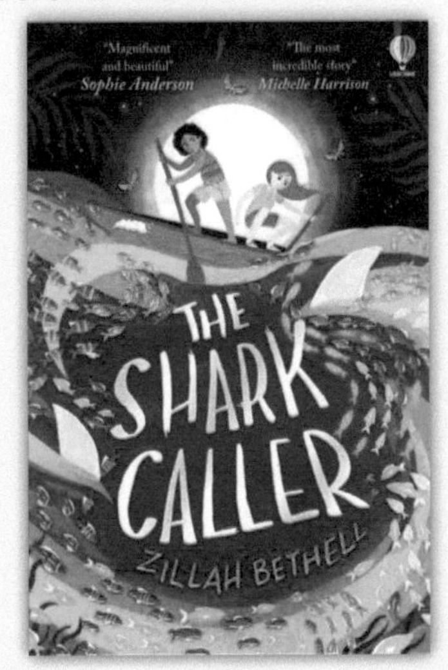

This is a story about friendship, grief, and growing up, set against the sharks and the ocean of Papua New Guinea. For ages 9 and up. A great read-aloud.

Writer's Workshop

Write a fish report. Include a drawing and a chart showing the taxonomic levels your fish belongs to.

Memorization Station

Vertebrate: an animal with a backbone and belonging to the phylum Chordata.

Fish: aquatic vertebrate animal that has gills and no limbs with digits

Amphibian: four-limbed cold-blooded vertebrate that starts life with gills and then develops lungs

Reptile: air-breathing cold blooded vertebrate that has a scaly body

Warm-blooded: ectotherm, an animal that uses metabolic processes to keep its body temperature stable, regardless of the environment

Gills: paired organs on fish and some amphibians that have the ability to extract oxygen from the water

Spinal chord: a bundle of nerve fibers that runs the length of the body in animals and connects the body to the brain

Cold-blooded: endotherm, an animal that has a body temperature that is regulated by the environment or varies with the environment

Step 2: Explore

Choose a few hands-on explorations from this section to work on as a family. They should be appealing activities that will create mental hooks so your kids remember the information in the unit. Save the rest of the explorations for the next time you do this unit in four years when your kids are older. You can also read the sidebars together and explore some little rabbit trails.

This unit includes printables. See the introduction for instructions on retrieving your Printable Pack.

EXPLORATION: Phylum Chordata Poster

For this activity, you will need:

- "Vertebrate Classification" from the Printable Pack plus the vertebrate animal cards that follow
- A large sheet of poster board
- Paints and paintbrushes
- Scissors
- Glue stick

The phylum Chordata includes all the **vertebrate** animals (plus a few others that have spinal chords but no bones). The next major level down is the class level. There are seven classes of vertebrates.

- Agnatha are jawless **fish** without scales, like lampreys.
- Chondrichthyes are fish that have cartilage instead of bones. Sharks and rays fit here.
- Osteichthyes are all bony fish. Seahorses, tuna, trout, salmon, and goldfish are some examples.
- Amphibia are animals that spend part of their lives in the water and part on land. They have to keep their skin moist and must lay their eggs in water. Frogs, toads, and salamanders are **amphibians**.
- Reptilia live their lives on land, have lungs that breathe air, and have thick, scaly skin. Snakes, alligators, lizards, and turtles are **reptiles** in this class.
- Aves have wings, feathers, and lay eggs. This includes all birds. Birds are **warm-blooded,** regulating their body temperature to keep it constant.
- Mammalia includes all the animals that have hair or fur and give birth to live young. Dogs, bats, humans, kangaroos, and lions are examples of mammals. Mammals are also warm-blooded.

All the Chordata together represent well under 10% of all animals on Earth. But chordates tend to be larger and more obvious so they get more attention than the invertebrates. It's important to remember that all the species of life are essential for a healthy planet.

1. Design a poster for the classes of vertebrates. Divide your poster into 8 sections: one for the title and one each for the classes. Paint your poster board background.
2. Cut apart the cards on the "Vertebrate Classification" sheet. Glue the pieces to the poster board, one in each section.
3. Cut apart the animal cards and sort them into the correct classes. Glue them to the poster board in the right places. Use the descriptions of each class, above, as a guide. If you still aren't sure, look up the animal and find out which class it is in.

Fish

A fish lives in water, uses **gills** to breathe, has a **spinal chord**, and is usually **cold-blooded**. There are more than 34,300 species of fish of all sizes, lifestyles, and aquatic habitats.

EXPLORATION: Fish Anatomy

For this activity, you will need:

- "Fish Anatomy" and "Fish Anatomy Answers" from the Printable Pack
- Colored pencils or crayons
- Pen
- A real fish to observe (optional)

Some of the unique features of fish include the swim

Additional Layer

In nature, it seems there are always exceptions. The six living species of lungfish are one of those exceptions.

Photo by George Berninger Jr., CC by SA 4.0, Wikimedia

These fish breathe air with lungs, similar to a land animals, instead of breathing water with gills. But they are still fish. Learn more about their bodies.

Memorization Station

Memorize the seven classes of vertebrates and which animals are in each class.

bladder, the lateral line, and the gills. These parts of a fish make it possible for the animal to thrive in an underwater environment. The fish in the printable is a generic fish. Think about what a generic mammal would look like and you get the idea that this is very general.

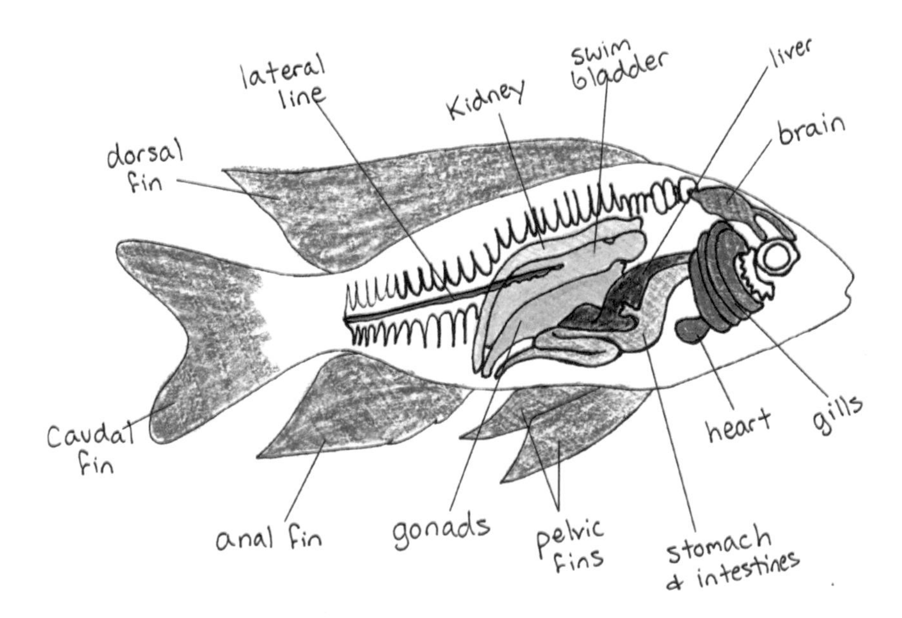

1. Color the gills and the heart red. Label them. Gills are respiratory organs. They contain thin filaments of tissue filled with blood capillaries. The fish bring in water with dissolved oxygen in it through their mouths and then push it out over the gills and out of their bodies.
2. Color the stomach and the intestines pink. Label them.
3. Color the liver brown. Label it. The liver removes toxins from the body.
4. Color the gonads green. If it is a male fish, it will make sperm in its gonads. If it is a female, it will make eggs. Female fish release eggs into the water and then a male fertilizes the eggs with sperm.
5. Color the lateral line orange. Label it. The lateral line is a specialized sense organ that allows the fish to feel vibrations or movement of the water. This is how fish keep track of the rest of their school, know when a predator is approaching, or follow fleeing prey. The lateral line lies just below the skin on both sides of a fish and is often visible on the outside of the fish.
6. Color the swim bladder light blue. The swim bladder is filled with air and allows the fish to adjust its buoyancy in the water so it can stay afloat without much effort.
7. Color the kidney yellow. Kidneys filter waste out of

Memorization Station

Memorize the features that define a fish:

- Backbone
- Gills for respiration
- Fins for movement
- Scales for body protection

Check out the Printable Pack for a sheet featuring these features.

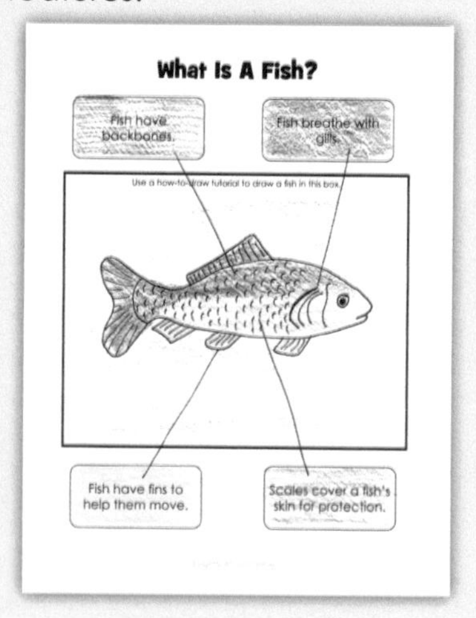

Bookworms

There are many excellent picture books about fish to read during this unit:

- *Swimmy* by Leo Lionni
- *The Rainbow Fish* by Marcus Pfister
- *The Pout-Pout Fish* by Deborah Diesen
- *A Fish Out of Water* by Helen Palmer
- *Fish is Fish* by Leo Lionni

the blood. In fish, the kidney is also very important in regulating the salt to water ratio in the fish's body. In freshwater fish, the kidney makes sure the body retains enough salt; in saltwater fish, the kidney works constantly to expel the excess salt.

8. Color the brain purple. Fish can taste chemicals in the water in the same way humans smell chemicals in the air. In some fish, like sharks, this olfactory portion of the brain is large. The brain is attached to a spinal column that runs the length of the fish.
9. Color the fins dark blue. Label each type of fin.
10. If you have access to a real fish, find some of the features that are visible from the outside. The lateral line, fins, and gill covers can all be seen in most fish.

EXPLORATION: Shark Trivia Golf Game

For this activity, you will need:

- A book or video about sharks
- "Shark Trivia" cards from the Printable Pack
- Large cardboard box, bigger than a shoe box
- Paint
- Tape
- Scissors
- Golf ball (or another small ball)
- Golf club - toy or a real one

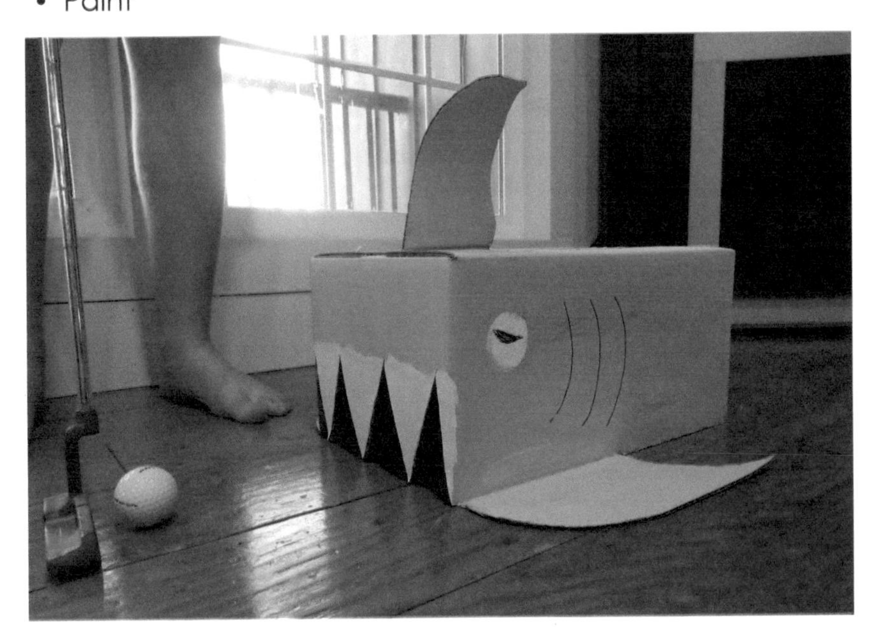

There are over 500 different species of sharks from gentle giant whale sharks to the tiny dwarf lanternsharks to the **predator** great white sharks. Each type of shark has unique life cycles, feeding habits, and territories. There is so much to learn about sharks!

Fabulous Fact

Fish are bilaterally symmetrical, they can be divided into two mirror image halves.

This is a pufferfish in the waters off Hawaii. Photo by Brocken Inaglory, CC by SA 3.0, Wikimedia.

But, as always in nature, there are exceptions. The flounder, which lies flat on the seafloor for camouflage, has both eyes on one side of its body.

Deep Thoughts

Many people are afraid of sharks. Sharks really do attack people sometimes, but shark attacks are even more rare than being struck by lightning.

Are you afraid of sharks? Why?

Some people say shark-phobia is a result of misunderstandings and a lack of education. Do you think this is true?

Does education often contribute to or eliminate fears? Think of examples.

Memorization Station

Predator: an animal that hunts and eats other animals

Oviparous: Eggs are fertilized and develop outside the mother's body

Viviparous: eggs are fertilized and develop inside the mother's body

Ovoviparous: eggs are fertilized inside the mother, but develop on their own with an egg sac

Bookworms

Misunderstood Shark by Ame Dyckman is about a shark who interrupts an underwater broadcast. Is he going to eat everyone? Is he just misunderstood?

Lots of cool shark facts are interspersed with this silly story for ages 3 to 8.

Writer's Workshop

An abstract is an introduction and summary of a scientific paper. Read an abstract from a scientific paper about sharks. Then, write your own abstract on pretend research.

1. Read a book or watch a video about sharks. As you learn, write questions and answers about sharks on the back of the "Shark Trivia" cards. You may need to read or watch more than once or pause frequently so kids can keep up.
2. Make a shark mouth out of a cardboard box. Cut large teeth from one end of the box. Paint the teeth white and the shark gray or blue. Add a cardboard fin to the top of the box and paint on eyes and gill slits.
3. Practice golfing (or rolling) a golf ball through the gaps between the teeth from a set distance away.
4. Play a trivia game. One player putts a ball, trying to get it between the shark's teeth. If they succeed, then they get to be asked a trivia question. If they get the question right, then they keep the card. The players take turns rolling the ball and answering questions. The player with the most cards at the end is the winner.

EXPLORATION: Shark Reproduction

For this activity, you will need:

- "Shark Reproduction" from the Printable Pack
- Scissors
- Colored pencils or crayons
- Glue stick
- Internet
- Video about sharks from the YouTube playlist for this unit

Sharks have three different methods of producing young:

Oviparous	Viviparous	Ovoviparous
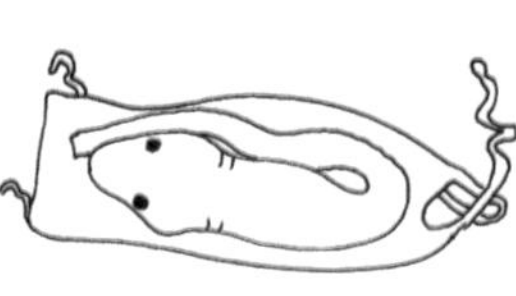		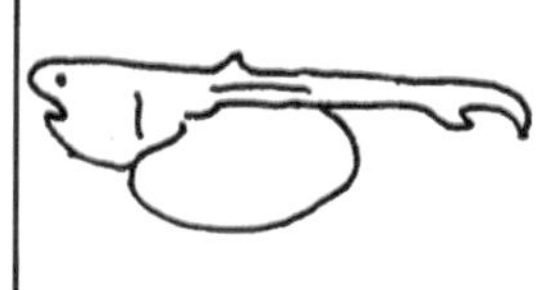
The shark lays eggs in a leathery egg sac called a mermaid's purse where the baby shark is protected and nourished until hatching.	The shark embryos develop inside the mother and are nourished through an umbilical cord.	The eggs are laid internally, the young hatch, and then they have an egg sac that nourishes them until they are ready to be born.

Some ovoviparous shark babies feed on unfertilized eggs or their weaker siblings instead of an egg sac.

1. Cut the labels off the bottom of the "Shark Reproduction" printable. Glue them next to the picture

they match. Color the printable.

2. Under each label, make a list of sharks that reproduce in that way. You will have to research information online.
3. Watch a video about sharks and take notes on the back of the printable.

EXPLORATION: Seahorse

For this activity, you will need:

- Book or video about seahorses
- How-to-draw a seahorse tutorial from a book or the internet
- Pencil
- Colored pencils
- Card stock
- Medium box, like a shoe box
- Paints and paintbrush
- Scissors
- Glue stick

Seahorse heads look sort of like a horse's head, which is where they get their name. They live in shallow oceans from about 45° N to 45°S latitude. There are 45 different species of seahorses.

1. Read a book or watch a video about seahorses.
2. Sketch and then color a seahorse on card stock paper using a how-to-draw tutorial from the internet or a book.
3. Cut out the seahorse, leaving a tab at the bottom to use for gluing the seahorse in the box.
4. Paint the inside of a medium box to look like an undersea background. You can use an image from the

Famous Folks

Eugenie Clark was a marine biologists who focused on sharks. She was known as the "shark lady."

She was the first to artificially inseminate fish, train sharks, and discover a species of sole that produces a chemical to scare sharks away. She wrote books for the public and promoted ocean conservation.

Bookworms

Shark Lady: True Adventures of Eugenie Clark by Ann McGovern is a biography of the most famous shark scientist in the world.

For ages 8 to 13.

internet for inspiration.

5. Add rocks, corals, sea weeds, and other sea life to the diorama, along with your seahorse.

☻☻☻EXPLORATION: Life Cycle of a Trout

For this activity, you will need:

- "Life Cycle of a Trout" from the Printable Pack
- Colored pencils or crayons

Most fish have similar **life cycles**. They lay **eggs** in the water which are **fertilized** after they are laid. Then, the fry hatch from the eggs and are on their own. They grow from fry to juveniles and from **juveniles** to **adults** in a few months. Then, once they are mature, they find a mate and lay and fertilize eggs once again.

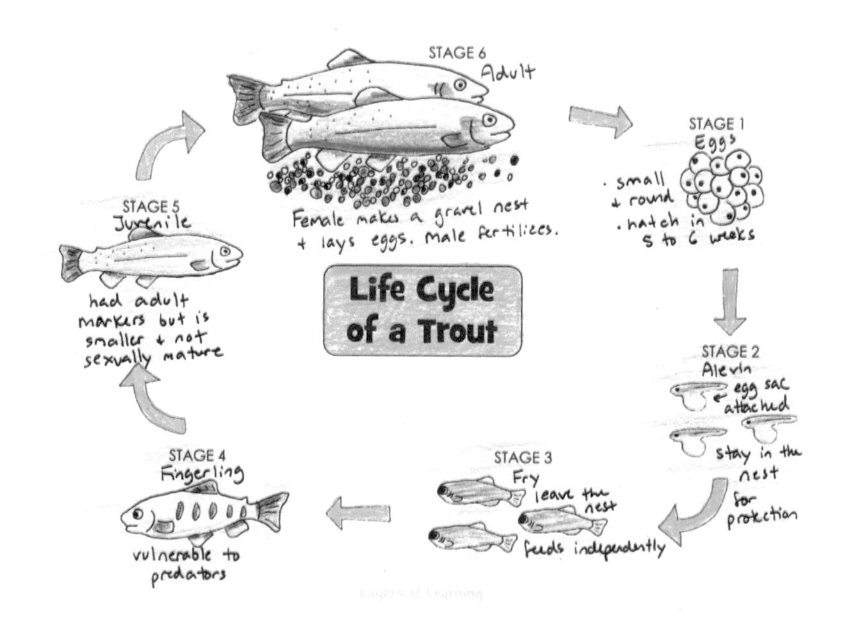

1. Color the stages of the "Life Cycle of a Trout" as you read about each stage. Take notes on the page. Stage one is the egg. Trout eggs are small and round, with a dark spot in them. They hatch in five to six weeks. If the water is warmer, the time period is shorter.
2. Stage two is the alevin. The newly hatched trout has an egg sac attached to its body to nourish it. It stays in the nest of pebbles for another two to three weeks.
3. Stage three is the fry. The yolk sac is absorbed into the baby fish's body and it leaves the nest. It begins to feed from the surface of the water.
4. Stage four is the fingerling. Once a fry has grown to about three inches long, it is a fingerling, but it is still very

Memorization Station

Life cycle: stages of development that a living thing undergoes, including reproduction

Egg: the female reproductive cell in eukaryotic animals and plants

Fertilization: the union of a male reproductive cell, the sperm, with the female reproductive cell, the egg

Fertilization results in the formation of a new individual of the species.

Juvenile: a young and sexually immature individual of a species

Adult: a fully grown and sexually mature individual of a species

Bookworms

Fish For Kids: A Junior Scientist's Guide to Diverse Habitats, Colorful Species, and Life Underwater by Kevin Kurtz begins with the classification and definitions of fish and their anatomy and then has full-page spreads on 35 different species of fish.

Full-color pictures and a fair amount of text on each page. For ages 5 to 10.

vulnerable to predators.

5. At stage five, the trout has become a juvenile. It looks exactly like its parents, with all the same markings, but is smaller and cannot yet reproduce.
6. At stage six, the trout is an adult. It is able to reproduce. When the trout is ready to mate, it turns bright colors. The female makes a nest in the pebbles at the bottom of a stream or lake. The nest is called a "redd." The female chooses a mate and lays her eggs in the nest. Then, the male releases his sperm over the eggs. The pair move on and make another nest further along the bed. A female lays about a thousand eggs a year.
7. Put the "Life Cycle of a Trout" into your Layers of Learning Notebook in the Science section.

Famous Folks

Austrian Hans Hass made underwater documentaries that made octopuses, stingrays, sharks, and coral reefs popular. He was one of the first to speak up for the environment.

EXPERIMENT: Fish Tank Tests

WARNING: Setting up a fish tank is a long-term commitment. Never release fish, amphibians, crustaceans, or aquarium plants into the wild. Even if the species is native, aquarium life often includes diseases that can infect and destroy native wildlife.

For this activity, you will need:

- Fish tank and supplies—a net and food and any decorations you like
- 2 fish bowls or large clear containers
- 10-12 goldfish, guppies, or another fish from a pet store
- Fish tank heaters (optional)
- Various optional supplies from each test below
- Science Notebook

Additional Layer

Turn your fish tank into an ecosystem.

Let the tank get natural light and grow some algae. Install a freshwater filter and keep the pH balanced to be close to natural water. Add (clean) real pebbles and sand to the bottom of the tank. Add live plants to your aquarium. Add microorganisms like daphnia and planarians. Include several species of fish and perhaps snails, frogs, clams, or crabs to the tank. Include fish that naturally clean the tank, like bottom feeders.

Ask an expert at a pet store for more help in finding the right species of plants and animals to create a balance in the tank.

Fabulous Fact

All vertebrates develop from eggs and all vertebrates sexually reproduce. Sexual reproduction means an egg and sperm from two different individuals unite. Both a mother and father are necessary for the next generation to be born.

A few animals can change their sex if the environment demands it. Clownfish do this regularly. When the dominant female fish dies, the highest ranking male fish changes its sex and becomes the dominant female.

But natural sex changes are rare and happen in only a few species.

Famous Folks

Georges Cuvier was a French naturalist who wrote a 22-volume compendium with illustrations and descriptions of all the fish known at the time (1828-1829). Nearly half of the fish in the volume were new to science.

Scientists have performed experiments that show fish can communicate, learn, hear, see as well as humans, respond to their environments, and feel pain. Still, fish brains are much less developed than a mammal brain, and there are still many questions about the capacity of fish to think and be self-aware. There are ways to test intelligence and responses of animals that do not hurt the animals. Below, you will find a series of ideas for tests you can set up and perform on fish. You will need two tanks for most tests so you can have a test subject and a control. You can choose among the tests or design your own.

1. Can fish learn and remember? Lower two different colored containers (like thimbles painted red and blue) into the tank at feeding time each day. One container should have food and the other should be empty. Repeat this for at least a week, lowering the food into different locations in the tank so the color is the only constant. Then lower the containers in without any food. Do the fish still respond to the color that had the food all along?

2. Does the color of light affect fish behavior? Place fish in two different tanks or bowls. Give each tank a different light bulb color (LED, incandescent, black light) Observe the behavior difference in the two fish tanks. Are the fish in one tank more active than the other? Do the fish in one tank swim more at the bottom of the tank than the other? Do fish interact more with each other in one of the tanks? Do the fish eat faster in one of the tanks? Keep feeding and temperature the same.

3. Do fish adjust their respiration depending on temperature? Place fish into two different tanks or bowls. The tanks should both be at room temperature. Test fish breathing rates. Count breaths per minute several times a day. Then raise the temperature of one tank 5 deg with a tank heater. Check breathing rates again. Does it change? Try raising the temperature 5 more degrees. Do fish breathing rates change again?

4. Is a fish aware of itself as an individual? Take one fish out of the tank and put it in a bowl by itself. Let it get used to its environment for several hours. Then attach a mirror to the outside of one side of its tank. Does it respond to itself? What kinds of behaviors do you notice?

5. Does your fish respond to its environment color? Take several fish out of the tank and put them in another tank or a bowl. Put a blue background on three sides of the bowl. Feed and care for the fish as normally for several days and see if the fish respond to the new environment.

Do they change color? Does their behavior change? Repeat with other colors.

6. Do fish respond to sound? Put several fish in another tank or bowl and move it to a quiet room where there is no sound. Leave the main tank in a room where it will have normal sound; you can even play music or an underwater soundtrack for the fish. Does the sound level affect the fish behavior?

☻☻☻EXPEDITION: Pond Trip at Night

WARNING: It is illegal in some places to use a light to attract fish for the purpose of harvesting them.

For this activity, you will need:

- Flashlight
- Zipper plastic bags
- Light rope
- A nearby pond or lake
- Guidebook for local fish
- Science Notebook

People who study wild animals, such as fish, spend much of their time outdoors doing original research. This activity is designed to help you learn some research techniques, like trapping or attracting wild animals so you can study them. Think about how you could use the light trap to do research on the wildlife in your local pond or stream.

1. Seal a flashlight in a double thick layer of clear zipper bags, then tie a rope around the light.
2. Go to a pond, lake, or river at dusk or night. Lower the light into the water. Wait for creatures to be attracted to the light, sketch what you see, and identify the creatures if you can.
3. Take notes and make sketches in your Science Notebook.

Famous Folks

Jacques Cousteau is the world's most famous marine biologist. He co-invented the aqua-lung SCUBA device, wrote popular books, and hosted a documentary television series called *The Undersea World of Jacques Cousteau.*

He brought the depths of the ocean to people for the first time and the world hasn't been the same since.

Additional Layer

Many fish are beautiful and often kept by people as a sort of living art.

This is a mandarinfish or Synchiropus splendidus. Photo by Luc Viatour CC by SA 3.0, Wikimedia

Find an image of a fish you think is beautiful and sketch it in your Art Sketchbook. Then go look at *Goldfish* by Henri Matisse.

Amphibians

Amphibians are born and live their youth in the water; in adolescence, they develop lungs and breath air. They always go back to the water to reproduce, and most live their lives near and in the water even though they are breathing with lungs.

Fabulous Fact

A frog's tongue is attached at the front of its mouth instead of at the back. This allows it to more easily catch its food. Frogs and toads will only eat food that is alive, so they have to be quick. They have sticky tongues to help with this. Their tongues are also really strong. Frog tongues can pull in insects that weigh more than the frog weighs.

Writer's Workshop

Frogs and toads are in the same taxonomic group. The are only separated in folk culture by the texture of their skin and main habitat. Frogs are more likely to be in water and toads on land, for example.

Make a Venn diagram that compares and contrasts frogs and toads. Write their distinct characteristics in the outer circles and the things they share in common in the center.

EXPERIMENT: Frog Dissection

WARNING: This activity uses sharp tools and includes powerful chemicals. Children should be taught to use tools correctly and to never eat during this lab. Thoroughly wash hands and all surfaces after the experiment.

For this activity, you will need:

- "Frog Anatomy" from the Printable Pack
- Frog dissection kit from a science supplier
- Disposable gloves
- Goggles

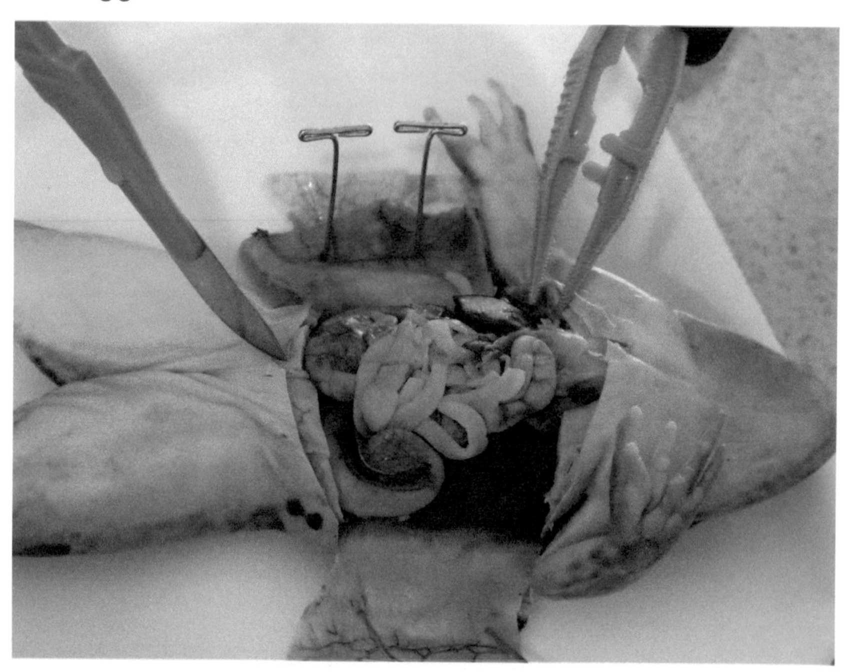

Frogs are amphibians with short bodies and no tails. There are over 6,300 identified species of frogs and probably many more that have never been described by science. They are in the order Anura, along with toads.

1. Label and color the"Frog Anatomy" printable using the "Frog Anatomy Answers" as a guide.
2. Examine the outside anatomy of your frog specimen including the typanic membrane, which allows the frog to hear, the eyes with their membrane coverings,

and the inside of the frog's mouth. How is the tongue attached?

3. Dissect a frog using the kit and guide from a science supplier. Identify as many parts as you can.

EXPLORATION: Life Cycle of a Frog

For this activity, you will need:

- "Frog Life Cycle" from the Printable Pack
- A book or video about frog life cycles
- Colored pencils or crayons
- Scissors
- Colored paper
- Glue stick
- A pond to search for frog eggs (optional)

Amphibians lay eggs, called spawn, in water. Their spawn are coated in a kind of jelly as a protection from bacteria and other animals. Soon, the eggs hatch into tadpoles. Tadpoles look very different from their adult parents and breathe underwater using gills. They feed on aquatic plants and then, later, on small aquatic animals as they grow. Front legs begin to form and their tails grow shorter and shorter. Eventually, their tails disappear. At the same time their outside appearance is changing, they are also changing on the inside. Internally, they are developing lungs to breathe with instead of the gills they were born with. They can live both in and out of the water at this point in their life cycle. All amphibians eventually come back to the water to lay eggs and continue this life cycle again.

1. Watch a video or read a book about the life cycle of a frog.

2. Color the "Frog Life Cycle" squares. Then, cut them out on the solid lines.

Bookworms

Frog and Toad Are Friends by Arnold Lobel is a childhood classic and a sweet story about friendship.

There are four books in this easy reader series for kids from ages five to seven.

Deep Thoughts

In order to study animals it is usually necessary to disturb them, capture them, and sometimes even kill them.

- What ethics do you think should be involved in studying animals?
- How do you balance the gathering of information with the well-being of the animals?
- Can you think of any kinds of research that might be worth it even if some animals are harmed?

Writer's Workshop

Try this story starter about a frog or a toad:

Everything was quiet at the pond until...

3. Glue each square to another piece of colored paper to make a backing. Sort the life cycle squares into the correct order and explain the frog life cycle. You can keep your cards in a paper pocket in your Science Notebook.

4. In the early spring, search at a local pond or a quiet place in a stream for frog eggs. Examine the eggs closely and listen and look for adult frogs in the area.

5. Every few days, or once a week, go check on the eggs and see how they are growing. You should be able to watch the different stages of the life cycle.

EXPLORATION: Frog Facts Beanbag Toss

For this activity, you will need:

- "Frog Beanbag Pattern" from the Printable Pack
- "Frog Facts" (2 pages from the Printable Pack plus a blank page to write your own facts)
- Green fabric or felt - .25 meters (1/4 yard)
- Sewing pins
- Googly eyes or fabric marker for eyes
- Hot glue gun and glue
- Sharp sewing scissors
- Needle and thread, sewing machine, or no-sew fabric glue
- Beans or rice
- Blanket for a "pond"

1. Cut out the "Frog Beanbag Pattern" and pin it to a piece of fabric. You will need two fabric frogs (top and

Bookworms

Fanatical About Frogs by Owen Davey is a child's guide to animals. Filled with wonderful illustrations and plenty of text to make it much more than a simple picture book. For ages 5 to 8.

Teaching Tip

If you want a beanbag shortcut, just fill a plastic zipper baggie with beans or rice, slip a picture of a frog into the front and zip it up.

Additional Layer

In the Middle Ages, the idea of a witch or magician being able to metamorphose people into frogs became a part of folklore.

Read the tale of the "Frog Prince" by the Brothers Grimm and think about why animal transformations were so horrible to medieval minds.

bottom), so you can either cut a double layer of fabric or cut the pattern out twice.

2. Sew the two fabric frog pieces together, wrong sides facing out until you have just a little gap left.
3. Turn the fabric frog right side out. Fill the frog with beans or rice. Then, sew up the last little gap. Hot glue googly eyes to the head.
4. Spread a blanket on the ground to make a pond. Spread the "frog facts" lily pads on the blanket.
5. There are two levels of play. Level one—kids try to hit a lily pad with the frog beanbag and the fact is read to them. Level two—instead of reading the fact, rephrase it as a question for the child to answer.
6. There are blank lily pads for older kids to come up with their own facts about frogs to be used in the game.

Fabulous Fact

Frogs and toads hibernate all winter in burrows. Their breathing slows way down and they enter a sleep state. They live off the fat from their bodies during this time.

This is an American toad.

EXPERIMENT: Thin-Skinned Amphibians

For this activity, you will need:

- 2 hard boiled eggs
- Food coloring
- Beaker or jar large enough to fit both eggs
- Water

Amphibians have thin skin that allows air and liquids to pass through so they can get more oxygen and water. This permeable skin makes amphibians sensitive to pollution, which is why some species are endangered.

1. Fill two bowls three quarters full of water. Put 5 drops of food coloring into each bowl.

Fabulous Fact

The axolotl, sometimes called the walking fish, isn't a fish; it's an amphibian. It reaches adulthood without losing its gills or its tail, but it does grow legs.

The axolotl is native to Mexico and is critically endangered in the wild.

Writer's Workshop

Frogs and toads drink water through their skin instead of their mouths. If you could absorb the house you lived in to eat or drink, what food or drink would you wish your house were made of? Do a 5-minute quick write about it.

Famous Folks

Tyrone Hayes did research into the herbicide atrazine and discovered it is an endocrine disruptor that makes male frogs turn into female frogs. He advocates for the reduction of herbicides and pesticides in agriculture both for frog and human health.

Photo by Earl Neikirk, CC 2.0

Fabulous Fact

Reptiles lay amniote eggs. This means the eggs are complex, having an amniotic and several other membranes that surround the developing fetus and take part in its development. Amphibians and fish do not have amniotic eggs, but reptiles, birds, and mammals do.

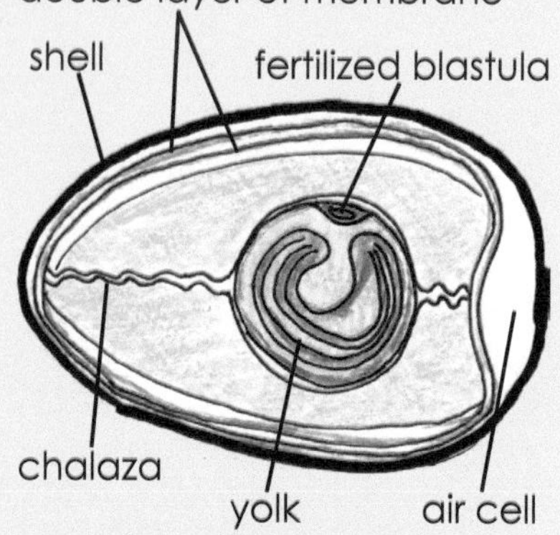

Reptile eggs can survive in the air because they have a rubbery outer shell that protects them from drying.

2. Peel one of the eggs, so only the thin membrane is left over the egg. Leave the other egg unpeeled. Put both eggs into the water. Leave them overnight.
3. Remove the eggs from the water and dry them off. Peel the second egg. Cut both eggs in half. Examine the eggs. The unpeeled egg is like the frog with thin, permeable skin. Water and air can get in, but so can pollutants.

EXPERIMENT: Toad Habitat

For this activity, you will need:

- Clay flower pot
- Hammer
- Acrylic paints (not water soluble)
- Paintbrushes
- Small spade
- Shallow dish, clay saucer, or pie pan
- Outdoor location in a yard or garden

Toads are really just frogs that like to spend more time on land and have warty skin. They like to live in gardens where there are lots of bugs and some water to drink and sit in. But they need to feel safe and sheltered, so they usually won't move in unless they have a home.

1. Paint a flower pot anyway you like it. Let the paint dry completely overnight.
2. Find a location near water and next to or underneath bushes or tall grasses where the toad will be shaded. Break a small hole in the rim of the pot with a hammer. This will be the entrance your toad will use to get into and out of his or her house. Place the pot upside down

on the soil. You might want to dig the rim of the pot into the soil just a little so it will stay put.

3. Dig another shallow hole near the pot for your saucer to fit in. The rim should be level with the surrounding soil. Fill the saucer with water. Refill it every few days so there is always water.
4. Keep an eye out to see if a toad moves in. Observe from a distance to see what the toad does, what he or she eats, how he or she sounds, and so on.

Reptiles

Reptiles live their entire lives as air breathers. They have thick skin that helps prevent moisture loss, and they lay eggs to reproduce. Most reptiles are quadrupeds with four legs, but snakes have no legs at all.

EXPLORATION: Reptile Orders

For this activity, you will need:

- Air dry clay
- Paints and paint brushes
- "Reptile Orders" from the Printable Pack
- Colored pencils

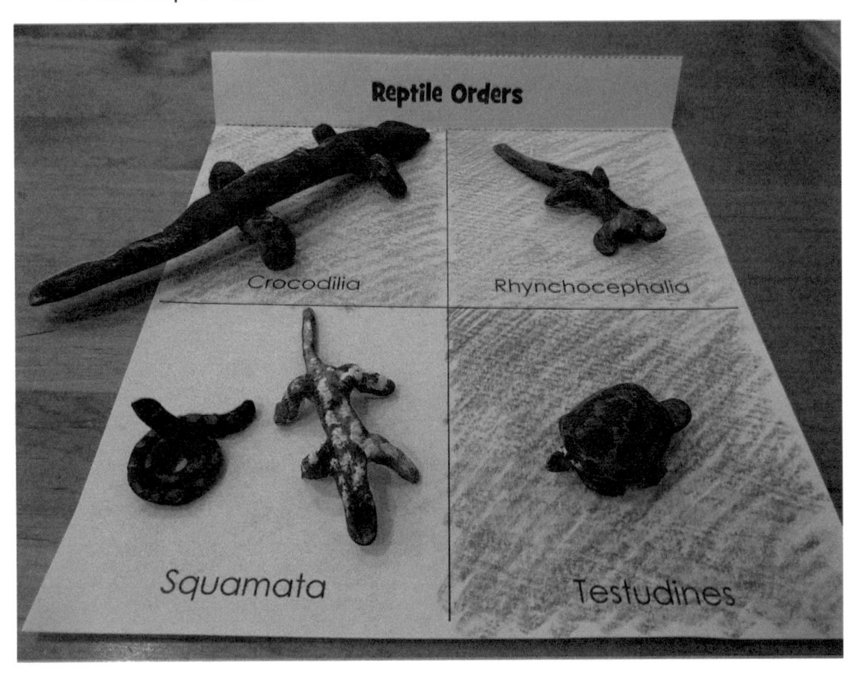

There are four orders of living reptiles:

- Crocodilia: crocodiles, alligators, caimans, gharials. These animals have four sprawling limbs, replace their teeth throughout life, and have strong jaws and a

Memorization Station

Memorize the things that make amphibians unique:

- Lay eggs in the water
- Are born in the water and breathe with gills when young
- Have moist skin
- Are cold-blooded
- Undergo metamorphosis to develop lungs for breathing air

Then memorize the things that make reptiles unique:

- Lay eggs on land
- Breathe air with lungs
- Have scaly skin or plates for protection
- Are cold-blooded

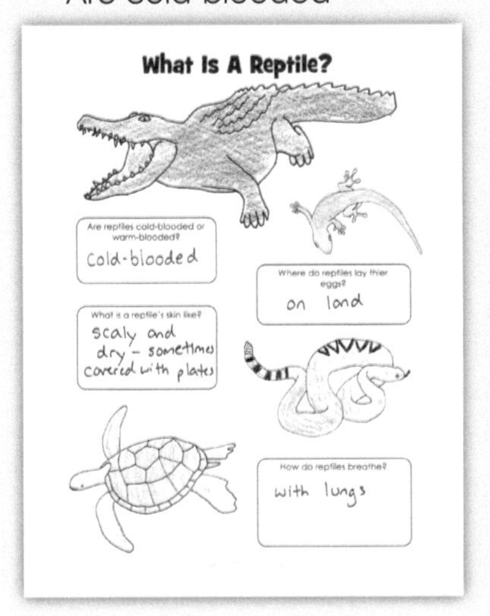

Use the coloring sheets from the Printable Pack.

Famous Folks

Grace Olive Wiley worked with venomous snakes, breeding and studying them. She was fired more than once for her habit of letting venomous snakes wander at will. She died of a snake bite at the age of 64, her snakebite kit being 30 years out of date.

Writer's Workshop

Compose a diamante poem about a reptile of your choice. A diamante poem has seven lines. The first and last lines only have one word in each of them and are usually nouns. The second and sixth lines each have two words that are adjectives. The adjectives you choose will describe the nouns from your first and last lines. The third and fifth lines each have three participles, or -ing words. The fourth line, right in the middle, is the longest line, with four nouns. When you put them all together, the poem creates a diamond shape.

Alligator

Fierce, Sneaky

Gliding, Hiding, Lurking

Eyes, Teeth, Claws, Tail

Hunting, Swimming, Biting

Fast, Cunning

Crocodylia

powerful bite.

- Rhynchochephalia: tuataras. There is only one species which lives in New Zealand. It has a primitive and small brain and a three-chambered heart.
- Squamata: lizards, snakes. These animals have scaly skin, can unhinge their jaws to eat things larger than their heads, use internal fertilization, and can have live young or eggs, though most lay eggs.
- Testudines: turtles, tortoises, terrapins. They have an external bony shell that protects them from predators. Some species live in water while others live on land. They all lay eggs on the land.

1. Make snakes, lizards, turtles, crocodiles, and tuataras out of clay. Let them dry overnight. Paint them.
2. Color each of the "Reptile Orders" boxes. Fold up the title end on the page so it stands up behind the boxes. Place the reptiles in their correct orders in the boxes.

EXPLORATION: Snake Anatomy

For this activity, you will need:

- "Snake Anatomy" from the Printable Pack
- Card stock
- Colored pencils
- Scissors
- String

Snakes have long, thin bodies which means all their internal organs must also be long and thin. The left lung of the snake, for example, is a small, vestigial organ because there isn't room for two full-sized lungs. You'll also notice that in snake bodies, paired organs like gonads and kidneys are staggered down the length of the body.

Snakes also swallow their prey whole, which means their internal organs must be able to shift around to accommodate large meals. Humans have a diaphragm and connective tissues that keep our hearts from shifting around in our bodies, but snakes have no diaphragm, allowing their hearts to shift around their big meals.

1. Print the "Snake Anatomy" onto white card stock. Color the organs in the snake body.
2. Cut the snake out along the outside and the inside of

the coil. Turn the paper over and color the skin of the snake. Look up a real snake and try to copy the pattern you see on its body.

3. Hang the snake up by a string from your ceiling.

EXPEDITION: Lizard Hunter

WARNING: Reptiles are wild animals and may bite. Some species of reptile are venomous. Before you go out looking, research the potentially dangerous species in your area so you can avoid them. After handling a wild animal, always wash your hands before eating or drinking. Do not keep a wild reptile as a pet.

For this activity, you will need:

- A large transparent jar or box with a lid
- Nail and hammer
- Reptile identification guide for your region

1. Punch a few holes in the lid of a jar or box with a hammer and nail.
2. Go outside with your jar and search for lizards. They tend to like small confined spaces like between logs or stones or under scrap wood or debris. Catch a lizard and identify it. Be careful! Salamanders can look like lizards, but a salamander will always be found in a wet zone.
3. Learn as much as you can about one species that lives where you are. Observe this species in captivity for up to a few hours. Let it go, observe it in the wild. Become an expert on that one species.
4. Keep notes and draw sketches of your chosen species.

Additional Layer

Chameleons are reptiles that can change color to match their surroundings.

Print "Chameleon" from the Printable Pack on card stock. Paint the circle in bright colors. Cut out the chameleon shape, except the eye.

Cut out both circles and use a brad to fasten the chameleon circle over the colorful circle.

Bookworms

Joan Procter, Dragon Doctor by Patricia Valdez is a picture book biography of Joan Procter, a British herpetologist (reptile scientist) who became the Curator of Reptiles at the British Museum in 1920 at the age of 23. This is her story.

Additional Layer

Tell the story of the tortoise and the hare from Aesop. You can find the story online.

Discuss the lessons that can be learned from this story.

Choose a reptile to be the star of a story you tell.

Deep Thoughts

Since at least the Middle Ages, governments have controlled human harvesting and handling of wild animals. This is called wildlife management. In the past, it was mainly so that wild populations could be reserved for the elite or so that over-harvesting wouldn't completely deplete populations.

Today, it is often for the protection of the species itself instead of for human usage of the species.

Discuss the difference in philosophy between the past and the present regarding wildlife conservation.

How does the new philosophy affect things like fishing laws?

Take photographs and write up a report. Your report should be based on observations you have made firsthand, not on books. Include information like habitat, social habits (Does it hang out with other lizards of the same species?), diet, behaviors, and anything else you observe.

EXPEDITION: Turtle Spotter

WARNING: In most places it is illegal to interfere with or touch sea turtles. If you live in a sea turtle zone, observe them from a distance. Some turtles, like snapping turtles, can be aggressive and dangerous. If you catch a turtle, don't keep it for longer than a few minutes, just long enough to sketch it. Let it go back to its home. Wash your hands afterward. Turtles can carry diseases harmful to humans.

For this activity, you will need:

- A local wild habitat likely to have turtles
- Reptile identification book for your area
- Science Notebook
- Pencils and colored pencils

1. Read a book or watch a video about turtles or tortoises.
2. Research which species of turtles or tortoises live in your habitat. Find out when and where they might be and go out to find some.
3. Observe their behavior and make a sketch in your Science Notebook.

EXPLORATION: Sea Turtle

For this activity, you will need:

- Book or video about sea turtles
- "Sea Turtle Notes"
- Bivalve sea shells
- Green craft foam
- School glue
- Scissors
- Googly eyes or a black permanent marker

There are seven species of sea turtles: green sea turtle, Kemp's ridley sea turtle, olive ridley sea turtle, hawksbill sea turtle, flatback sea turtle, and leatherback sea turtle. They live all over the oceans except in the Arctic. Sea turtles take decades to mature into adults, and all that time is spent at sea. Only the females return to shore to lay their eggs on sandy beaches. Because of the long time it takes for them to mature, the specific nesting requirements, and human predation, many sea turtles are in danger of extinction.

1. Read a book or watch a video about sea turtles. Use the "Sea Turtle Notes" to write down some interesting facts like the habitat, diet, and life cycle.
2. Sketch a sea turtle outline onto some green craft foam, using a sea shell as the turtle's shell. Cut out the outline. Add googly eyes to the turtle and glue the sea shell to the craft foam.
3. Tell an audience the things you learned about sea turtles.

EXPLORATION: Life Size Crocodilian

For this activity, you will need:

- Book or video about crocodiles or alligators

Writer's Workshop

Make an accordion crocodile book with things you have learned about crocodiles.

Draw a crocodile head at one end and a crocodile tail at the other.

Additional Layer

In 2015, a video of a group of marine biologists removing a plastic straw from a sea turtle's nose went viral and convinced many people, including several corporations, to stop using plastic straws and other single-use plastics.

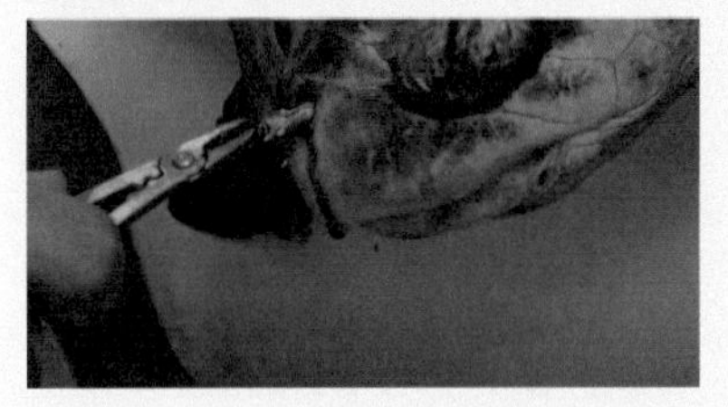

Plastic is an amazing product that can be instrumental in human health and safety, but it is also one of the worst pollutants on the planet.

Find out how plastic pollution is harming wildlife. Come up with a balanced plan to reduce unnecessary plastic use. What materials can be used instead? How can plastic safely be disposed of?

Writer's Workshop

Write about the life cycle of a turtle using sequencing words like first, next, and last.

Famous Folks

Gustave is a famous crocodile hatched in the Nile River in about 1955 in the country of Burundi. He is thought to have killed as many as 200 people. He is legendary and feared by the locals.

Various herpetologists have tried to capture Gustave to study him, but none have been successful.

Deep Thoughts

American alligators were once endangered, but wildlife agencies and laws protecting the species have brought them back to prolific populations. Today, they are often a nuisance or a danger to people.

Some people want their populations reduced through hunting and others say it is always wrong to hunt an animal.

What do you think? Does harvesting always harm the environment or is controlled harvesting sometimes helpful?

- Long sheet of butcher paper
- Markers
- Paints and paintbrushes

There are 27 species of crocodilians. Crocodilian species from dwarf caimans to huge saltwater crocodiles range in size from 1 meter to 7 meters long (3-23 feet). They are the biggest reptiles alive today and can weigh up to 2000 kg (about 4400 pounds). They also grow throughout their lives and can live up to 75 years in the wild.

1. Measure out a sheet of butcher paper so it is 4 meters long if you want to learn about alligators or 5 meters long if you want to learn about crocodiles. Draw a freehand outline of an alligator or crocodile on a long sheet of butcher paper. Make it fill the whole paper—tip of the nose at one end and end of the tail at the other.
2. Paint in your outline drawing with the colors and patterns of the animal.
3. Read a book or watch a video about alligators or crocodiles. As you learn, jot down some interesting facts. Once the paint has dried, write facts on or around the crocodilian's body with a marker.

Step 3: Show What You Know

During this unit, choose one of the assignments below to show what you have learned during the unit. Add this work to your Layers of Learning Notebook. You can also use this assignment to show your supervising teacher or your charter school as a sample of what you've been working on in your homeschool, if needed.

There are more ideas for writing assignments in the "Writer's Workshop" sidebars.

Coloring or Narration Page

For this activity, you will need:

- "Cold-Blooded Vertebrates" from the Printable Pack
- Writing or drawing utensils

1. Depending on the age and ability of the child, choose either the "Cold-Blooded Vertebrates" coloring sheet or the "Cold-Blooded Vertebrates" narration page from the Printable Pack.
2. Younger kids can color the coloring sheet as you review some of the things you learned about during this unit. On the bottom of the coloring page, kids can write a sentence about what they learned. Very young children can explain their ideas orally while a parent writes for them.

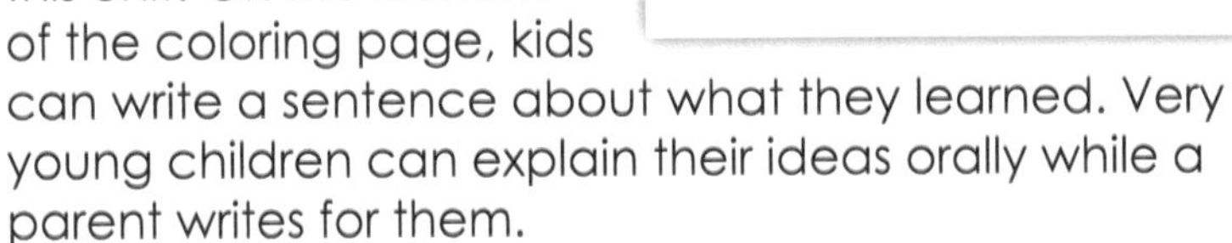

3. Older kids can write about some of the concepts you learned on the narration page and color the picture.
4. Add this to the Science section of your Layers of Learning Notebook.

Science Experiment Write-Up

For this activity, you will need:

- The "Experiment" write-up or "Experiment Report Template" from the Printable Pack

1. Choose one of the experiments you completed during this unit and create a careful and complete experiment write-up for it. Make sure you have included every specific detail of each step so your experiment could be repeated by anyone who wanted to try it.
2. Do a careful revision and edit of your write-up, taking it through the writing process, before you turn it in for grading.

Cold-Blooded Vertebrate Salad Bowl

For this activity, you will need:

- Bowl

Famous Folks

Jairo Mora Sandoval was a Costa Rican environmentalist who was murdered in 2013 while trying to protect sea turtle nesting sites from poachers. The sea turtle eggs are valuable in Costa Rica where people eat them, and the drug cartels had begun doing a black market side business.

Photo by Christine Figgener CC 3.0, Wikimedia

Jairo's death was meant to be a warning. Instead, it galvanized international protections, policing, and laws surrounding sea turtles and their nest sites.

Bookworms

Wild Survival: Crocodile Rescue by Melissa Cristina Marquez is about young Adrianna and her family who travel the globe rescuing animals. Ages 8 to 12.

Unit Trivia Questions

1. True or false - Cold-blooded is an official taxonomic division.

   False. Cold-blooded is useful for dividing animals into two groups, but it is not part of taxonomy.

2. Name the three groups of animals that are cold-blooded.

   Fish, amphibians, reptiles

3. Explain the unique features of the life cycle of an amphibian.

   Amphibians are hatched in the water. The young have gills, then as they grow, the gills metamorphose into lungs and the pollywogs grow legs. The adults live on land and breathe air.

4. Which of these animals are considered fish:

   a)Trout

   b)Sting ray

   c)Shark

   d)Whale

   e)Lamprey eel

   A, b, c, and e are all fish. Whales are mammals.

5. A vertebrate is an animal with a ____________.

   Backbone/spinal chord

6. True or false - All vertebrates develop from eggs.

   True. Eggs can be microscopic or they can be larger than your fist.

7. Scientists who study animals are called zoologists. A scientist who specializes in fish is called an ______________.

   Ichthyologist

- Strips of paper
- Pencils
- Science Notebook
- Timer

This game is easy to set up and play, but it is best with at least four players.

1. Have each person write 3 to 5 words, each on its own strip of paper. The words should be related to the things you learned about cold-blooded vertebrates in this unit. It could be words like "tiger shark" or "gills" or "life cycle." Allow the players to use their Science Notebooks for inspiration.
2. Fold the strips of paper in half and put them all in a bowl. Divide into two teams.
3. Set the timer for 1 minute. Start the timer, then one player for the first team will draw a slip of paper from the bowl. Without saying the word on the paper, give clues to get your team to guess the word. Keep drawing slips of paper until the timer goes off. Then, the play goes to the other team. Keep playing until the slips of paper are gone. Each team gets a point for every word guessed correctly.
4. Play round two the same, with the same slips of paper, but instead of giving verbal clues, act out the words on the papers.
5. For round three, use the same papers again, but this time, you may only say one word as your clue. The team with the most points at the end wins.

Writer's Workshop

For this activity, you will need:

- A computer or a piece of paper and a writing utensil

Choose from one of the ideas below or write about something else you learned during this unit. Each of these prompts corresponds with one of the units from the Layers of Learning Writer's Workshop curriculum, so you may choose to coordinate the assignment you choose with the monthly unit you are learning about in Writer's Workshop.

- **Sentences, Paragraphs, & Narrations:** Use the "Simple Sentence Mix and Match" printable as inspiration to make your own mix and match sentence parts about cold-blooded vertebrates. Swap sentences with a partner and match up their crazy sentences!
- **Descriptions & Instructions:** Pick a cold-blooded

vertebrate animal and describe it scientifically, being as accurate as possible about limbs, size, coloring, markings, and habitat.

- **Fanciful Stories**: Think of a fictional frog, toad, snake, crocodile, or fish. If the animal behaved more naturally, how that would that change the story? Rewrite an episode with more realistic behavior.
- **Poetry:** Photocopy a page about a fish, frog, or reptile. Make a black out poem, crossing out all the words you don't want.
- **True Stories:** Tell a story of a time you had an encounter with a fish, frog, or reptile.
- **Reports and Essays:** Fill out a K-W-L chart about one thing you learned about during this unit. In the Things I Wonder section, think of something you could research.
- **Letters:** Write a friendly letter to an animal you studied. Thank it for letting you learn about it.
- **Persuasive Writing:** Would you rather be a frog or a snake? Explain your reasoning, including facts about your preferred animal's lifestyle that you appreciate.

Big Book of Knowledge

For this activity, you will need:

- "Big Book of Knowledge: Cold-Blooded Vertebrates" printable from the Printable Pack, printed on card stock
- Writing or drawing utensils
- Big Book of Knowledge

1. Color, draw on, write on, or add to each of the Big Book of Knowledge pages you are using. Only add the printables if you learned these concepts during this unit. If there are other topics you focused on, feel free to add your own pages to your Big Book of Knowledge.

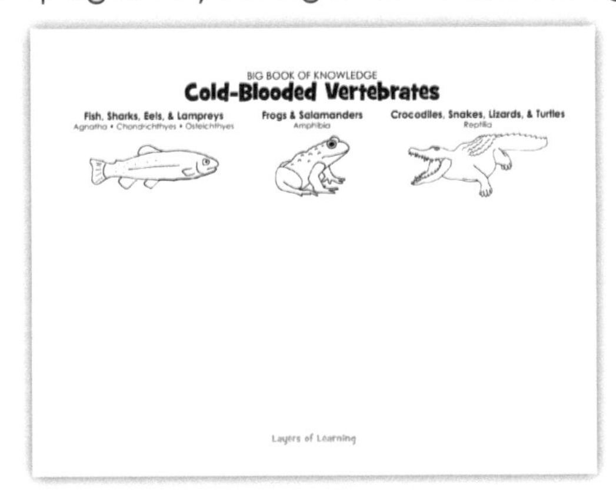

2. Use your Big Book of Knowledge regularly to help you review, quiz, or create games that will help you commit the things you've learned to memory.

Big Book of Knowledge

The Big Book of Knowledge is a book for you, the mentor, to use as a constant review of all of the things you're learning about. You can use it to quiz your kids or prepare tests or review games. Whenever you learn something in Layers of Learning that you want your kids to remember, add it to your Big Book of Knowledge.

Assemble your Big Book of Knowledge in a binder or with binder rings. Divide it into sections for each subject.

In the Printable Pack for this unit you will find a "Big Book of Knowledge" sheet. You can add this sheet to others you collect or create yourself as you progress through the Layers of Learning curriculum. Customize the Big Book of Knowledge to your family by adding facts and topics that you enjoyed exploring as you were learning.

Visit Layers of Learning online to find more information on how to assemble and use your own Big Book of Knowledge.

You will also find cover and section pages to print along with creative games to play with your Big Book of Knowledge to keep school, even the tests, fun!

Unit Overview

Key Concepts:
- Warm-blooded animals maintain their body temperatures regardless of the environment.
- Birds and mammals are warm-blooded.
- Mammals have a complex reproductive process with internal fertilization, placental nourishment of the young, and feeding with milk after birth.

Vocabulary:
- Bird
- Beak
- Feather
- Diurnal
- Nocturnal
- Crepuscular
- Camouflage
- Mammal
- Uterus
- Placenta
- Ovary
- Fallopian tube
- Embryo
- Fetus
- Birth
- Marine mammal

Scientific Skills:
- Knowing the scientific names for animal body parts helps in identifying them in the wild.
- Experiments on living creatures need to be carefully planned and many trials run to be sure you are getting true results.
- Close observations of animals and their behavior in the wild can lead to both answers and to more questions.

WARM-BLOODED VERTEBRATES

Warm-blooded vertebrates are animals with a backbone that regulate their own internal temperature to keep it constant. Warm-blooded means the animal uses energy from its food to produce heat. The body must maintain a specific internal temperature. If the body falls below or rises above the proper temperature, then the animal may die. Birds and mammals are the two groups of warm-blooded vertebrates.

Birds have wings and feathers and lay eggs. They live in nearly every environment on Earth from the polar regions and mountains to grasslands, forests, and deserts. There are even birds that spend a lot of time under the water. Most birds can fly but not all of them; penguins, ostriches, emus and kiwis are examples of flightless birds. Not only the wings, but the lightness of the bones, the muscle structure, the digestive tracts, and the respiratory systems of birds are especially constructed for flight.

This is a female superb fairywren. It is a passerine, or songbird, from Australia. Many female birds are drab browns, grays, and whites while the males are often colorful.

Mammals bear live young, feed their babies with milk, and have hair or fur. Most mammals spend their entire lives on land but a few—like whales, dolphins, and manatees—live in the water full-time. All mammals must breathe air. There are two special groups of mammals: marsupials, which carry their developing young in pouches on their bodies, and monotremes, which lay eggs instead of nurturing young inside their bodies.

This is a harbor seal, a mammal that lives along coastlines in temperate and Arctic zones in the Northern Hemisphere.

Step I: Library List

Choose books from your library that go with this topic. Here's a list of some favorites and also a list of search terms so you can utilize what your library offers. Read the books with your kids and/or assign them some to read independently. It is from these books your kids will learn most of the facts they need from this unit.

Search for: birds, mammals, and individual types you are interested in like wolves or parrots or farm animals

☻ ☻ ☻ *Encyclopedia of Science* from DK. Read "Birds," "Mammals," and "Primates" on pages 332-336 and "Sexual Reproduction" on page 367.

☻ ☻ ☻ *Kingfisher Science Encyclopedia*. Read "Birds," "Mammals," "Animal Reproduction," and "Animal Behavior" on pages 84-91.

☻ ☻ ☻ *The Usborne Science Encyclopedia*. Read "Flying and Gliding," "Moving on Land," "Feeding," "Teeth and Digestion" on pages 306-313, and "Creating New Life" on pages 324-327.

☻ ☻ ☻ *National Geographic Field Guide to the Birds of North America* by Jon L. Dunn and Jonathan Alderfer. This guide has lot of pictures, range maps, and tips for identifying birds.

Family School Levels

The colored smilies in this unit help you choose the correct levels of books and activities for your child.

☻ = Ages 6-9
☻ = Ages 10-13
☻ = Ages 14-18

On the Web

For videos, web pages, games, and more to add to this unit, visit the Biology Resources at Layers-of-Learning.com.

You will find a link to video playlists, web links, and more.

Bookworms

If you're looking for a family read-aloud, we'd like to suggest this one.

Mrs. Frisby and the Rats of NIMH by Robert C. O'Brien is about a sick young mouse and his mother's struggle to save his life when the family must move in the spring. Can the odd and frightening rats who live in the rosebush help?

Memorization Station

Memorize the features that are specific to birds:

- Two wings and two legs
- Feathers
- Lay eggs on land
- Breathe air with lungs
- Warm-blooded

You can use "What Is A Bird?" from the Printable Pack to learn and remember these features.

Bookworms

The Burgess Bird Book for Children by Thornton W. Burgess is a childhood classic. Peter Cottontail interviews birds and the reader learns about their appearance, nesting, song calls, and more. For ages 6 to 9.

Draw 50 Animals: The Step-by-Step Way to Draw Elephants, Tigers, Dogs, Fish, Birds, and Many More by Lee J. Ames. Beginning with simple shapes, learn to draw recognizable animals. Great for beginners. Use this to gain drawing skills so when you go out in the field you can make better sketches in your notebooks.

Amazing Mammals by DK. Full-color guide to lots of interesting mammals with awesome facts to learn.

DK Eyewitness: Bird by David Burnie. An awesome guide to introduce your whole family to the world of birds.

A Mammal Is an Animal by Lizzy Rockwell. For the youngest children, this book explains what is a mammal and what is not. Filled with just enough facts for little ones.

The Big Book of Birds by Yuval Zommer. Adorable color illustrations and just the right amount of bird facts for young kids.

Ultimate Explorer Field Guide: Mammals by National Geographic Kids. This book teaches kids how to go outside and find wild mammals to observe. Includes tracks, signs, and places to actually watch for the animals plus basic info on species. Applies mostly to North American animals.

The Atlas of Amazing Birds by Matt Sewell. Charming illustrations paired with sparing text describing one bird per double-page spread, all sorted geographically.

Mammals by Donald F. Hoffmeister and Herbert S. Zim. A field guide that helps you identify the mammals you spot in North America. Look for a guide for your region.

Birds of North America by Paul D. Hess. This is a large book with full-color photographs of 657 different species of birds along with facts about each species.

The Complete Illustrated Encyclopedia of Birds of the World by David Alderton. A browsable book about 1,600 different species of birds from all over the world. It is filled with photographs and illustrations.

What It's Like to Be a Bird: From Flying to Nesting, Eating to Singing—What Birds Are Doing and Why by David Allen Sibley. Excellent basic information about all sorts of aspects of bird life from a life-long ornithologist and bird watcher.

How Birds Work: An Illustrated Guide to the Wonders of Form and Function—from Bones to Beak by Marianne Taylor. Explains all kinds of things about birds, both general and specific. Why does a heron stand on one leg? Why are an owl's ears asymmetrical? So much more!

Step 2: Explore

Choose a few hands-on explorations from this section to work on as a family. They should be appealing activities that will create mental hooks so your kids remember the information in the unit. Save the rest of the explorations for the next time you do this unit in four years when your kids are older. You can also read the sidebars together and explore some little rabbit trails.

This unit includes printables. See the introduction for instructions on retrieving your Printable Pack.

Birds

Birds are **vertebrates** that are organized into class Aves which includes animals with beaks, feathers, and hard-shelled eggs. There are more than 10,000 species of birds, about half of them passerines—small songbirds like robins, finches, and sparrows.

EXPLORATION: Bird Orders

For this activity, you will need:

- "Bird Orders" from the Printable Pack (10 pages)
- Watercolor paints
- Paintbrushes
- Scissors
- Transparent tape

Birds are in the class Aves. Within that class, there are 39 orders of birds. We will focus on 18 of the most familiar and numerous, plus one superorder of flightless birds.

Additional Layer

Not all birds fly. Learn about penguins, ostriches, and kiwis—all flightless birds.

Photo by Christopher Michel, CC by SA 2.0, Wikimedia.

Memorization Station

Bird: warm-blooded vertebrate with feathers, wings, and a beak

Vertebrate: an animal with a backbone and belonging to the phylum Chordata

You may want to take some time together to look up lots of examples of vertebrates to get an idea of just how diverse this group is. Can you think of some animals that don't have a backbone?

Writer's Workshop

Learn about bird migration by visiting the Cornell Lab of Ornithology online and searching for "migration."

Write a 5-paragraph essay about migration. Include an introductory paragraph; 3 body paragraphs on how migration works, why birds migrate, where birds migrate; and a conclusion paragraph.

Bookworms

Charlie's Raven by Jean Craighead George is about a boy who finds and cares for a baby raven that imprints on him. Can Charlie protect his new friend from the murderous neighbors?

Additional Layer

You can dissect a chicken wing to see how a bird's bones, muscles, and blood vessels are arranged in the wings.

You just need chicken wings from the grocery store, scissors, a sharp knife or scalpel, and your Science Notebook.

Start by sketching the outside. Then, carefully cut the skin off the outside with scissors.

Cut and tease apart the skin, fat, and muscles to see the structure of the parts. Push and pull on the bones and muscles to see how they work. Find tendons, cartilage, and blood vessels.

Make sketches as you go.

1. Paint the birds from the "Bird Orders" pages. You can look up images of the birds to find out what they look like.
2. Once the paint has dried, cut the birds out on the dashed lines. Hang them in a window with transparent tape.

EXPLORATION: Parts of a Bird

For this activity, you will need:

- "Parts of a Bird" and "Parts of a Bird Answers" from the Printable Pack
- Pen and colored pencils

When you use a guidebook for identifying birds, it will explain the bird's features and coloration using certain anatomy terms. For example, the guidebook might say, "The Baltimore Oriole has an orange breast and a white flank." It is important to know these anatomy terms so you can identify birds properly and understand their parts.

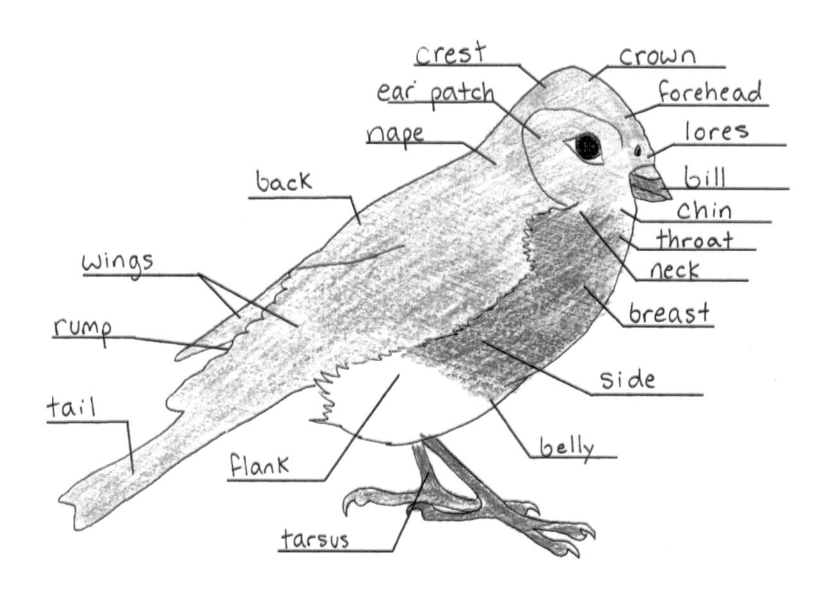

1. Use the "Parts of a Bird Answers" as a guide to filling in the blanks on the "Parts of a Bird" printable.
2. The bird in the diagram is a passerine, a songbird. You can color it however you'd like. All birds, no matter which order they belong to, have similar anatomy, so the "Parts of a Bird" sheet will be helpful for identifying all birds.
3. Add your diagram to your Science Notebook and use it to review the bird parts regularly.

EXPERIMENT: Bird Feeder

For this activity, you will need:

- Pine cone - large ones work better
- Peanut butter
- Bird seed
- String
- Scissors
- Paper plates or trays
- 2 small bowls
- Butter knife
- Science Notebook and colored pencils (optional)

One of the easiest ways to observe birds is to make a bird feeder and attract them to you.

1. Cut the string into 30 cm-long pieces. Pour bird seed into one bowl and scoop peanut butter into another.
2. Put your pine cone onto a paper plate to catch all the mess. Then, spread peanut butter over your pine cone.
3. Over the plate, sprinkle the peanut butter covered pine cone with bird seed.
4. Tie your piece of string around the top of your pine cone and then around a tree branch or from a porch. Hang it where you can see it out of a window.
5. Keep a record of the birds you spot at your bird feeder. Pay special attention to what time you see the birds and the bird behaviors you observe. You might also want to make some sketches in your Science Notebook.

   It may take several days or even a couple of weeks for birds to find your feeder, so don't give up!

EXPEDITION: Bird Watching

For this activity, you will need:

- A bird identification guidebook for your region

Additional Layer

Make a bird call. You need a small 2"x1"x1" piece of hardwood and a round screw eye.

Drill a hole in the end of the hardwood, slightly smaller than the screw. Work the screw in and out of the hole to loosen it up.

Twist the screw back and forth quickly to make it squeak. Hold still for fifteen, minutes while using your bird call and see if you can attract some birds.

Additional Layer

Make a handprint bird craft. Cut an oval body, circle head, and triangle tail. Add on a beak and feet. Use your handprint for the wing.

Pair this craft with a book about birds.

- Binoculars (optional)
- Cornell Ornithology Lab bird calls - you can access these online or purchase books that incorporate audio bird calls from the Cornell Lab
- Science Notebook
- Pencils and colored pencils

Birdwatching involves observing birds in the wild. It can be the wild of your backyard, a park, a nature preserve, or the wilderness. You are most likely to see lots of birds in locations near water sources—like a stream or pond—and in the early morning or late evening. But birds are everywhere and active at all times of the day and night.

Part of bird watching is seeing the birds, but another part of it is hearing the birds. To "spot" a bird, you need only hear it. With some practice you can get good at identification through sight and sound.

1. Choose a location and time where you think you will be able to spot lots of birds. Take along a bird identification book and, if you have them, some binoculars.
2. Spend time just sitting quietly, listening and watching. Record what you see and hear.
3. Go home and listen to the bird calls of the species you identified on the Cornell Lab of Ornithology website. Just search for the species with the search function on the website to get details and hear the bird calls. Memorize three or four bird calls. You can also try their game "Bird Song Hero."
4. Go out bird watching at least one more time. This time,

Writer's Workshop

Choose an animal to learn more about and write a report on.

Use the "Animal Report Outline" from the Printable Pack for help in writing the report.

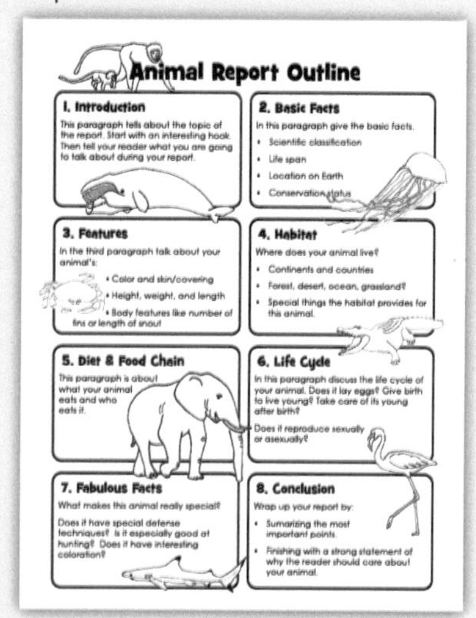

Animal Report Outline

1. Introduction
This paragraph tells about the topic of the report. Start with an interesting hook. Then tell your reader what you are going to talk about during your report.

2. Basic Facts
In this paragraph give the basic facts.
- Scientific classification
- Life span
- Location on Earth
- Conservation status

3. Features
In the third paragraph talk about your animal's:
- Color and skin/covering
- Height, weight, and length
- Body features like number of fins or length of snout

4. Habitat
Where does your animal live?
- Continents and countries
- Forest, desert, ocean, grassland?
- Special things the habitat provides for this animal.

5. Diet & Food Chain
This paragraph is about what your animal eats and who eats it.

6. Life Cycle
In this paragraph discuss the life cycle of your animal. Does it lay eggs? Give birth to live young? Take care of its young after birth?
Does it reproduce sexually or asexually?

7. Fabulous Facts
What makes this animal really special?
Does it have special defense techniques? Is it especially good at hunting? Does it have interesting coloration?

8. Conclusion
Wrap up your report by:
- Sumarizing the most important points.
- Finishing with a strong statement of why the reader should care about your animal.

You can shorten the report assignment for younger kids by removing some of the paragraphs from the middle of the report or by having the child write one sentence on each topic instead of an entire paragraph.

Expedition

An aviary is a large enclosure where birds are kept and can be viewed. This is the National Aviary in Pittsburgh, Pennsylvania, United States.

See if there is an aviary nearby that you can visit.

see if you can identify some of the birds by their calls as well as by visual sightings.

5. As you watch birds, take notes and make some sketches in your Science Notebook of the things you see. It is interesting to note the time of day, where the birds were seen (do they favor a certain tree?) and their behavior (were they feeding, fighting, nesting?).

EXPLORATION: Bird Beaks

For this activity, you will need:

- "Bird Beaks" and "Bird Beak Answers" from the Printable Pack
- Pictures of birds from magazines or online
- Internet connection or bird identification handbook
- Index cards
- Scissors
- Glue stick

Bird **beaks** are used for eating, fighting, climbing, boring holes, preening, grasping, singing, courtship, building nests, regulating body temperature, and feeding young. A bird beak has to be extremely useful for the bird to survive and thrive. Beaks come in many different shapes for many different food sources and lifestyles birds have.

1. Complete the "Bird Beaks" worksheet by matching the bird beak shape to the food the bird eats. Use the "Bird Beaks Answers" as a guide. Note that only a few bird beak shapes are shown, there are many more variations in the animal world. These are generalizations and the worksheet introduces the idea that bird beaks are, in part, shaped for the food the bird eats.

2. Cut out pictures of birds from magazines. Glue the pictures to index cards. Look up each bird online or in a bird identification book to find out what it eats. Write down its diet on the card. Does its beak shape match what you expected its food to be? What might affect the bird's beak shape besides just the food it eats?

Bookworms

The Triumphant Tale of the House Sparrow by Jan Thornhill is about the little bird from the Middle East that spread through Europe, Africa, and Asia with the spread of human cultivation of grain, becoming one of the most despised birds in history.

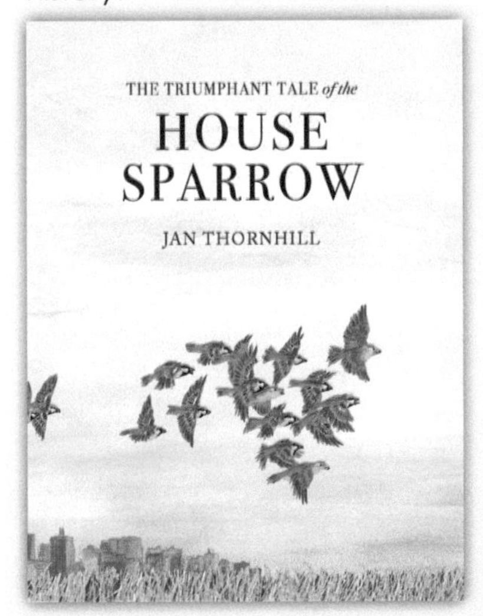

This books melds history with zoology. For ages 9 to 12.

Memorization Station

Beak: a bird's bony projections from the upper and lower mandibles of the face

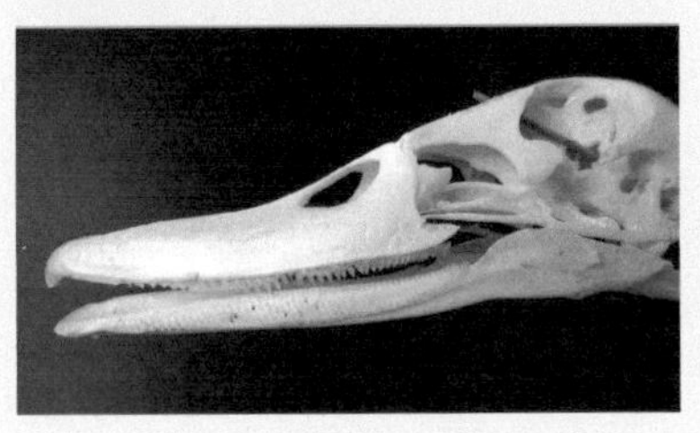

Deep Thoughts

Many birds migrate, but no one knows how they navigate. Think about possibilities and then how you might do an experiment to test your hypothesis.

Memorization Station

Feather: outer covering of a bird's skin that is made of hollow shafts fringed with barbules

Famous Folks

A scientist who studies birds is called an ornithologist. One of the most famous ornithologists ever was John James Audubon.

He collected and drew every species of bird he could get his hands on, making a collection of paintings that could be used as a tool for scientists as well as art lovers.

3. Older students can take this further by making a database of bird beak shapes, bird foods, and bird behaviors that seem to relate to beaks. See if you can find any correlations. For example, a recent study at Cornell found that birds with longer beaks sang slower, which can relate to mating and socialization among birds. Try to focus the study down to make it manageable. For example, study just the various species of wrens or only birds of prey or compare two birds with the same basic beak shape.

EXPERIMENT: Feathers

For this activity, you will need:

- "Feathers" and "Feathers Answers" from the Printable Pack
- Pencil and colored pencils
- Feather from a bird
- Magnifying glass
- Microscope (optional)

Feathers are formed in the epidermis, the outer layer of a bird's skin. They are made of complex twisted and crossed strands of keratin, a protein that also makes up hair, finger nails, horns, and hooves. There are two main types of feathers: vaned and down. Vaned feathers cover the outer part of the bird and aid in flight, in shedding water, and in decoration. Down feathers are close to the skin and used for warmth and sometimes to line a nest.

1. First, label the parts of an individual feather on the "Feathers" sheet, using "Feathers Answers" as a guide.

   The vane is the plumed part of a feather. The rachis

is the shaft, the long slender part of a feather that runs up the center. The barbs are the long, thin shafts that grow out from the center rachis. They have little barbules on them that hook one rachis to another to make a smooth, continuous surface on the feather. The calamus is the large hollow potion of the central shaft that attaches the feather to the bird's body. Finally, the afterfeathers are the downy lower barbs.

2. On the wing diagram on the lower half of the "Feathers" page, color the primaries red. These feathers attach to the bird's hand bones and are long and strong to provide the forward thrust when the bird flaps its wings.
3. Color the secondaries orange. These feathers attach to the lower arm bone and generate lift during flight.
4. Color the primary coverts yellow. Covert feathers cover other feathers. They help to smooth the air flow over the wing. Color the greater secondary coverts green, the median coverts light green, and the lesser coverts blue.
5. Color the alula light blue. Alula feathers, usually just 3-5 feathers, are attached to the bird's thumb. The bird lifts them when flying slowly or when landing to keep lift and avoid stalling at slow speeds.
6. Color the tertials purple. The tertials act as a covering for the flight feathers when the bird's wing is folded against its body. They are attached to the skin and not to any bone and so do not have the strength to actually aid in flight.
7. Now, examine a real feather. Identify the vane, rachis, barbs, calamus, and afterfeathers. Run your fingers up and down the feather, zipping and unzipping the barbs from one another. When a bird preens, part of what she is doing is smoothing the barbs together to make a nice smooth surface. Notice how the calamus is hollow. It is both strong and light.
8. Look at the feather with a magnifying glass or a microscope. On each of the barbs there are tiny hooks that hold one barb to the next. This is how they zip.

EXPERIMENT: Bird Bones

For this activity, you will need:

- A chicken thigh bone, clean and dry
- Sheet of paper
- Transparent tape
- Paper plate
- Coins, nails, or other small heavy objects

Fabulous Fact

Many species of animals—from spiders to ducks to humans—exhibit sexual dimorphism. This means that you can tell it is a female or a male based on obvious physical characteristics such as size, coloring, or behavior.

Image by Francis C. Franklin / CC-BY-SA-3.0 Wikimedia

The brightly colored mandarin duck on the left is a male. The more drab duck on the right is a female. They are the same species, but different sexes.

Why do you think the females are less showy than the males?

Fabulous Fact

Many birds and mammals are synanthropic, which means they like to live around people because humans improve their lives.

This is a rufous hornero bird from South America. It likes to hang out in parks and yards where people have cut the grass and made it easier to find bugs to eat.

Additional Layer

In the Middle Ages, the nobility and commoners hunted with trained hawks or falcons. In Britain, birds of prey were so popular that one could hardly walk down a street without seeing a goshawk perched on a wrist.

Image by Jamain, CC by SA 4.0

Learn more about medieval falconry. You might enjoy the Layers of Learning *High Middle Ages Europe* and *Medieval Britain* units for more about the Middle Ages.

Bookworms

An Anthology of Intriguing Animals by DK includes many kinds of vertebrates, not just birds and mammals.

Each page highlights a specific animal like "elephants" or "orca" and has facts and colorful illustrations. For ages 3 to 9.

Bird bones are hollow but strong. The hollow spaces inside the bones are connected to the respiratory system, so air can travel through them and around the bird's body.

1. Cut a chicken thigh bone in half so you can see inside. The bone is hollow. Hollow tubes are stronger than solid ones. This means that bird bones are stronger than mammal bones. Bird bones are also more dense, with fewer air spaces in the cells that make up the actual bone. Flying and landing require strong bones.

   On the downside, a bird's bones tend to shatter and splinter when they break whereas a mammal bone breaks more cleanly. It is easier for a mammal to heal from a broken bone than a bird.

2. Make a cylinder out of paper. Tape it so it stays in place. Set a paper plate on top of the cylinder. Add coins to the paper plate until it crushes the paper cylinder. Weigh the maximum amount of coins you fit on your cylinder before it broke. Were you surprised by how much weight it could hold?

EXPERIMENT: Life Cycle of a Quail

For this activity, you will need:

- "Life Cycle of a Quail" from the Printable Pack
- Colored pencils
- Scissors
- Glue stick
- String
- Hole punch
- Quail hatching kit with incubator (optional)

Each species of bird has a specific life cycle. There are

variations in the number of eggs laid, the length of incubation, how much care is given to the young, how long it takes a bird to leave the nest, and at what age the young bird learns to fly and fend for itself. In addition, many birds also migrate as part of their life cycle. They fly to the warmer part of the globe as the seasons change.

Quail are ground birds and most species of quail do not migrate. The male internally fertilizes the female in the spring. Then, the female lays clutches of up to 20 or so eggs (depending on the species). The male and female both tend the nest and incubate the eggs. After a few weeks, the quail chicks hatch. They are running around the forest floor within a few hours of hatching, finding food. The parents and chicks join up with other quail in the area and they all live together, searching for food and watching out for predators. It takes the chicks just a few months to mature into adults.

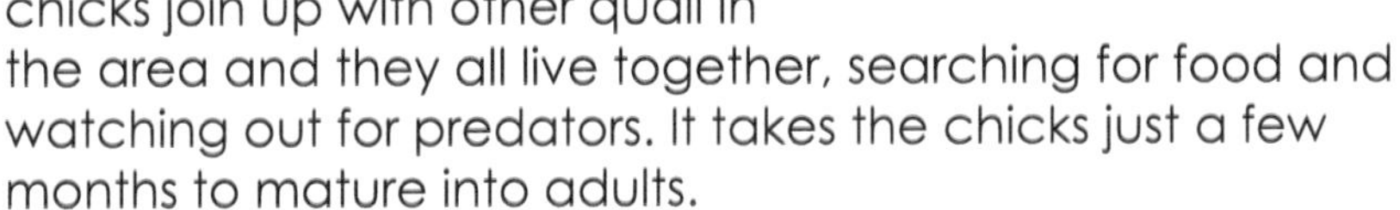

1. Color the "Life Cycle of a Quail" printable.
2. Fold the sheet on the gray dashed line. Then, with it still folded, cut along the solid black lines in the center.
3. Unfold the sheet and glue it into a cylinder with the words and images on the outside.
4. Punch a few holes along the top edge and tie strings on. You can hang the life cycle from the ceiling.
5. If you'd like, purchase a quail hatching kit from a science supplier. It comes with an incubator, fertilized quail eggs, and instructions. After the quail have hatched, you will be responsible for their care and they will need a large enclosed cage to live in, so be sure you want new pets before you hatch quail eggs.

EXPERIMENT: Owl Pellet Dissection

WARNING: Wear gloves and take care to clean everything thoroughly following this dissection.

Bookworms

Where Are You From? by In-Sook Kim is about how all animals come from eggs, but some are born oviparous, nurtured and developed inside the egg, and others are viviparous, nurtured by a placenta inside the mother.

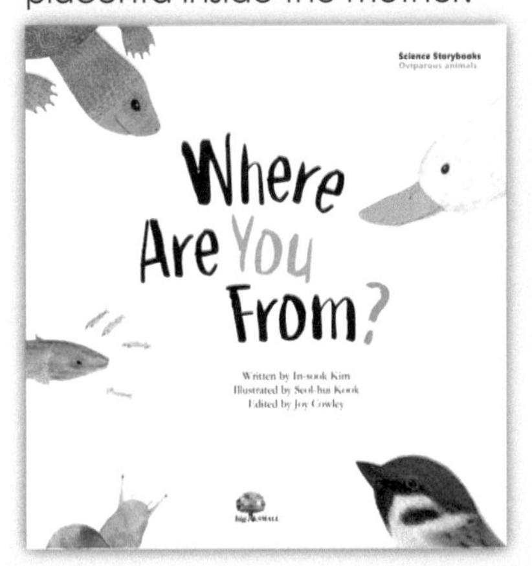

This is a picture book for ages 5 to 10, but it contains biology principles that are for all ages and makes a good intro to a lesson.

Fabulous Fact

Cormorants were once trained to fish in rivers in China, Peru, England, Greece, and France. Today, it is a dying practice.

For this activity, you will need:

- Owl pellets dissection kit
- Gloves
- School glue
- Paper
- Probe (toothpicks or skewers work too)

There are over 200 species of owls. Most birds are **diurnal**, active during the day, but nearly all owls are **nocturnal**. The rest are **crepuscular**, hunting at dawn and dusk. Owls hunt for small birds, mammals, amphibians, and reptiles. Their hunting strategy relies on stealth, which is achieved by their **camouflage** coloring and serrated leading edges on their flight wings, which cut the sound down to near silence. Like other birds, owls have no teeth to chew their food. After swallowing their prey whole, they digest the soft parts and spit back up the bones and fur in oval pellets.

The owl pellets tell the scientist what the owl eats, where it is roosting, and what small animals are available for hunting.

1. Dissect an owl pellet with a dissection kit to guide you. Wear gloves and make sure to cover your work surface with something disposable.
2. Assemble the bones of the skeleton as well as you are able. Glue it to a sheet of paper in the correct order. It's a bit like putting together a puzzle to deduce what animal your owl ate. Identify the prey the owl ate if you can.
3. When you are finished, put your tools and owl pellet debris in the trash. Wash your hands and any surfaces thoroughly.

Memorization Station

Diurnal: active during the day

Nocturnal: active at night

Crepuscular: active at twilight

Camouflage: coloring or patterns meant to blend in with surroundings

Mammal: warm-blooded vertebrate with hair, milk production, and the birth of live young

Fabulous Fact

This is a picture of the leading edge of an owl's flight feather. Notice the serrations that reduce noise during flight.

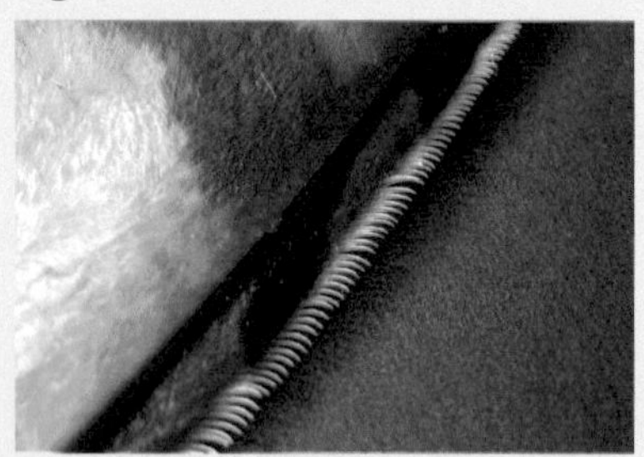

Image by Kersti, CC by SA 3.0

Fabulous Fact

Naked mole rats are a unique mammal because they are cold-blooded.

Mammals

EXPLORATION: Mammal Orders

For this activity, you will need:

- "Mammalia" from the Printable Pack, including the 2 pages of orders and the 8 pages of animal cards.
- Scissors
- Colored pencils or crayons (optional)
- Long sheet of paper like from a roll of freezer paper
- Meter stick or ruler
- Glue stick

Mammals are animals that have hair and feed their young with milk. There are 15 orders of mammals that we will define here (the marsupials are actually one level up at infraclass, but we will group them together for simplicity). Biologists hotly debate mammal classification, so you'll see different systems than the one we present in this unit. We choose these orders because they are up to date and widely accepted. However, the system we show here may change as more information is learned and biologists refine the system. Also, we left out several obscure orders with fewer members to simplify things.

1. Cut apart the animal cards in the Printable Pack. Color the animals if you'd like. Spread the animals out on a table and group them in a way that makes sense to you. Remember that sometimes it is features you can't easily see, like teeth or what an animal eats, that puts it into a certain group.

2. Cut out the order names from the first two pages. Each order has a short description of the features that animals

Additional Layer

Mammals have the most highly developed brains of any living things. In particular, mammals have a neocortex, the outer layer of the brain which controls sensory perceptions, spacial reasoning, motor commands, conscious thought, and in humans, language. Lower order mammals have a smooth neocortex. Primates and a few other intelligent mammals have a wrinkled neocortex, increasing the surface area. Humans have a much thicker neocortex than any other animal.

Memorization Station

Use the "What Is A Mammal?" from the Printable Pack to learn and memorize the definition of a mammal. Glue the correct mammal description tags to the page on the dashed rectangle spaces.

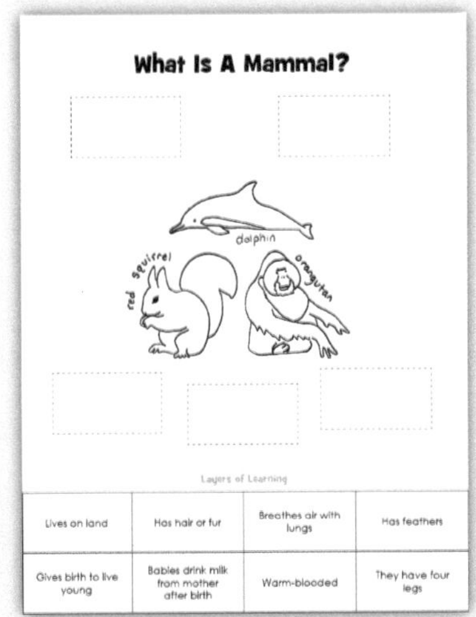
What Is A Mammal?

Lives on land	Has hair or fur	Breathes air with lungs	Has feathers
Gives birth to live young	Babies drink milk from mother after birth	Warm-blooded	They have four legs

The answers are:

- **Has hair or fur**
- **Breathes air with lungs**
- **Gives birth to live young**
- **Babies drink milk from mother after birth**
- **Warm-blooded**

Writer's Workshop

Choose a favorite mammal and write a "fact and fiction" project. The "fact" is a factual and researched report about the animal. The "fiction" is a fun fictional story using your mammal as the main character.

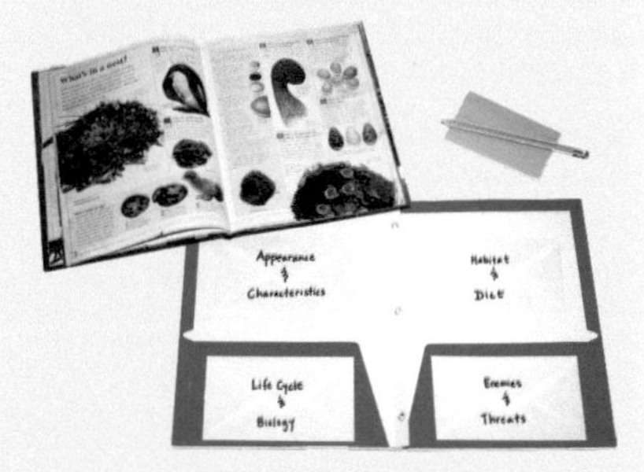

Use a two pocket folder and label one pocket "fact" and the other "fiction." Place your writing into each pocket accordingly. Decorate the front cover of your folder with pictures of your mammal.

Bookworms

The Incredible Journey by Sheila Burnford is about three pets who face huge distances and great dangers to make it back to their family. A charming story that makes a great read-aloud.

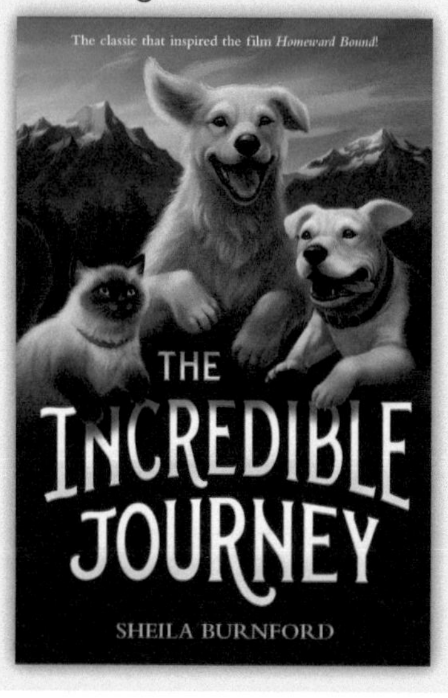

in that order possess. Regroup your animals into the orders as well as you can.

3. Cut a sheet of paper 3 meters long. Use a pencil and a ruler to lightly draw lines dividing the paper every 20 cm. You should end up with 15 even spaces, for the 15 orders of animals. (There are more than 15 orders, but we are only learning about 15 of them.)
4. Glue the order names along the top of the paper, one in each section. Use two sections for Artiodactyla and Carnivora, the two with the most animal cards. Put two orders, one above the other, for the orders with the fewest animal cards: Chiroptera, Pilosa, Cingulata, Monotremata, Proboscidea, and Sirenea.
5. Then, use the key below to put each animal in the correct order. If you'd like, older students can learn about the families and sort the animals further. This is especially useful in the Artiodactyla, Carnivora, Primates, and Rodentia orders.

Order	Members of the Order
Artiodactyla	giraffe, deer, moose, elk, cow, American bison, yak, springbok, sheep, bighorn sheep, dromedary camel, llama, gazelle, hippopotamus, musk ox, pig, warthog, Indian boar, water buffalo
Perissodactyla	horse, donkey, zebra, tapir, rhinoceros
Carnivora	wolf, dog, jackal, fox, grizzly bear, giant panda, raccoon, coati, red panda, marten, skunk, river otter, weasel, cat, lion, cheetah, mountain lion, hyena, civet, mongoose, walrus, monk seal
Cetacea	narwhale, blue whale, harbor porpoise, dolphin, orca
Chiroptera	bat
Pilosa	giant anteater, sloth
Cingulata	armadillo
Eulipotyphly	mole, hedgehog, shrew
Lagomorpha	rabbit, hare, pika
Marsupialia (infraclass)	wallaby, kangaroo, koala, bandicoot, numbat, quoll
Monotremata	echidna, platypus
Primates	aye-aye, chimpanzee, gibbon, gorilla, lemur, macaque, orangutan, howler monkey, tarsier, vervet monkey, humans, mandrill baboon

Probiscidea	African elephant
Rodentia	beaver, rat, jerboa, hamster, muskrat, marmot, gopher, chipmunk, field mouse, red squirrel
Sirenia	manatee

EXPLORATION: Life Cycle of a House Mouse

For this activity, you will need:

- "Life Cycle of a House Mouse" from the Printable Pack
- Paper
- Glue stick
- Scissors

A house mouse, *Mus musculus*, is a mammal of the order Rodentia. It is a wild animal, but moves into people's homes and buildings to find shelter and food. Because of this habit, it half domesticated itself and *M. musuculus* is often kept as a pet by people as well.

The house mouse life cycle is similar to other mammals. It is fertilized internally, the fetus develops inside the mother's uterus, and then is born alive, but it remains in need of nourishment and protection from its mother until it reaches maturity.

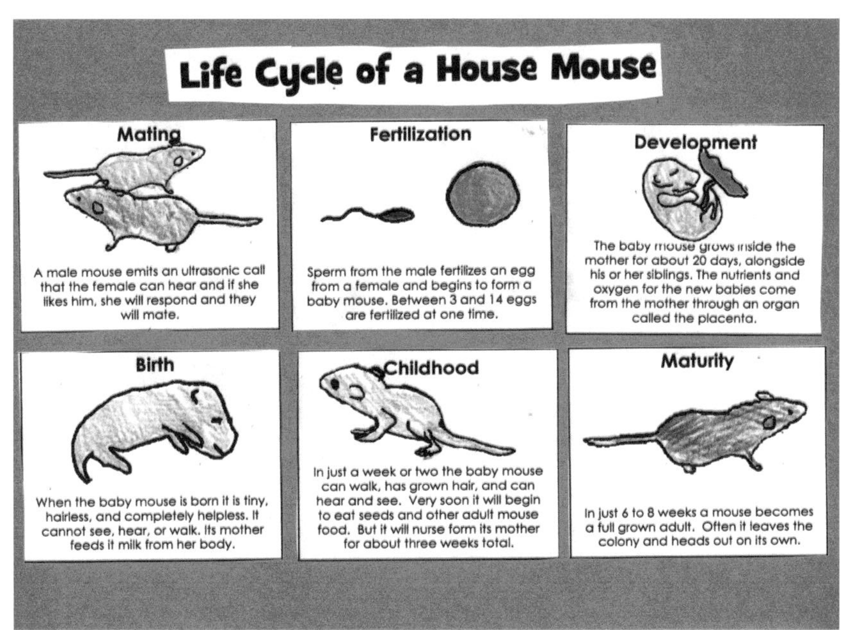

1. The panels on the "Life Cycle of a House Mouse" sheet are out of order. Cut them apart and glue them to another sheet of paper in the correct order.

   The panels should go in this order: mating → fertilization → development → birth → childhood → maturity.

Writer's Workshop

Write a paper that compares and contrasts the life cycles of two different organisms, like a quail and a mouse.

Explain the stages in each life cycle in the first two paragraphs and then point out the important differences in the final paragraph.

Additional Layer

Mus musculus is a model organism in biology, which means it is highly studied as a type of other organisms. If we study the effect of a sugary diet on a mouse, then we know a little bit more about the effect of a sugary diet on humans as well.

Other model organisms include *E. coli* bacteria, which is used to study disease and evolution, *Drosophila melanogaster* (fruit fly), which is used to study genetics, and *Saccharomyces cerevisae* (baker's yeast), which is used to study cancer. There are many more model organisms and, without the experiments that have been and are being done on them, very few of the medical advances of the last 200 years would have been possible.

Learn more about model organisms and their crucial role in medical research.

Deep Thoughts

From a scientific viewpoint, which only deals with the material or physical universe, humans are a type of animal. Smarter and more developed perhaps but still just one animal among many. This viewpoint is extremely useful in understanding biology, physiology, and ecology of humans and how we affect and are affected by the natural world in a dispassionate way.

But a scientific, materialistic viewpoint is not the only way to look at the world. Remember, science can tell us nothing about philosophy, beauty, morality, or realms beyond the purely physical.

Do you think humans are one animal among many or something set apart from animals? Why? What is your evidence?

Here is a cat and its pet human.

Additional Layer

Learn the names for animal groups. Brainstorm some together and then look up ones you don't know.

Murder of crows, parade of elephants, charm of hummingbirds, flock of flamingos, pack of wolves, pride of lions, skulk of foxes, crash of rhinos . . .

EXPERIMENT: Placental Mammals

For this activity, you will need:

- "Development of Placental Mammals" from the Printable Pack.
- Colored pencils or crayons

Most mammals are placental. The mother has a **uterus** where the young develop with the aid of a **placenta** organ until a relatively late stage of development. The developing baby takes up a great deal of room and the mother's abdomen becomes distended to accommodate the young. All mammals go through the same basic stages of development as newly developing life. The differences are mostly in the timing. It takes higher animals like humans and chimpanzees longer to develop than lower animals like mice and rabbits.

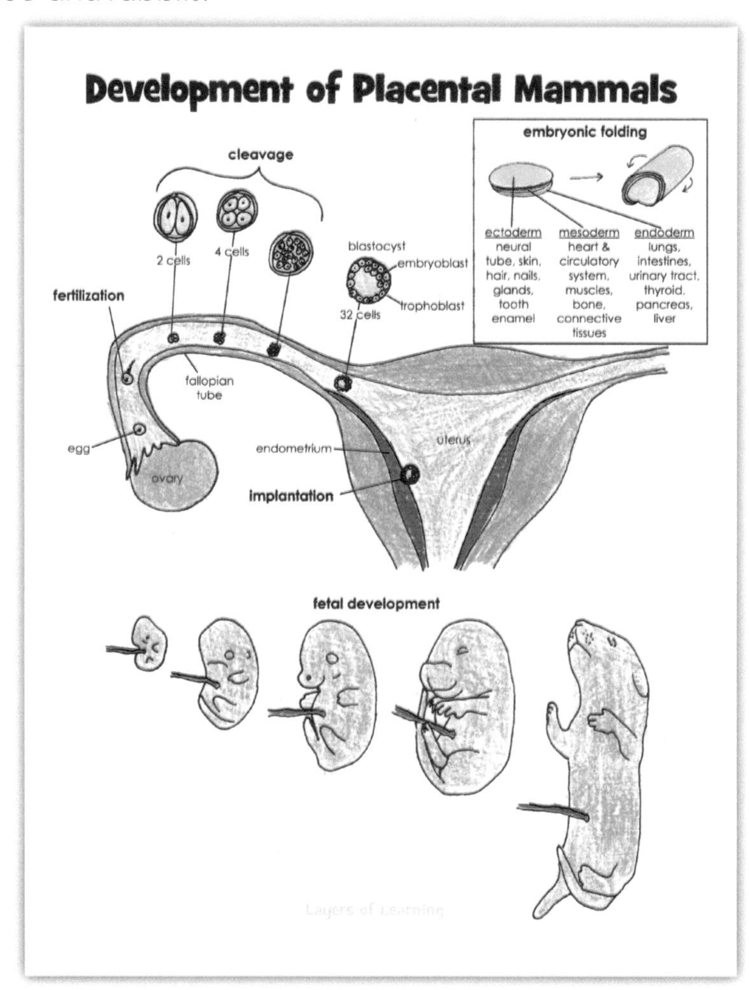

1. Read and follow the directions below to complete the "Development of Placental Mammals" sheet.
2. Fertilization is the event that begins the development of a new life. An egg from the mother joins with a sperm from the father inside the mother's body. Color the

mother's **ovary**, where the eggs are stored, pink. Color the **fallopian tube** where the fertilization occurs pink as well. Color the egg and the fertilized egg yellow.

3. After just a few hours, cleavage (which means to split) begins. The egg cell splits into two cells and then keeps splitting without the overall size changing.

   Once there are 32 cells, the cells rearrange themselves to form a sphere called a blastocyst. The cells on the outside of the sphere are called the trophoblast. The trophoblast will become the placenta. The cells inside the sphere are called the embryoblast. This will become the **embryo**. Embryo is the name given to the earliest stages of development, before organs have begun to form.

   Color the cleavage stages and the blastocyst yellow.

4. Next, at implantation, the blastocyst travels into the uterus and implants in the edometrium. Endometrium is the lining of the uterus, which thickens and become rich with blood and nutrients in preparation for implantation.

   Color the uterus pink. Color the endometrium red. Color the implanted embryo yellow.

5. The uterus and the blastocyst begin to connect. Trophoblast cells line up with the mother's cells to exchange nutrients and become the placenta and umbilical cord. The embryoblast becomes the body of the new life, the embryo. The uterus begins to nourish the embryo with sugars and oxygen. The cells in the embryo are pluripotent—they can turn into any kind of body cell that is needed. They are called stem cells.

6. At this point, the embryo cells differentiate into three flat layers, one on top of the other. The top is the ectoderm, the middle is the mesoderm, and the bottom is the endoderm. The three layers fold in on themselves to make a tube, with the endoderm on the inside.

   Color the ectoderm blue, the mesoderm red, and the endoderm yellow in the "embryonic folding" box.

7. Next, the embryo differentiates further into organs. The ectoderm becomes a neural tube, including the brain, skin, hair, nails, glands, and tooth enamel. The mesoderm becomes heart and circulatory system, muscles, bone, and connective tissue. The endoderm becomes lungs, intestines, urinary tract, thyroid, pancreas, and liver.

8. As soon as organs have begun to develop, the new

Memorization Station

Uterus: female organ where fertilization occurs and the offspring develop before birth

Placenta: organ that nourishes the fetus in a pregnant female through an umbilical cord

Ovary: female reproductive organ that produces eggs

Fallopian tube: one of a pair of tubes that carry the egg from the ovary to the uterus

Embryo: earliest stages of development of a mammal before organs have begun to develop

Fetus: after organs have begun to develop, offspring that grows within the uterus of a mammal

Birth: emergence of the offspring from the mother's body

Bookworms

The Elephant's Girl by Celesta Rimington is about little Lexington Willow who was swept away by a tornado, saved by an elephant, and adopted by a zookeeper. A charming and magical story for ages 8 to 13.

life is called a **fetus**. In mice, this is at about 3 days. In sheep, it happens after about 15 days. In humans, it takes 21 days. The timing varies by species. The rest of the development of the fetus, both the exact process and the speed, is different for each species. Generally for mammals, the fetus' head and brain develop early and look proportionally large compared to the rest of the body. Limbs develop as buds first and then gain definition as time goes on. The fetus can feel sensations and begins to regulate its own body in the very early stages of fetal development. All through this time, the fetus is nourished by the mother through the placenta and umbilical cord.

Color the umbilical cord of the developing mammal fetus red. Color the fetus pink.

9. When the fetus is ready to survive outside the mother's body, **birth** occurs. For mice, this is 21 days. For sheep, it is about 150 days. For humans, it is about 270 days. The newborn keeps developing after birth, again varying by species. A mouse pup is born hairless, blind, and deaf. A human infant is born with hair, color vision, and hearing fully developed. A deer fawn is born able to walk within 20 minutes of birth.

EXPLORATION: Marsupial Mammals

For this activity, you will need:

- "Life Cycle of a Kangaroo" from the Printable Pack
- Colored pencils or crayons
- Video about the kangaroo life cycle

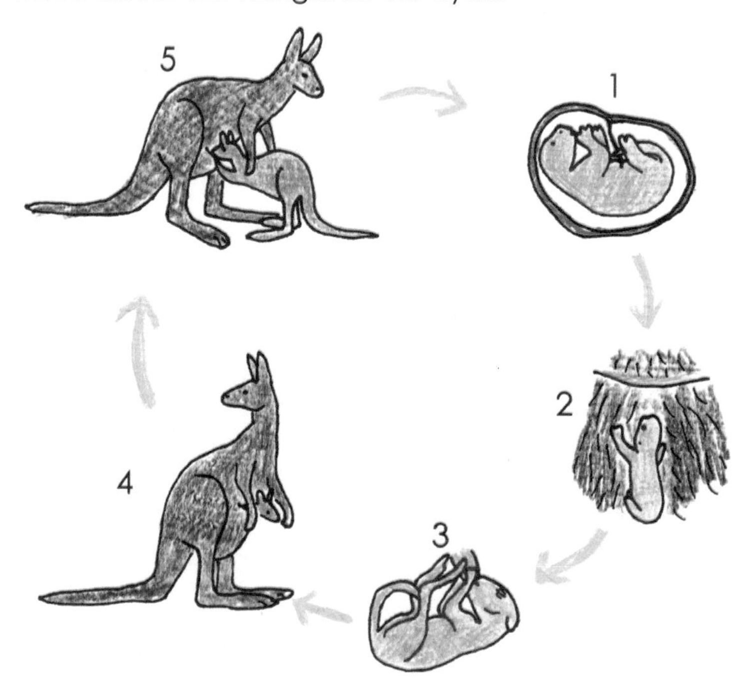

Writer's Workshop

Write a report about your favorite animal.

For young children there is a printable "My Favorite Animal Is" in the Printable Pack. Cut apart the pages, draw, color, write, then assemble the pages in any order you like.

Older kids can use the *Writer's Workshop: Reports & Essays* unit for more direction.

Fabulous Fact

Australia and New Guinea have hundreds of marsupial species. They are divided into six orders. The two largest orders are the dasyuromorphia—which includes small shrew like animals, quolls, and the Tasmanian devil—and the diprotodontia—which includes kangaroos, wallabies, possums, koalas, and wombats.

Dasyurus maculatus or tiger quoll

Dendrolagous goodfellowi or tree kangaroo

Marsupials of the Americas, such as opossums, are in completely different orders.

Marsupial mammals give birth very early in the development of the fetus and nourish their young in a pouch for the remainder of their development. The whole developmental process is different from placental mammals.

1. Read the descriptions below and follow along with the "Life Cycle of a Kangaroo" printable. Color each stage as you go. The numbers correspond to the numbers on the sheet.

   The placenta develops from the yolk and the amniotic sac, not from a layer of the blastula as in placentals. Development inside the mother is completed within 5 weeks, for some species it's as short as 12 days. But marsupial mothers can put the development of their babies on hold for months, if needed, until conditions are right for birth.

2. The newborn joey is blind, hairless, deaf and about the size of a jelly bean or smaller. The newborn marsupial's front limbs and face are the most highly developed parts at birth, so it can climb up its mother and suck from her teats.

3. The newborn marsupial crawls up its mother's body and into the pouch where it attaches itself to a teat where it is nourished with milk as its development continues.

   Early birth is more risky for the young, but it protects the mother from the dangers of long pregnancies, birthing large young, and famine during pregnancy.

4. The joey continues to develop inside the pouch for several months. At the end of this period, it begins to emerge from its mother's pouch to feed and learn from its mother.

5. After a few more months, it will spend most of its time outside its mother's pouch but still feed from her until it is about 18 months old.

6. Watch a video about kangaroo life cycles. Discuss how marsupial mammals differ from placental mammals.

☻EXPERIMENT: Fetal Pig Dissection

WARNING: This activity uses sharp tools and includes powerful chemicals. Children should be taught to use tools correctly and to never eat during this lab. Thoroughly wash hands and all surfaces after the experiment.

For this activity, you will need:

Fabulous Fact

There are two orders of marsupials in the Americas. The Didelphidae are opossums, which live in both North and South America. The Pauciturberculata are shrew opossums and only live in South America.

This is *Caenolestes sangay* the eastern caenolestid, a type of shrew opossum from the eastern side of the Andes in Ecuador.

Photo by Jbritomolina CC by SA 4.0, Wikimedia

Deep Thoughts

Discuss the differences and similarities between bats and birds or between sharks and dolphins.

There are good reasons for animals being put in the class they are in. But, could you argue for a different definition? Come up with a convincing argument for why birds and bats should be considered the same class.

The purpose of this thought exercise is to get kids to consider carefully all the possibilities and work out the reasons for themselves.

Bookworms

Pigology by Daisy Bird is an illustrated encyclopedia about pig behavior, pig breeds, pigs in popular culture, and the history of pigs. For ages 6 to 13.

Pair animal dissections with books that teach about the animal to foster wonder and reverence instead of callousness.

Writer's Workshop

Choose a mammal and become an expert on it. Read books, watch videos, and observe the animal in person if possible. Record body features, behaviors, eating habits, habitat, and life cycle.

When you are done researching, write a report and create a project about your animal. Your project could be a:

- Poster
- Diorama
- Power point
- Video
- Paper maché model
- Conservation or service project

- "Pig Anatomy" and "Pig Anatomy Answers" from the Printable Pack
- Colored pencils
- Fetal pig dissection kit
- Disposable gloves
- Goggles
- Rubber bands - larger size

Mammals all have similar internal anatomy with the digestive tract, circulatory system, lungs, and so on in roughly the same locations. If you study the internal anatomy of the pig, you know mammal anatomy.

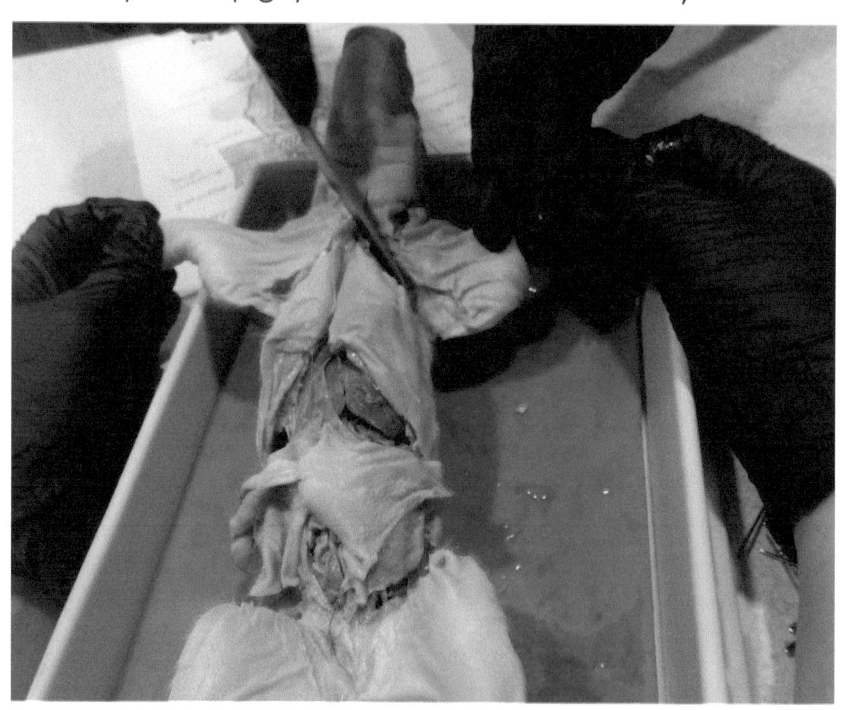

1. Color and label the "Pig Anatomy" using the "Pig Anatomy Answers" as a guide.
2. Dissect the pig using the guide that came with your dissection kit. We also recommend videos to guide you through this dissection.

EXPERIMENT: Mammal Intelligence

For this activity, you will need:

- Dog - if you don't have a pet dog, borrow a friend's dog
- Treat or toy - this should be a favorite of the dog
- Can large enough to hide the treat or toy
- Other materials as needed

Mammal brains are roughly twice the size of a bird brain in a similarly-sized animal and ten times as large as a reptile brain. Mammals also have a six-layered neocortex and other regions of the brain are more highly developed version of other vertebrates. Mammals are intelligent.

1. Give your dog an intelligence test. First, the dog needs to be comfortable and relatively calm. The environment where the test is conducted should be familiar to the dog and free of distractions.
2. Place a treat or toy in front of the dog, then place a can over the treat or toy. Give the dog three points for immediately getting the treat, two points for nosing at the can but failing to get the treat, and one point for ignoring the can.
3. Design two more tests for dog intelligence. The tests should show whether a dog can problem solve. The tests should not include any risks of injury to the dog.
4. Show the dog lots of love when you finish the tests.
5. If you like, you can give other dogs the same tests. Are some dogs smarter than others?

EXPERIMENT: Respiration Rates
For this activity, you will need:

- Pet mammal - rabbit, hamster, dog, cat, etc.
- Stopwatch

Mammals breathe with lungs which have many compartments, giving mammals a large surface area to take oxygen from the air.

1. Watch your pet at rest. Can you see it breathing? Count the number of breaths in one minute. Repeat your count three times and then average the numbers to get

Writer's Workshop

There are many different species of wild dogs around the world. Pick one species, read about it, then write about it. Include information on habitat, hunting, and pack structure.

Here are some ideas:

- Wolf
- Fennec fox
- Dingo
- Bush dog
- Dhole
- Coyote
- Jackal
- Maned wolf
- Cape hunting dog

Additional Layer

A classic experiment on animal intelligence is the mouse in a maze. You can try this if you have a rodent for a pet. Build a maze out of wood with a plexiglass top so you watch the mouse learn. Put some food, like peanut butter at the goal of the maze. Time how long it takes the mouse to get from one end of the maze to the other. Run the maze a second time. Does the mouse get faster? In other words, does the mouse learn?

Deep Thoughts

We know that some species of birds are highly intelligent. Crows, ravens, and parrots have all tested highly on problem solving, reasoning, and communication tests, often beating out primates like chimpanzees and gorillas.

How can that be if their brains are missing a neocortex and are so much smaller? What do you think we are missing when we study intelligence?

a resting breathing rate.

2. Repeat the experiment with a different mammal. You could use another pet or a human.
3. Compare the resting breathing rates of the two mammals. Are the breathing rates different? Why do you think they might be different?
4. Think of some more tests you could run with breathing rates. Are children different from adults? How does breathing rate change when exercising?

EXPLORATION: Marine Mammals

For this activity, you will need:

- Book about a marine mammal
- Science Notebook
- Internet how-to-draw tutorials (optional)
- Colored pencils or crayons
- Art paper
- Ultra fine-tipped permanent marker
- Watercolor paints and brushes

All mammals breathe air with lungs, so most mammals live on land. But some mammals spend most or all of their lives in the water. Walruses, seals, sea lions, polar bears, whales, dolphins, manatees, and otters are **marine mammals** that spend all or most of their lives in the sea.

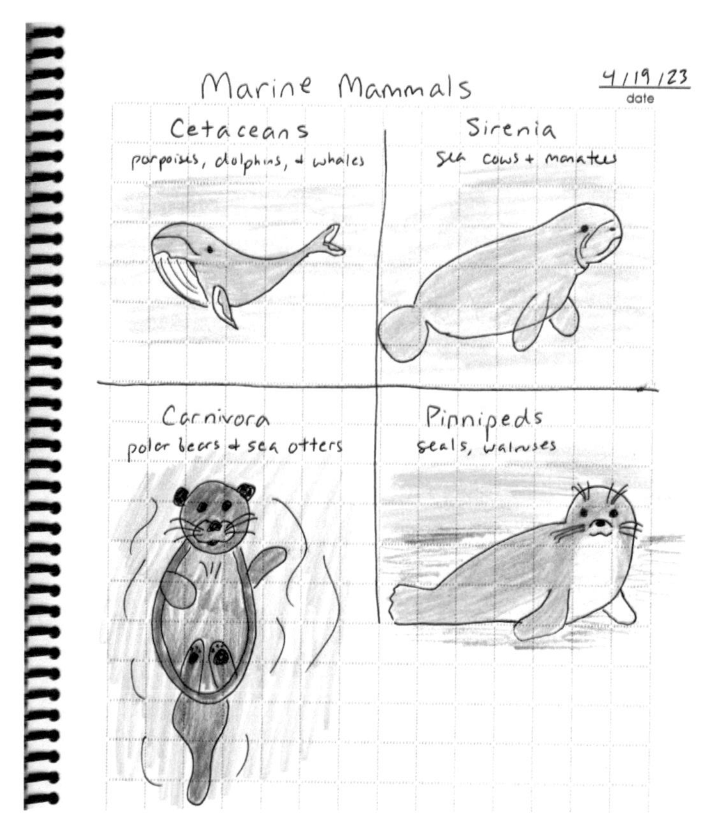

Additional Layer

There are two main types of whales: toothed and baleen.

Toothed whales have teeth and eat fish, squid, and other sea creatures.

Baleen whales have sheets of baleen, which is made of the same stuff as fingernails, just inside their mouths. Baleen filters tiny creatures out of sea water as the whale swims.

There are 14 species of baleen whales and 76 species of toothed whales, which include dolphins and porpoises.

There are many more differences between the two types of whales. Go learn more.

Color "Whales" from the Printable Pack.

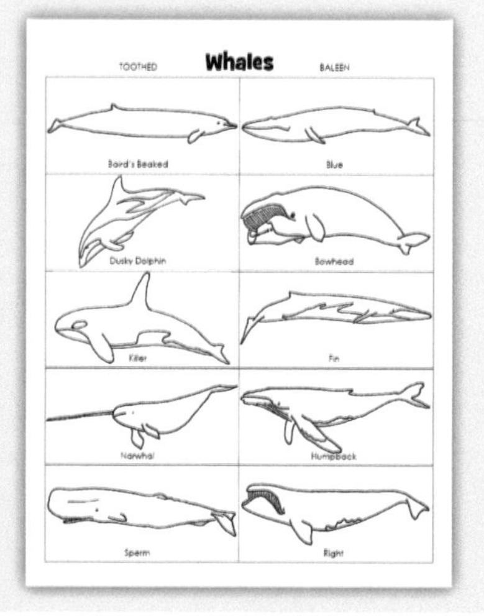

Memorization Station

Marine mammal: a mammal that spend most or all of its life in the sea and is dependent on the sea

1. Read a book about a marine mammal.
2. Discuss the marine mammal you learned about:
   a. Why is this animal considered a mammal?
   b. What body parts and behaviors does the animal have that helps it live in the sea?
   c. What similarities are there between this mammal and mammals that live on land?
   d. What foods does the animal eat?
3. Head a page in your Science Notebook "Marine Mammals." There are four groups of marine mammals:
   - Cetaceans: whales, dolphins, porpoises
   - Sirenia: sea cows, manatees
   - Carnivora: sea otters, polar bears
   - Pinnipeds: seals, walruses

   Which group is your animal in? Draw a four square chart in your Science Notebook of the four groups. Sketch an example of each group in its box. You can use online how-to-draw tutorials if you like. Color your pictures.
4. Sketch the marine mammal you read about really big on a piece of art paper. You can use an online how-to-draw tutorial.
5. Once you have it how you like, ink the lines with an ultra fine-tipped permanent marker. Paint the animal and its background with watercolor paints.
6. Show off your art to an audience and present three interesting facts about your animal to the audience.
7. Display your paintings until the end of this unit.

EXPEDITION: Mammals in the Wild

WARNING: Wild animals can be dangerous. Do not approach wild animals or feed them. Even animals like deer, squirrels, and rabbits can bite, kick, or otherwise harm you. Keep your distance!

For this activity, you will need:

- A wilderness area where you can observe wildlife
- Mammal identification guide
- Science Notebook
- Pencil and colored pencils
- Binoculars (optional)
- Camera trap (optional)

Expedition

Take a trip to a farm, a veterinarian, a pet store, a wildlife rescue center, or a wildlife refuge.

Go prepared with information or questions and take at least one good picture while you are there.

When you come back home, print your picture and put it in your Science Notebook along with notes about the trip.

Famous Folks

David Attenborough is a biologist and nature presenter who is famous for his BBC wildlife shows.

Photo by Department of Foreign Affairs and Trade website – www.dfat.gov.au

You might like watching *Wild Isles, Life In Color, Planet Earth* or *Planet Earth II* during this unit.

Bookworms

Here are some more of our favorite books with animal characters:

A Beatrix Potter Treasury by Beatrix Potter.

Redwall by Brian Jacques.

An Elephant in the Garden by Michael Morpurgo.

The Trumpet of the Swan and *Stuart Little* by E.B. White.

Brighty of the Grand Canyon and *Misty of Chincoteague* by Marguerite Henry.

Summer of the Monkeys and *Where the Red Fern Grows* by Wilson Rawls.

Rascal by Sterling North.

The Story of Doctor Doolittle by Hugh Lofting.

Black Beauty by Anna Sewell.

The Mouse and the Motorcycle by Beverly Cleary.

The One and Only Ivan by Katherine Applegate.

Pax by Sara Pennypacker.

Billy and Blaze by C.W. Anderson.

1. Go to an observation point where you are likely to see mammals. The best places are near water. You can also look for animal tracks in the mud on a riverbank.
2. Use an identification guide to determine what species you have seen. Draw a sketch and make notes in your Science Notebook.
3. If you can, set up a camera trap in a location where mammals are likely to come. A camera trap takes a picture when it detects motion. You can purchase a camera trap set up for less than $50.

EXPEDITION: Zoo

For this activity, you will need:

- A zoo near you
- Camera to take photos
- Poster board
- Crayons, markers, or colored pencils
- Other poster-making supplies as needed

Most of the animals kept by zoos are vertebrates.

1. Visit a zoo with all your new knowledge about animals. Pay attention to whether an animal is a reptile, amphibian, fish, bird, or mammal.
2. Choose an animal you really like at the zoo. Make sure you get a good photograph of it. Back at home, research more about your animal and make a poster about it, including the photo you took at the zoo. The poster should have information about diet, habitat, life cycle, and what the animal is really good at.

Step 3: Show What You Know

During this unit, choose one of the assignments below to show what you have learned during the unit. Add this work to your Layers of Learning Notebook. You can also use this assignment to show your supervising teacher or your charter school as a sample of what you've been working on in your homeschool, if needed.

There are more ideas for writing assignments in the "Writer's Workshop" sidebars.

Coloring or Narration Page

For this activity, you will need:

- "Warm-Blooded Vertebrates" from the Printable Pack
- Writing or drawing utensils

1. Depending on the age and ability of the child, choose either the "Warm-Blooded Vertebrates" coloring sheet or the "Warm-Blooded Vertebrates" narration page from the Printable Pack.
2. Younger kids can color the coloring sheet as you review some of the things you learned about during this unit. On the bottom of the coloring page, kids can write a sentence about what they learned. Very young children can explain their ideas orally while a parent writes for them.
3. Older kids can write about some of the concepts you learned on the narration page and color the picture as well.
4. Add this to the Science section of your Layers of Learning Notebook.

Science Experiment Write-Up

For this activity, you will need:

- The "Experiment" write-up or "Experiment Report Template" from the Printable Pack

1. Choose one of the experiments you completed during

Famous Folks

In the 1960s, Louis Leakey recruited three young women scientists to study primates in Africa. The women went on to distinguish themselves as the foremost experts in the world.

Jane Goodall studied chimpanzees in Tanzania.

Dian Fossey studied mountain gorillas in Rwanda.

Birutė Galdikas studied orangutans in Borneo.

Photo by Simon Fraser University

Unit Trivia Questions

1. Warm-blooded means:

   a) The blood is heated by the heart.

   b) The animal maintains its body heat by using energy from its food.

   c) The animal has a higher body temperature than cold-blooded animals.

2. What are the two groups of warm-blooded animals?

   Birds and mammals

3. True or false - All animals come from eggs.

   True, all eukaryotic organisms begin as fertilized eggs. Sometimes the eggs are microscopic and sometimes they are quite large.

4. A bird's beak is made of:

   a) Keratin, like fingernails and horn

   b) Cartilage

   c) Bone

5. Name the features of a bird.

   Birds have feathers and wings, are warm-blooded, and lay eggs.

6. Name the features of a mammal.

   Mammals have hair, are warm-blooded, give birth to live young, and feed their young with milk.

7. Name the stages of development of a mammal.

   Fertilization, embryo, fetus, birth, juvenile, adult

8. A manatee is called a ________ mammal because it lives in the sea.

   Marine

this unit and create a careful and complete experiment write-up for it. Make sure you have included every specific detail of each step so your experiment could be repeated by anyone who wanted to try it.

2. Do a careful revision and edit of your write-up, taking it through the writing process, before you turn it in for grading.

Game Show

For this activity, you will need:

- Five pieces of colored paper cut into quarters
- Permanent marker
- Masking tape
- Wall
- Big Book of Knowledge

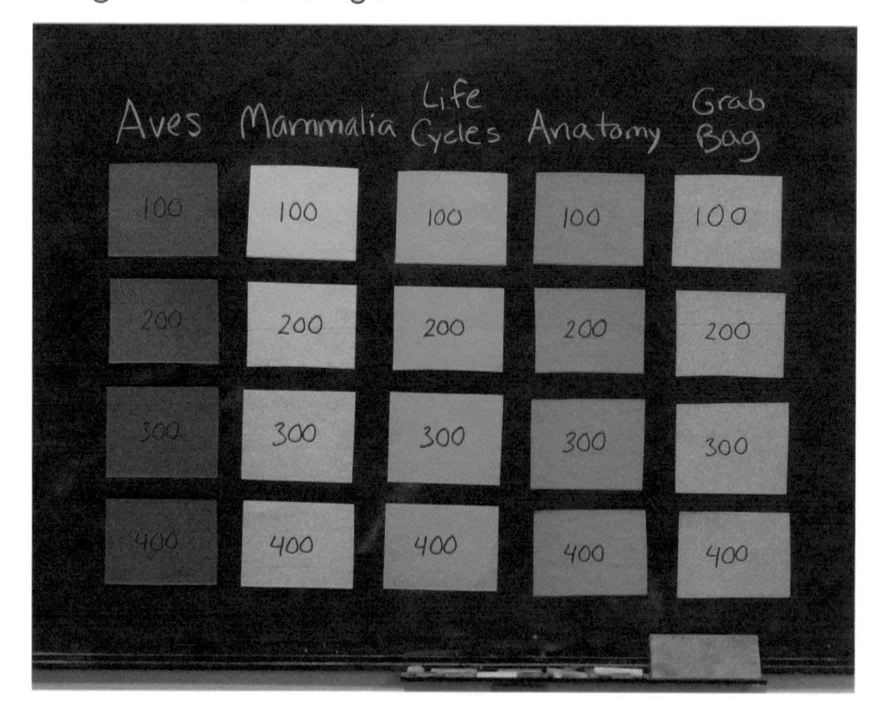

1. Number each set of colored paper rectangles 100, 200, 300, and 400. Tape them up on the wall in a grid with the colors in columns. Above each column, write the category on another piece of paper or on a strip of masking tape. Use categories like these: Aves, Mammalia, Life Cycles, Anatomy, and Grab Bag.

2. The first player chooses a category and a point amount. The mentor asks a question from the Big Book of Knowledge—easier questions for lower point amounts and younger children, harder questions for higher point amounts and older children.

3. Keep track of the points as you go. Highest points wins.

Animal Order Expert

For this activity, you will need:

- Slide show maker
- Internet
- Book or books about the order you choose
- Screen to use for presentation

1. Pick a bird or mammal order to become an expert on. Read a website, watch a video, or read books about the order. Find images online and make a slideshow presentation that includes images and text.
2. If possible, observe animals from the order firsthand. Include your original observations in your presentation.
3. Present your information and slideshow to an audience.

Big Book of Knowledge

For this activity, you will need:

- "Big Book of Knowledge: Warm-Blooded Vertebrates" printable from the Printable Pack, printed on card stock
- Writing or drawing utensils
- Big Book of Knowledge

1. Color, draw on, write on, or add to each of the Big Book of Knowledge pages you are using. Only add concepts you learned during this unit. If there are other topics you focused on, feel free to add more of your own pages to your Big Book of Knowledge.

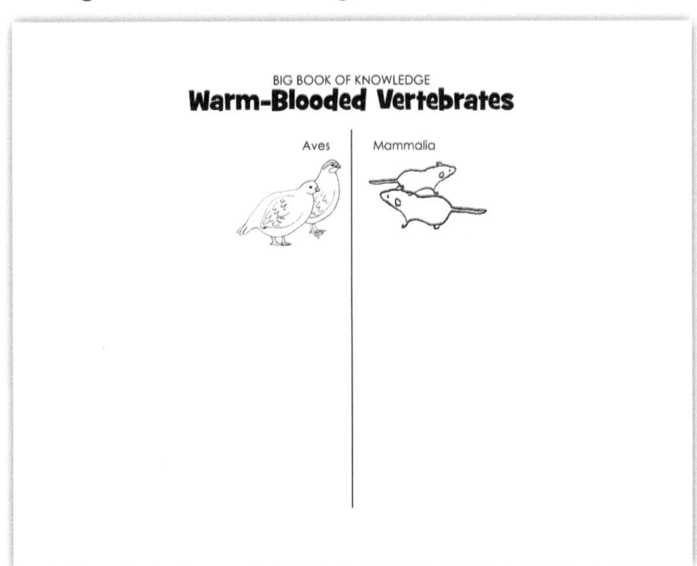

2. Use your Big Book of Knowledge regularly to help you review, quiz, or create games that will help you commit the things you've learned to memory.

Big Book of Knowledge

The Big Book of Knowledge is a book for you, the mentor, to use as a constant review of all of the things you're learning about. You can use it to quiz your kids or prepare tests or review games. Whenever you learn something in Layers of Learning that you want your kids to remember, add it to your Big Book of Knowledge.

Assemble your Big Book of Knowledge in a binder or with binder rings. Divide it into sections for each subject.

In the Printable Pack for this unit you will find a "Big Book of Knowledge" sheet. You can add this sheet to others you collect or create yourself as you progress through the Layers of Learning curriculum. Customize the Big Book of Knowledge to your family by adding facts and topics that you enjoyed exploring as you were learning.

Visit Layers of Learning online to find more information on how to assemble and use your own Big Book of Knowledge.

You will also find cover and section pages to print along with creative games to play with your Big Book of Knowledge to keep school, even the tests, fun!

ECOLOGY

Ecology is the study of interactions between organisms and the environment they live in. Ecologists think about the water, soil, air, microorganisms, plants, and animals in a place. It's a big-picture science that steps back to see how all living things interact with each other and all the non-living things to make life possible. No organism can thrive by itself.

Ecologists study everything from the lichen on tree bark to the deer population of the forest and see it as a whole system.

All ecosystems are in a state of change all of the time. Populations of animals or plants migrate, fluctuate, and compete with each other. The weather shifts, storms happen, climate adjusts, volcanoes erupt, and living things move to survive or to fill a void. Ecologists study how ecosystems develop and change over time. They explore what makes an ecosystem strong and resilient and what causes damage or even destroys one. The goal is partly just to understand but also to learn how we, as humans, can help the ecosystems of our planet to be healthy.

Unit Overview

Key Concepts:

- Ecology is a science that studies interactions between living things in their natural environment.
- The study of ecology is a new science and we are just at the beginning of understanding the complexities of the natural world.
- Human beings can have a large effect for good or for ill on the living world.

Vocabulary:

- Ecology
- Environmentalism
- Conservation
- Biome
- Ecosystem
- Ecological niche
- Habitat
- Energy
- Energy flow
- Producer
- Consumer
- Decomposer
- Food chain
- Apex predator
- Food web
- Trophic level
- Community
- Biodiversity
- Ecological succession
- Population
- Pollution
- Contaminant
- Natural resources
- Introduced species
- Extinction
- Endangered species

Theories, Laws, & Hypotheses:

- Keystone species hypothesis
- Food web theory
- Energy flow theory
- Succession theory
- Intermediate disturbance hypothesis
- Extinction theory

Step I: Library List

Choose books from your library that go with this topic. Here's a list of some favorites and also a list of search terms so you can utilize what your library offers. Read the books with your kids and/or assign them some to read independently. It is from these books your kids will learn most of the facts they need from this unit.

Search for: ecology, ecosystems, biodiversity, food chains, conservation, desert biome, forest biome, ocean biome, coral reef biome, grassland biome, arctic biome, hibernation, migration

Encyclopedia of Science from DK. Read the entire "Ecology" section on pages 369-400.

Kingfisher Science Encyclopedia. Read "The Natural Balance" through "Endangered Species" on pages 434-445.

The Usborne Science Encyclopedia. Read "Ecology," "Food and energy," "Balance in nature," and "Conservation" on pages 330-337.

Over and Under the Pond by Kate Messner. A boy and his mother go out on a pond in the Adirondack Mountains and notice all the living things. Also look for *Over and Under the Snow* by the same author.

Migration: Incredible Animal Journeys by Mike Unwin. Follows the travels of 20 different animal species to see how they cover vast distances with the changing seasons.

Human Footprint by Ellen Kirk. A highly visual book with sparse words that highlights how much we (in the wealthy western world) consume. Makes you think about our impact on the planet. Great book to discuss with kids of all ages.

Who Eats What? Food Chains and Food Webs by Patricia Lauber. An excellent, simple introduction to food chains.

Secrets of the Garden by Kathleen Weidner Zoehfeld. A family plants a garden in the spring and then watches as the plants grow, insects and animals eat the plants, animals eat other animals, and the people eat the plants.

What If There Were No Bees? A Book About the Grassland Ecosystem by Suzanne Buckingham Slade. So much more than bees. Explains keystone species, the importance of a single organism, and how ecosystems are interrelated parts.

My First Book About Backyard Nature: Ecology for Kids by Patricia J. Wynne and Donald M. Silver. Takes you on a tour of a backyard that might be similar to yours. Animals, plants, bugs, soil, mushrooms, and more are put in context of an ecosystem.

Frog Heaven: Ecology of a Vernal Pool by Doug Wechsler. Takes the reader on an intimate tour of a vernal pool in New England, USA. What lives there? How do the organisms interact?

Bringing Back The Wolves: How A Predator Restored

Family School Levels

The colored smilies in this unit help you choose the correct levels of books and activities for your child.

= Ages 6-9
= Ages 10-13
= Ages 14-18

On the Web

For videos, web pages, games, and more to add to this unit, visit the Biology Resources at Layers-of-Learning.com.

You will find a link to video playlists, web links, and more.

Bookworms

If you're looking for a family read-aloud, we'd like to suggest this one.

Hoot by Carl Hiassen is about Roy who just moved to south Florida, runs into a bully, and gets caught up trying to save some endangered burrowing owls from a pancake project. Hilarious and moving, this book is full of chaos and growing up.

Fabulous Fact

Ecology is the study of how living things interact. Ecologists have four aims:

1. Understand how and why life does what it does.
2. Study energy transfer through living systems.
3. Trace successional development of environments.
4. Study the abundance and distribution of life.

This ecologist works for the National Park Service at Joshua Tree NP. She is teaching children how to replant a species to help restore a damaged ecosystem.

Expedition

During this unit, take a trip to a wilderness area near you. Spend time smelling the clean air, watching ants, and enjoying the beautiful sights of nature.

When you get back home, pick a picture from your trip to put in your Science Notebook with a caption.

an Ecosystem by Jude Isabella. A picture book about the eradication and then restoration of the wolves of Yellowstone Park.

☺ ☻*Tracking Trash* by Loree Griffin Burns. A scientist near Seattle began to track where the trash that washed up on beaches came from. Explores the trashy impact humans are having on our oceans.

☺*Ecology* by Brian Lane and Steve Pollock. This is a DK Eyewitness book full of colorful images and excellent information about populations, food webs, earth cycles, diversity, and so much more.

☺*Biodiversity* by Laura Perdew. A mini-textbook about life on Earth, how it evolved into the many species we have now, and how it is interconnected. Includes many projects kids can do to learn more.

☺*The World of Food Chains With Max Axiom, Super Scientist* by Liam O'Donnel and Bill Anderson. A graphic-style book where our hero goes into nature to find out about energy and its path through the ecosystem.

☺*Exploring Ecosystems with Max Axiom* by Agnieszka Jòzefina Biskup. Our hero goes exploring with an ecologist to learn about ecosystems in this graphic-style book.

☺*Inside Ecosystems and Biomes* by Debra J. Housel. This book does a great job explaining what an ecosystem is, how they are interconnected and the role of organisms.

☺ ☻*The Wondrous Workings of Planet Earth* by Rachel Ignotofsky. Teaches ecology concepts like populations, energy flow, and communities. This takes the reader on a tour of specific ecosystems. Focuses pretty hard on how humans are negatively impacting the earth. Read aloud to younger kids as well.

☺ ☻*The Ecology Book* by Tom Hennigan and Jean Lightner. This book is aimed at Christians who want to understand the diversity and design of the world in terms of God and creation. It is best used by a parent guiding a child.

☻*The Ecology Book* by DK. This is a general book that covers all the basic concepts of ecology on a beginner adult level but with lots of images and graphs.

☻*The Forest Unseen: A Year's Watch in Nature* by David George Haskell. The author, a naturalist, repeatedly visits a small patch of forest near his home through the course of an entire year. This book is the story of that little piece of Earth—a small patch that gives a window into the whole world and how it lives season by season.

☻*A Sand County Almanac* by Aldo Leopold. A classic in the world of ecology, this book takes readers on a tour of the North American wilderness, its beauty, and what we stand to lose if we don't learn to appreciate it.

☻*Messages From Islands: A Global Biodiversity Tour* by Ilkka Hanski. This book takes a close look at several different island ecosystems to learn how biodiversity develops through evolution, how biodiversity changes over time, and what happens when habitats are changed or lost.

☻*Nature's Best Hope: A New Approach to Conservation that Starts in Your Yard* by Douglas W. Tallamy. This is a book about how every single family can improve the natural world by changing our yards from grass mono-cultures and ornamental oases into natural corridors for native wildlife.

☻*Niche*, *Ecological Succession*, and *Aposematic Coloration* from Bozeman Science on YouTube. Use these videos as lectures for your high schooler. Have your student take notes from the videos. On the Bozeman Science site you'll also find a Biology unit completely focused on Ecology with lots of great topics explored. You can assign your high schooler to complete this series of videos alongside this unit.

Step 2: Explore

Choose a few hands-on explorations from this section to work on as a family. They should be appealing activities that will create mental hooks so your kids remember the information in the unit. Save the rest of the explorations for the next time you do this unit in four years when your kids are older. You can also read the sidebars together and explore some little rabbit trails.

This unit includes printables. See the introduction for instructions on retrieving your Printable Pack.

Ecosystems

Before you begin, make sure you understand the differences between **ecology**, **environmentalism**, and **conservation**.

☻ ☻ ☻**EXPLORATION: Biomes**

For this activity, you will need:

- "Biomes" and "Biomes Instructions" from the Printable Pack
- Colored pencils

Memorization Station

These terms are often confused:

Ecology: the science of how living things interact

Environmentalism: a political movement that seeks to protect nature though the force of government

Conservation: using natural resources responsibly so they are still available for the future

Writer's Workshop

Find out what you have to do to become an ecologist or wildlife biologist. Write yourself up a resume for your dream wildlife job as though you have all the education and training you would like.

You might find the *Writer's Workshop: Letters* unit useful in learning how to write a resume.

Fabulous Fact

Different ecologists categorize biomes differently. There is no set standard defining which biomes exist and where precisely they are on Earth. Some scientists name five biomes while others say there are twenty.

Also, biomes are very generalized. You can look at nearly any map of biomes and realize that it's not entirely accurate. You'll be aware that there's a desert or a temperate forest or even a tundra climate that is not shown on the map. But the concept of a biome is only useful if it is generalized.

Memorization Station

Biome: a geographical region with dominant plant and animal populations due to the climate

Ecosystem: a community of plants and animals that interact with the non-living things in a specific place

Additional Layer

Print out the "Habitat Sort Cards & Mats" (8 pages), then cut apart the cards and have your little ones sort the animals into their habitats.

Why do animals live in some places and not in others?

What makes this animal's body perfect for this habitat?

Bookworms

Alexander von Humboldt: Explorer, Naturalist, & Environmental Pioneer by Danica Novgorodoff is a picture book biography of the founder of ecology.

- Book or video about biomes
- Medium box, like a shoe box
- Natural items from your area, along with miscellaneous craft supplies

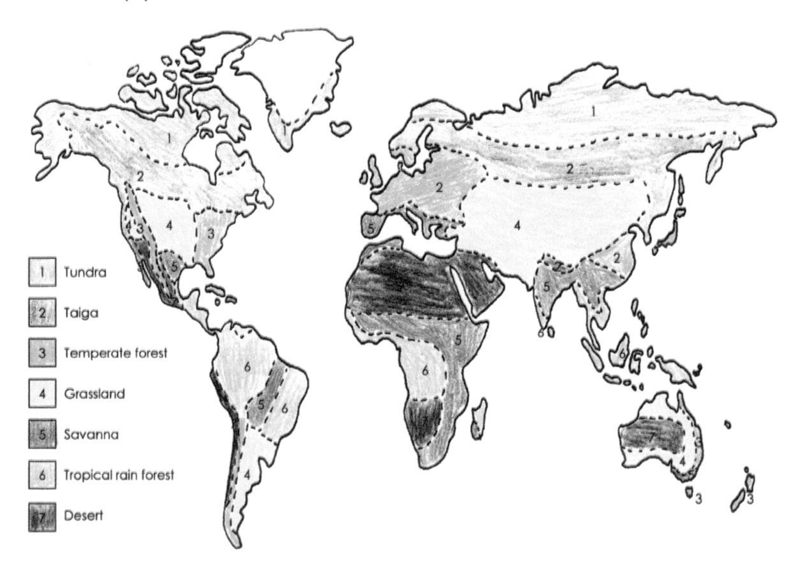

A **biome** is a large ecological area on earth with plants and animals that require similar conditions. Biomes are dependent on the amount of rainfall, combined with the average yearly temperature. They are subjective, so different people define them in different ways. In this unit, seven different biomes are explored—tundra, taiga, temperate forest, grassland, savanna, tropical rain forest, and desert.

1. Complete the "Biomes" map using the "Biomes Instructions" sheet.
2. Read a book or watch a video about the biome you live in to learn more.
3. Go outside and collect some specimens that are typical of your biome. For example, you might collect a pine cone and a wildflower that are native to where you live. Make a display box of your biome using the items and craft supplies.

EXPERIMENT: Pop Bottle Ecosystem

For this activity, you will need:

- Clear soda pop bottle
- Small annual plant like pansies
- Potting soil
- 2-3 small fish like guppies or gold fish
- Aquarium rocks
- String
- Paper coffee filter
- Scissors
- Water at room temperature
- Fish food
- Science Notebook

An **ecosystem** is an interconnected zone where life exists. It includes the air, the water, the soil, the animals, the plants, and the bacteria in that zone. The space an ecosystem takes up depends on how a person defines it. The entire Amazon Rainforest could be an ecosystem, but so could the life cradled in the water trapped in a single leaf high in the canopy of the rainforest.

1. Cut the top off a clear 2-liter pop bottle just below the shoulder, where the bottle stops curving and goes straight down.
2. Add aquarium rocks, water, and a few small fish to the bottom section of the pop bottle.
3. Get two coffee filter papers and cut two small holes in the center. Cut a piece of string long enough so it can reach from the top of the bottle down to the water with the goldfish. Tie a knot in one end of the string. Thread the unknotted end through the top of the filter papers, through the

Additional Layer

The organization of life begins with the cell, the smallest unit of life, and extends up to the biosphere. The top six levels, starting with organism, are considered the levels of ecology.

- Cell
- Tissue
- Organ
- Organ system
- Organism
- Population
- Community
- Ecosystem
- Biome
- Biosphere

Make a circle-shaped step book that shows each of the levels of ecology.

You can learn or review all of the levels by labeling foam cups with the level names and gluing pictures representing each level to the cups. Mix them up and have the kids put them back in order for a quick quiz.

The "Levels of Organization of Life" pictures you see on the cups, above, are in the Printable Pack.

little holes you made. Set the filter paper down inside the inverted top of the bottle that you already cut off.

4. Put your plant into the coffee filter inside the inverted top of the pop bottle. You may need a little extra gardening soil to fill up the space.
5. Place the plant and top of the bottle into the lower part of the bottle with the fish so the string dangles into the water.
6. Poke a small hole in the side of the lower part of the bottle, above the water line, but low enough that your child can put a flake or two of fish food through the hole to feed the fish each day.

   The water, rich with nutrients from the fish, will wick up the string to water your plants. You have made a little ecosystem with a plant, animals, water, soil, rocks, air, sunlight, and food that you replenish every day.
7. Draw a labeled diagram of your pop bottle ecosystem in your Science Notebook. Explain how the parts of the ecosystem are connected to one another.

EXPEDITION: Ecological Niche

For this activity, you will need:

- A park, cemetery, yard, or forest with large mature trees
- String - cut to about 3 meters long
- Compass or compass app on a phone
- Tree identification guide
- Moss and lichen identification guide (optional)
- Hand lens or magnifying glass
- Science Notebook

Famous Folks

John Muir was a naturalist who advocated for wilderness preservation in the United States in the late 1800s. He is the father of the National Parks.

You can do an exploration about the him and the National Parks in the Mainland North America unit.

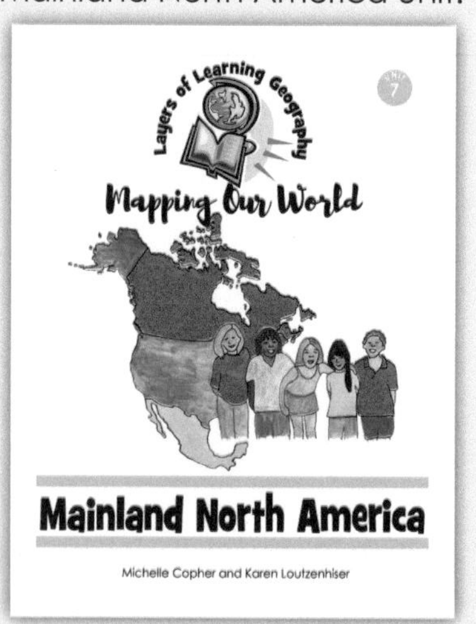

Writer's Workshop

Go outside and observe something beautiful in nature. It could be the clouds above a busy city or the blades of grass in your yard. Imagine it could talk to you. Write a dialogue between you and nature.

An **ecological niche** is the exact role an organism plays in its ecosystem. The living thing is shaped by the ecosystem and, in turn, it shapes the ecosystem it lives in. In a temperate forest, a tawny owl nests and roosts in large trees and preys on small rodents. So the niche of the tawny owl is as a flying deciduous forest predator of small rodents that hunts at night.

1. Choose a large, mature tree to study. It can be in your yard, a park, a cemetery, or in a wild place. Deciduous trees are best but conifers work well too. Make sure there are things growing on the trunk.
2. Using a tree identification guide for your area, identify the species of tree. Make a sketch of the tree, including a close up sketch of the bark and leaves in your Science Notebook. Take notes about the roughness of the bark, the amount of light, and moisture.
3. Wrap a piece of string around the trunk of the tree at about eye level. Tie the string off with a shoelace-style knot that will come undone easily. Using a compass, find north; note which side of your tree is north. Draw two small circles, one in the top half and one in the bottom half, in the center of a page in your Science Notebook. Each circle represents the tree at the point where the string is tied. Mark your circle with the N, S, E, and W points of the compass.
4. Look for mosses and lichens that are growing on the tree at the point where the string is wrapped around the tree. Sketch each organism around the circle you have drawn in your Science Notebook. Use a magnifying glass or hand lens to look at the moss or lichen up close. Use a moss and lichen identification guide to identify which species you have. Identifying species of mosses and lichens is difficult because they are so small and, at first glance, look the same. Do your best. Write down the species names where you have sketched them in your Science Notebook.

   If you don't have a moss and lichen identification guide, you can still do this. Just take careful notes and careful observations of the lichens and mosses you come across, sketch them in your notebook, and name or number each one yourself.
5. What do you notice about the places where the moss and lichen grow? Where does each species seem to grow best? Is one side of the tree shadier than another? Does one side seem to get more water than another?
6. Use the string you wrapped around the tree to measure

Memorization Station

Ecological niche: the role an organism plays in its ecosystem

Additional Layer

If you change one element in an ecosystem, you can cause drastic changes to the whole system. A good example of this are the wolves of Yellowstone National Park.

This is a wolf in Yellowstone Park.

The wolves were intentionally eradicated in the early 20th century and then reintroduced in 1995. Detailed studies have been done to show the far-reaching effect this one species has on the whole ecosystem of Yellowstone. Look up more information.

Deep Thoughts

John Muir wrote:

And into the forest I go, to lose my mind and find my soul.

What do you think this quote means?

Expedition

Most modern zoos do research and work on conservation efforts in breeding and returning animals to wild places where they have gone extinct.

Visit a zoo or marine park, enjoy looking at the animals, and look for information about breeding or conservation programs.

Memorization Station

Habitat: the place an organism lives, including its food, water, shelter, and space

A biome is broader than a habitat and may include several habitats within it. For example, the rainforests of Africa are a biome, but the habitat of the okapi is only above 1,600 feet in elevation within the rainforest.

So a biome is a geographical area that can be mapped for the whole globe, but a habitat is specific to every individual species.

the circumference by measuring the length of the string that wrapped around the tree.

7. Repeat the process on the same tree, but this time at knee level on the tree. Are there any differences or do your new findings confirm what you have already seen?
8. Repeat the entire process on another tree in the same area. You can repeat this as many times as you need to get a good survey of the moss and lichen growing patterns in your location.
9. Analyze your results. Does each species of moss or lichen seem to prefer a certain location on trees? Do they prefer certain tree species or types of bark over others? What is the niche of each of the species you studied?

EXPLORATION: Four Things Every Organism Needs

For this activity, you will need:

- "Habitat" from the Printable Pack
- Pencil and colored pencils or crayons
- Books or videos about crickets
- 4-6 crickets (Buy them at a pet shop where they are sold as food for other animals.)
- Large box to keep the crickets in
- Mesh lid for the box to keep the crickets from escaping
- Cricket food, moist bread, vegetables, or dry cereal, etc.
- Sponge or cotton ball soaked with water for crickets to drink from
- Several different colors of nail polish or model paint (one for each cricket)
- Thermometer
- Stopwatch
- Two gallon-size bags filled with dry rice
- Science Notebook
- Other materials, as determined by your research

Every living thing needs food, water, shelter, and space. It gets these from its **habitat**, the place an organism lives. Consider a date palm in an oasis in the Sahara Desert. Where does it get its food, water, shelter, and space? What about the white-winged guan that lives in the cloud forests of Bolivia? What does the wobbegong carpet shark of the coral reefs of Indonesia need to live? They all need the food, water, shelter, and space that their specific habitats give them.

1. On the "Habitat" sheet, fill in the blanks with: food, water, shelter, and space. In each box on the lower part of the page, draw a different animal in its habitat.

Include all four things the animal needs.

2. Learn about crickets by reading books and/or watching videos about them. Write down information in your Science Notebook about the food crickets eat, how they drink water, what their habitat is like, what temperature they like, if they need hiding places, and so on.
3. Design a cricket habitat inside a large box. Add a lid made of mesh material so they can't escape. Be sure to provide food, water, shelter, and space for the crickets. Draw a sketch of your cricket habitat in your Science Notebook. Label it.

4. Mark each of your crickets with a small dab of model paint or nail polish so you can keep track of which is which.
5. Experiment with your crickets to see what kind of food they like best. Place four different kinds of food in each of the four corners of their shelter. Mark how many times each cricket visits each kind of food. They are more likely to go straight for the food if they are hungry and haven't eaten in a couple of days. It won't hurt them. Record your results in your Science Notebook.
6. Put one bag of rice in a freezer for several hours to cool it down. Microwave the other rice bag for 3 minutes to warm it up. Place the cold bag under one end of the cricket habitat and the hot bag under the other end of the cricket habitat. Observe the crickets to see what kind of temperature they prefer. Use a thermometer to test the air temperature in the part of the habitat they

Fabulous Fact

Male crickets start to chirp when they are about 6 weeks old. Females do not chirp. Once your crickets begin to chirp, use a stopwatch to record how many times the males chirp in 25 seconds. Divide that number by 3 and add 4 to get the temperature in degrees Celsius.

Fabulous Fact

The keystone species hypothesis says that some animals have a bigger than expected effect on their environments.

A typical keystone species is a predator that prevents a herbivorous species from eating all the plants in a particular area.

Wolves, ochre sea stars, sea otters, American bison, elephants, prairie dogs, gnus, and beavers have all been called keystone species.

Bookworms

Return of the Sea Otter by Todd McLeish is about the people along the Pacific Coast who studied and recovered the sea otter.

An easy read for 14+ years.

Additional Layer

Chemotrophs are a group of organisms that get their energy from inorganic chemicals like iron and hydrogen sulfide. They can be found at hydrothermal vents on the ocean floor.

Learn more about these living things, how they get food, and how they fit into the energy flow of the planet.

Bookworms

Rotten! Vultures, Beetles, Slime, and Nature's Other Decomposers by Anita Sanchez is totally gross and completely awesome.

For ages 8 to 12, this book is 83 pages long, stuffed with facts, and absolutely entertaining to read and view. The illustrations are great! This makes an excellent read-aloud as well.

move to. Record your results in your Science Notebook.

7. Design an experiment to determine whether crickets prefer light or dark spaces. Before you run the experiment, write your hypothesis down in your Science Notebook. Record your results as well.
8. Design an experiment to see what kind of shelter, if any, crickets prefer. You can use toilet paper tubes, sections of egg cartons, or other items of different sizes, some with one opening, some with two, and so on.

Energy Flow

All life needs **energy** to survive and the amount of energy in an ecosystem is a major factor in determining how plentiful the life there is. Nearly all energy comes from the sun, is changed into food by plants, and then moves through animals as they eat plants and each other.

EXPLORATION: Producers, Consumers, Decomposers
For this activity, you will need:

- A large basket, like a clothes hamper
- Rolled up socks
- Book or video about producers, consumers, and decomposers from the Library List or the YouTube playlist
- Science Notebook

Energy is necessary for all living things. Living things get their energy from either sunlight or from eating other things that were once living. **Energy flows** from the sun to **producers**, then from producers to **consumers** and **decomposers**.

Living things can be categorized into three different roles—producer, consumer, and decomposer. Plants, algae, and bacteria are producers who make energy from sunlight. Animals and bacteria are consumers who eat plants or animals. Fungi and bacteria are decomposers who break down dead material.

1. Read a book or watch a video about producers, consumers, and decomposers.
2. Head a page in your Science Notebook "Producers, Consumers, and Decomposers" and add it to your table of contents. Divide your page into three parts and label the three parts—producers, consumers, and decomposers.
3. Set up an empty basket at one end of a room and stand back a few feet (further if your kids are older and closer if they are little). Call out a living thing. The child then has to declare the living thing a producer, consumer, or decomposer before shooting a sock into the basket.
4. Each time a successful shot is made, draw the living thing in the correct spot in your Science Notebook. So, if "oak tree" was called, the child will answer "producer," take a shot at the basket, and then draw an oak tree in his or her Science Notebook with the other producers.
5. Play until you have at least two living things in each section of your notebook page.

EXPLORATION: Food Chains
For this activity, you will need:

- A book or video about food chains from the Library List or the YouTube playlist
- "Food Chains" from the Printable Pack
- Internet (optional)

Animals eat food to get energy and nutrients so they can grow and reproduce. The food they eat is part of a chain of energy stretching back to the sun. A wolf eats a rabbit. The rabbit eats grass. Grass captures energy from the sun. The grass, rabbit, and wolf are part of a **food chain**.

1. Read a book or watch a video about food chains.
2. Complete the "Food Chains" printable by drawing plants and animals in each empty square to complete the food chains. If you aren't sure what eats what, look it up online.

Memorization Station

Energy: the ability to do work such as move, grow, or metabolize

Energy flow: the movement of energy through an ecosystem as living things photosynthesize or eat

Producer: an organism that makes its own food from sunlight energy

Consumer: an organism that eats other living things in order to get energy

Decomposer: an organism such as fungi, invertebrate, or bacteria that breaks down dead animals and plants

Food chain: a series of organisms each dependent on the next for a food source

Apex predator: an animal at the top of the food chain without natural predators of its own

Writer's Workshop

Think of the last time you were away from the city and out in nature. Maybe you went camping or visited a nature trail. Write what you loved about it. Use all of your senses. How did it smell, feel, look, and sound?

Famous Folks

Charles Elton was a British zoologist. He developed the ideas behind ecological niche, food chains, trophic levels, population cycles, and invasive species.

Additional Layer

Animal interactions between predator and prey aren't all one-sided. Prey animals often fight back with camouflage, poison, or other protection.

Choose one of these poisonous butterflies or moths. Learn more about it, then create a craft that shows off its warning markings.

- Monarch butterfly
- Cinnabar moth
- Giant African swallowtail
- Pipevine swallowtail
- Tiger moth

First, paint your hands in the colors of the butterfly you chose. Press your hands on to a sheet of white paper. Paint the butterfly's body. After the paint has dried, add feelers, legs, and the name of your butterfly.

Memorization Station

Food web: a system of interlocking food chains

This is a simplified Arctic marine food web. All food webs actually contain thousands of species.

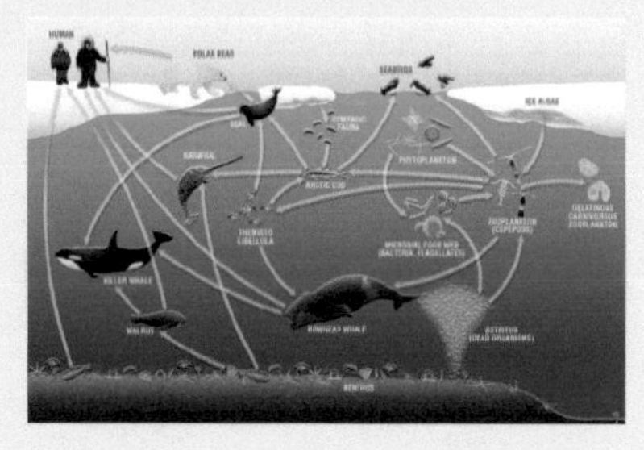

3. Play a food chain game. One person names a producer. The next person must name a consumer that eats that producer. Keep naming organism in the food chain until you get to an **apex predator** and the food chain ends. Repeat as many times as you like, trying to make your food chains as long as you can.

4. Put the Food Chains printable in your Layers of Learning Notebook in the Science section.

☺ ☺ ☺ EXPLORATION: Antarctic Food Web

For this activity, you will need:

- Books or videos about whales and dolphins
- "Cetacean Anatomy" and "Cetacean Anatomy Answers" from the Printable Pack
- "Antarctic Food Web" from the Printable Pack
- Scissors
- Colored pencils or crayons

A **food web** is a system of interlocking food chains. In real life, animals don't always eat one kind of food or get eaten by one kind of organism. There are usually multiple interactions taking place.

Cetaceans are whales, dolphins, and porpoises. There are two main body types: toothed and baleen. Toothed whales have different feeding habits and occupy a different position on the food chain than baleen whales do.

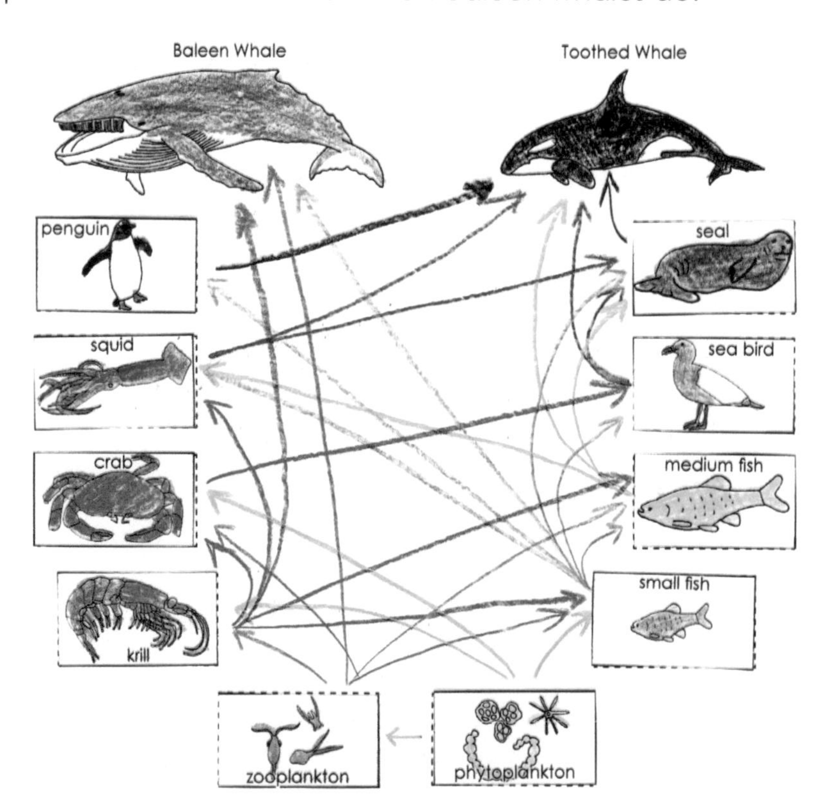

1. Read a book or watch a video about whales and dolphins. Pay attention to the differences between toothed and baleen cetaceans.
2. Label the "Cetacean Anatomy" using the "Cetacean Anatomy Answers" as a guide.

   Baleen whales feed by filling their mouths with water that contains fish or krill or zooplankton (microscopic animals that drift in the open ocean). The throat pleats are folds of skin that can expand to hold thousands of gallons of water. The whales open their mouths, fill them with food-filled water, then close their mouths and expel the water through the baleen, keeping the food inside their mouths.

   Toothed whales feed by sending out echolocation signals using their melon, an organ in their heads that can focus and change a sound made by the animal. Once they find prey, they grasp it with their teeth and eat it.
3. Cut the organisms off the bottom of the "Antarctic Food Web" sheet. Cut them apart. Color the organisms. Glue them into the spaces on the sheet as seen in the image in this Exploration. Draw arrows showing the direction the food moves in the food web.

   Phytoplankton is eaten by krill, crabs, small fish, and zooplankton. Zooplankton is eaten by small fish, medium fish, krill, crabs, and baleen whales. Krill is eaten by small fish, baleen whales, medium fish, squid, and crabs. Crabs are eaten by sea birds. Small fish are eaten by baleen whales, penguins, medium fish, sea birds, squid, and seals. Medium fish are eaten by squid, seals, and toothed whales. Sea birds are eaten by seals, and toothed whales. Squid are eaten by seals and toothed whales. Penguins are eaten by toothed whales and seals. Seals are eaten by toothed whales. Baleen whales and toothed whales are the top of their food chains.

EXPLORATION: Trophic Levels

For this activity, you will need:

- Video about trophic levels from the YouTube playlist
- "Trophic Levels" from the Printable Pack

When energy is passed along the food chain only about 10% reaches the next level. The remaining 90% is used by each organism to grow, stay warm, move about, and reproduce. That is why there will always be more producers than consumers. And there will always be more primary

Fabulous Fact

Food web theory says that dynamic interactions between predator and prey are defined by the flow of energy between different trophic levels.

Bookworms

What If There Were No Sea Otters? by Suzanne Buckingham Slade is about the difference one species can make in an ecosystem.

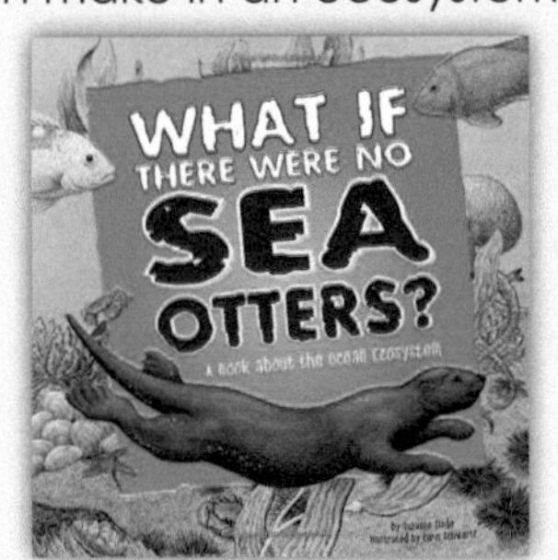

Fabulous Fact

Energy flow theory says that energy moves from producers to consumers, and that as it moves up the food chain, the total energy available decreases.

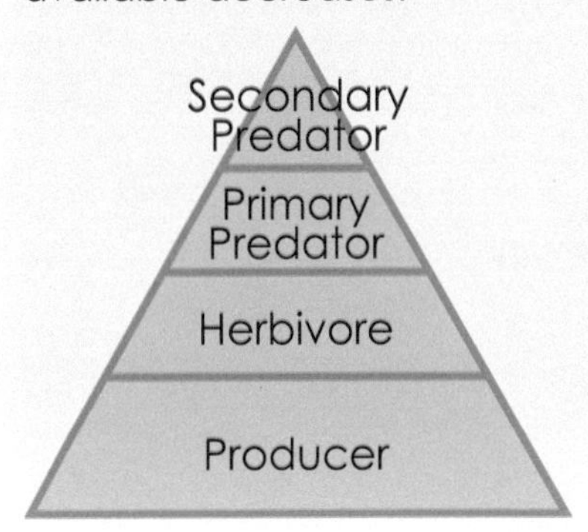

consumers than apex predators. There just isn't enough energy to support more animals at the top of the food chain.

The position an organism occupies in the food chain is called its **trophic level**. A plant that makes energy from sunlight is on trophic level one. An animal that eats plants is on trophic level two. An animal that eats animals that eat plants is on trophic level three. And an animal that eats animals that eat other animals is on trophic level four. The most trophic levels that can exist in any food chain is five because there isn't enough energy for more.

1. Watch a video about trophic levels.
2. Complete the "Trophic Levels" worksheet according to the directions. Add the page to your Layers of Learning Notebook in the Science section.

Biological Communities

In ecology, a community is a group of living things in a particular area. We study communities to learn how creatures interact with each other and the natural world.

EXPERIMENT: Backyard Community

For this activity, you will need:

- Outdoor area like a backyard or a park
- Species identification guides for your area
- Science Notebook
- Pencil
- Poster board
- Markers or crayons

A **community** is two or more species that live together in the same place. A community is similar to an ecosystem except that it only includes living organisms and not factors like water, soil, or air. Also, a community can be narrowed to the way a scientist wants to use it. You may speak of the community of Windermere Lake, England, which would involve all living organisms in and around the lake, or you could speak of the fish community of Windermere Lake, which would include only the various fish species.

1. Choose an area that has a simple community structure, like your back yard or a small park. Make a list of all the species, animal and plant, that you can find. Use species identification guides.
2. What do you know about this community? Can you make a food web for this community? Are there any

Memorization Station

Trophic level: the position an organism occupies in the food chain

Trophe is a Greek word that means food or nourishment.

Fabulous Fact

Biodiversity has increased over the history of the planet. There are more species now than at any previous time.

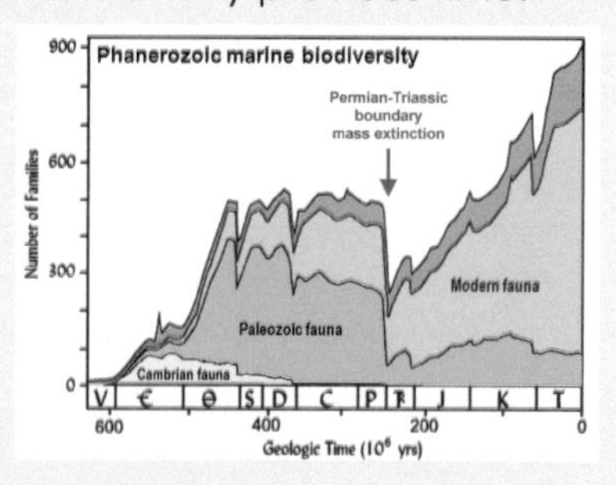

In this image by Thomas Alego, you can see the increase in biodiversity across geologic time among marine animals. Today is at the far right of the image.

Bookworms

Nature's Best Hope (Young Reader's Edition) by Douglas W. Tallamy is an argument that if individuals will make a few simple changes to their own backyards, we can turn around habitat loss.

organisms that are part of this community that are not part of the food web (humans probably fit here)? What job does each species do in this community? Does it seem to you that this is a healthy community? If not, what would you do to make it better?

3. Make a poster to show the community you studied. Draw a picture or map of the community in the center. Draw or print pictures of many of the species and write a caption for each one, explaining its job in the community.

Bookworms

One Small Square: Backyard by Donald Silver examines all the living things in just one tiny patch of backyard. Packed with text and pictures for ages 6 to 12.

Memorization Station

Community: a group of species that live together in a specific place

Biodiversity: the variety of living things on the earth or in a specific ecosystem

Deep Thoughts

Over time, biodiversity on Earth has increased. We know that mankind has shaped the earth in ways no other species ever has before. We have polluted it, overused it, and changed the atmosphere, water, and plant cover. We clear cut forests and cut the tops off of mountains. As we do these things, we have an impact on wildlife and biodiversity as species go extinct.

There is a lot of debate about how much human beings are changing the earth's living systems. Do you think the impact is big or small?

What is your evidence and reasoning for your position?

EXPEDITION: Biodiversity

For this activity, you will need:

- Two different habitats like your backyard and an empty lot or piece of wild land
- Measuring tape
- Microscope or magnifying glass
- Small spade
- Paper and pencil
- Calculator
- Nutrient broth (optional)
- Nutrient agar (optional)
- Petri dishes (optional)
- Science Notebook

Biodiversity is the variety of species in an ecosystem. Biodiversity is highest is the tropics and lowest in very cold or very dry places. Biologists estimate that there are more than 8 million different species of living things on Earth right now. We've only documented about 1.2 million of them. Some of the undocumented are in the soil in your backyard, deep in the sea, in the pond down the road, in rainforests, in coral reefs, and deep in deserts that appear barren.

High biodiversity is a good thing because living things play off of one another, buffer one another, and rely on one another in incredibly complex relationships. High biodiversity means that if one food source falls off, another is there to rely on. It also means that harmful diseases or organisms are kept in check.

1. Measure a 3 by 3 meter square in your first habitat.
2. Count the number of different species in the habitat square. Consider plants, animals, insects, and worms. Dig up a bit of soil and sift it carefully to find tiny creatures. You can use a microscope or magnifying glass to look closely. Record your data in your Science Notebook.
3. ☻Older students can also count bacteria species. Take a small dirt sample and put it into a nutrient broth solution where it can grow for several days. It needs to stay nice and warm. Then, swab some of the nutrient broth over a petri dish filled with agar. Put a lid on it and let it grow in a warm place for several days. Each bacteria species will form a colony that can be differentiated by color and texture. We often ignore bacteria when we consider the living parts of an ecosystem, but they are vital for healthy ecosystems.
4. Count the total number of organisms in your habitat. Some will be easy to count, like trees or birds. Others you will have to estimate by counting a small number and then multiplying that number by the larger space. For example, if you count 90 individual grass plants in a 1 meter square then you can multiply that number by the number of square meters that contain grass to get an

Fabulous Fact

Biodiversity bestows the knowledge of survival under myriad circumstances that species have developed over thousands of years. That survival knowledge can help humans through medicine. For example, polar bears can go months without urinating. If a human tried that, she would be poisoned by her own blood. If we can figure out how polar bears do it, we might be able to cure kidney failure.

This is a rare animal called the pink fairy armadillo.

Biodiversity is also beautiful. From plankton in the sea to flowers to snail shells, the living things of Earth are beautiful and they make the world around them wonderful, bizarre, and sometimes grotesque, but always fascinating. Each species is a unique work of art with unique traits and structures. Each species is irreplaceable.

Deep Thoughts

When you hear the word "nature," what comes to mind?

Is there a sharp divide between humans and nature or are humans just another part of nature?

estimated total number of individual grass plants.

5. Calculate the diversity index by dividing the number of species by the total number of organisms in your habitat. The closer your number is to 1, the better your diversity.

6. Now repeat the process at a second site. Which site had higher biodiversity? What were the differences between the sites? Which site do you think had the most complete and in tact food webs?

EXPEDITION: Pollinators

For this activity, you will need:

- A book or video about pollination
- "Flower Characteristics For Attracting Pollinators" from the Printable Pack
- Flower identification guide
- Flowers outside in a garden, park, or meadow
- Science Notebook
- Pencil and colored pencils

All flowers need to move their pollen from the stigma to the style, sometimes of the same flower, but more often to another flower of the same species. Flowers move their pollen with the help of wind or animals and, in some rare cases, water.

1. Read a book or watch a video about pollination.

2. Go outside and find 3-4 different kinds of flowers in full bloom. Look carefully for the pollen inside the flower.

Bookworms

You Wouldn't Want to Live Without Bees! by Alex Woolf is about what would happen if all the bees disappeared. Not only would you miss honey, but you would miss all the plants the bees pollinate.

For ages 8 and up.

Famous Folks

Aldo Leopold was an American naturalist during the first half of the 20th century. He was influential in developing outdoor ethics and conservation.

He wrote *A Sand County Almanac*, which is in the Library List.

Writer's Workshop

Would you rather live deep in the forest or on a private island? Write about your choice and why you picked it.

3. Identify each of the types of flowers you are studying. Sketch each one and write a list of its traits such as size, flower shape, color, location of the pollen inside the flower, and anything else you notice.
4. Use the "Flower Characteristics For Attracting Pollinators" to predict how the flowers you are examining are pollinated.
5. Observe the flowers for a little time and see if you notice any pollinators moving around the flowers. Draw the pollinators and flowers interacting in your Science Notebook.

   If you don't see evidence of pollinators, you can look up information on your flowers to find out how they are pollinated.

Additional Layer

Symbiosis is when organisms live together and affect one another in some way. There are three types of symbiosis.

- Mutualistic organisms help one another.
- Parasitic organisms feed off of and harm another organism.
- In commensalism, one organism benefits while the other is unaffected.

Think of examples of each kind of symbiosis. It is easy to find information online about symbiosis.

Writer's Workshop

A parasite is a species of animal or plant that lives off of another species. It does not help its host in any way and usually does great harm to the host. Research information about parasites. Create a notebooking page about a particular parasite.

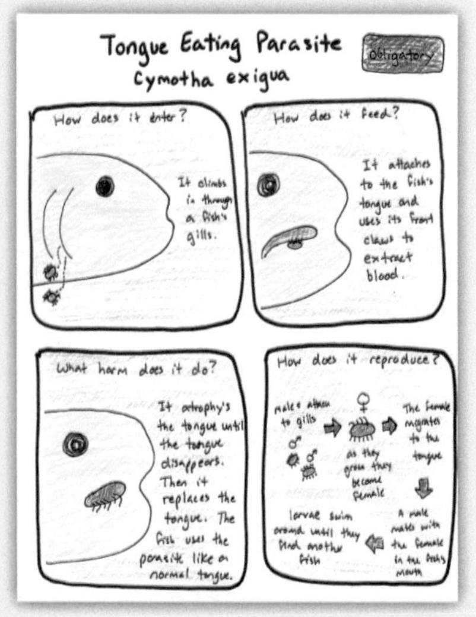

Include how it enters the host, how it feeds, what harm it does, and how it reproduces.

☺ ☺ ☻EXPEDITION: Comparing Freshwater Communities

For this activity, you will need:

- Two different sites with freshwater - streams, lakes, rivers, or ponds
- Meter stick or metric tape measure
- Stopwatch
- Jar lid
- Black and white spray paint or acrylic paint plus paintbrush
- String
- Nail and hammer
- Thermometer
- Small ball or an orange (needs to be able to float)
- Dip net or kitchen strainer lined with a cloth
- Large, wide plastic or metal bin - like a wash basin
- Hip boots, waders, or wading shoes to get wet
- "Freshwater Communities comparison" from the Printable Pack
- Clipboard and pencil
- Plant, reptile, amphibian, fish, and insect identification guides for your area

The freshwater communities you investigate will include all of the animals and plants that live in the water. We will not include animals and plants that live near the water or periodically visit the water.

1. Before you head outside, you will need to construct a Secchi disc. It will be used to calculate the turbidity (murkiness) of the water. Divide your jar lid into 4 quarters. Paint it alternating black and white sections, then punch two holes in the center of the lid near one another. Thread a long string through the holes and tie it off. You will be able to lower the disc into the water to test how long it is visible.
2. Pack up all of your materials and head to the first

freshwater site. Write down the name of the site and whether it is a stream, river, pond, or lake in the box that says "Site 1" on the "Freshwater Communities Comparison" chart.

3. Take the temperature of the water at the surface. Hold your thermometer bulb just under the water and shade it from the sun with your hand. Then take the temperature about 3 feet down. Record your measurements on the chart.

4. Next, determine the turbidity. If you can see the bottom, estimate the depth of the water. If you can't, use the Secchi disc. Lower your Secchi disc into the water until it just disappears from view. Measure the depth by measuring the length of the string from the surface to the Secchi disc. Then, lower the Secchi disc further and slowly raise it up until you can just see it. Measure again. Take the average of the two measurements to get turbidity. Record your measurements on the chart.

5. Next, measure the velocity of the water. A lake or a pond has a velocity of zero. If you are at a stream or river, measure out a distance of ten meters along the bank. Have one person stand upstream and one person stand downstream so you are ten meters apart. The upstream person tosses a small ball or orange just upstream. As it passes, start a stopwatch. The person lower down catches the ball and you stop the watch. Divide the time by ten to get the meters per second that the stream is flowing. Record your data on the chart.

6. Identify any plants or animals that are easily visible.

Deep Thoughts

No one knows why some animals herd together. Is it for safety, mates, comfort, group foraging and hunting, saving energy when traveling, or to access the experience of the older animals?

Discuss your ideas and what you think is most likely. Can you think of any ways to test your hypothesis?

Famous Folks

Rachel Carson was a writer and marine biologist who is best known for her game changing book, *Silent Spring*.

The book argued against chemical pesticides, claiming they were harming birds' eggs. It led to the banning of DDT (a chemical pesticide) across most of the world, ignited a national debate, and led to the establishment of the Environmental Protection Agency in the United States.

Fabulous Fact

Succession theory says that, over time, a natural environment will recover in stages from a disturbance.

Intermediate disturbance hypothesis predicts that the highest species diversity occurs when ecological disturbances are neither too frequent nor too rare. In other words, some disturbance is desirable.

Additional Layer

The time it takes to go from bare ground to a mature environment depends on the habitat. A redwood forest could take 500 years to grow back while a grassland may only take one year to recover.

This scientist is studying a bleached coral reef in Hawaii.

After a bleaching event, it takes about nine to twelve years for the reef to reach a mature stage.

Different plants and animals thrive at each of the stages of ecological succession as their habitat changes.

Change in habitats is normal and constant. The changes can be big, like a massive windstorm that blows over thousands of acres of trees, or they can be small, like one tree that falls and opens the forest floor to sunlight.

Record your findings on the chart. If you can't identify everything, don't worry, just do your best.

7. From the shore or from a wading depth near the bank, reach out into the water with a dip net or reach down with a kitchen strainer and pull it toward you through the water. Gently drag across the bottom. Sweep several times and then empty your net into a bin that has just a few inches of clean water in it. Avoid getting large globs of mud or dirt. You might also want to turn over rocks that are in the water to see what is attached.
8. Identify as many of the tiny animals as possible. You may see insect larvae or nymphs, tadpoles, minnows, snails, and leeches or worms.
9. Return all the creatures to the water. Carefully clean all of your equipment. Then repeat the process at another freshwater site. Be careful not to contaminate the second site with life from the first site.
10. Compare your findings between the two sites. Was the ecosystem of the second site significantly different from the ecosystem of the first site? How were the two communities different?

EXPLORATION: Ecological Succession

For this activity, you will need:

- Video or book about ecological succession
- "Ecological Succession" and "Ecological Succession Answers" from the Printable Pack
- Colored pencils or crayons

Ecological succession is the process of a habitat recovering from a disturbance over time. The disturbance could be natural—a volcanic explosion, a mudslide, a flood, or a wildfire. A disturbance could also be man-made—logging or clearing land for farming.

There are two types of succession: primary and secondary. Primary succession is when the land is completely barren, like after a volcanic explosion, a glacier, or a sand dune growing grasses for the first time. The first things to grow back are lichens, then grasses, then small plants and bushes, then trees, and finally, a mature forest.

In secondary succession, the soil is not completely destroyed. This would be after an event like a forest fire, a flood, or a drought. In this case, the first species to recover are grasses and small plants, then small trees, and finally, a mature forest. The first plants to come back after a disturbance are called "pioneers." The mature environment is called a climax community.

1. Watch a video or read a book about ecological succession. If you can find some that specifically talk about forests or coral reefs, even better!

2. Complete the "Ecological Succession" sheet using the "Ecological Succession Answers" as a guide. The bottom portion of the sheet is a storyboard to illustrate how ecological succession happens in two different ecosystem types.

EXPEDITION: Counting Populations

For this activity, you will need:

- A garden where you can dig up a section of soil
- Spade
- Measuring tape
- Paper and pencil
- 4 pieces of 60 cm long x 5 cm wide wood (approximately)
- Long brad nails
- Wood glue
- Hammer
- String or twine
- Scissors
- Push pins
- "Plant Survey" from the Printable Pack

Ecologists often want to count **populations** so they know if a species is struggling or overpopulating, and so on. It is usually impossible to physically count every member of a species so, instead, ecologists estimate numbers using sampling or mark and recapture. In sampling, you count a small portion of the total and then multiply by the remaining sections of space (in the forest, in the field, in the pond, and so on). In mark and recapture, scientists capture

Memorization Station

Ecological succession: the process of a habitat recovering from a disturbance

A forest fire, logging clear cut, volcano, mud slide, wind storm or hurricane, and river bottom dredging can all be the causes of a disturbance that nature will take time to recover from.

This is a clear cut in Norway where the loggers left a few seed trees, so the area will more quickly repopulate.

Population: the individuals of one species in a particular location

Famous Folks

One of the pioneers of ecology was Henry Chandler Cowles who studied ecological succession at the Indiana Dunes.

Fabulous Fact

If a population is increasing or declining in numbers, biologists try to find out why. They try to learn what optimal population numbers are and, in some cases, they try to regulate that number to prevent population crises.

Some species go through dying off stages where individuals starve. Often these cycles are predictable and regular.

Research information on the lemming population cycles. In your Science Notebook, write down the stages and why biologists think it happens.

Additional Layer

Governments and communities try to control and manage the harvesting of game animals such as deer and fish.

Play "Game Warden" with the three pages of board and cards from the Printable Pack.

certain animals, mark them with a tag of some kind, and then later go back and recapture members of the species. The portion that have marks from the previous capture compared to the new individuals creates a mathematical proportion that can be solved to find an estimate of the total number of individuals.

1. Head a new page in your Science Notebook "Counting Populations" and add it to your table of contents. You will be sketching your experiments and taking notes as you go.
2. Dig up 1 square meter of soil to a depth of 30 centimeters. Count all the worms you find in your sample. Multiply that by the size of your entire garden to get a population count. You can substitute worms for oak trees on your block, robins in the park, or squirrels in your neighborhood if a garden isn't available.
3. Make a plant survey grid. Build a square frame, each side about 60 cm long. Attach the sides together at the corners with wood glue and brad nails. Then, create a grid of push pins and pieces of string to make a 4x4 grid inside the frame.
4. Set the plant surveyor on any piece of weedy or wild ground. Draw sketches of each of the plant species inside your plant surveyor on the "Plant Survey" sheet. Identify each species with a plant identification guide.
5. Count how many plants of each species are present inside your plant surveyor. Determine the total area of the field or forest you are surveying and multiply the number of plants in your survey by the number of plant surveyors that would fit in the space. This will give you an estimate of the number of plants in the whole ecosystem. This works best if you take several plant surveys and average your results.

EXPLORATION: Where Do Animals Go In Winter?
For this activity, you will need:

- "Where Do Animals Go In Winter?" from the Printable Pack (2 pages, printed on card stock)
- Scissors
- Glue stick
- Metal brad
- Colored pencils or crayons
- Internet
- Science Notebook

In the winter, food is harder to find and temperatures are too cold for comfort. So what do animals do to survive? Some migrate, or move to warmer places. Others hibernate, or curl up in a den and sleep out the deepest part of winter. But many animals stay put, search for food, and keep warm with thick fur or feather coats.

1. Color the "Where Do Animals Go In Winter?" pages.
2. Cut the circles out on the heavy lines. Fasten the circle with a window on top of the complete circle using a metal brad.
3. Discuss hibernation, migration, and staying put. Hibernation is when an animal sleeps out the worst part of the winter in a hibernaculum (den). Migration is when an animal moves to a warmer location for the winter. And staying put means storing food or being able to find food under the snow and putting on a heavier coat to stay warm.
4. Cut the animals out. Paste them into the correct section of the lower circle. The second page has a table telling each animal's strategy for the winter.
5. Pick an animal from each section to research together online to discover more details.
6. Glue your circle chart into your Science Notebook.

Additional Layer

Make a model hibernacula for an animal.

Additional Layer

Use "How Hibernation Works" from the Printable Pack to create a page inside your Science Notebook about this topic.

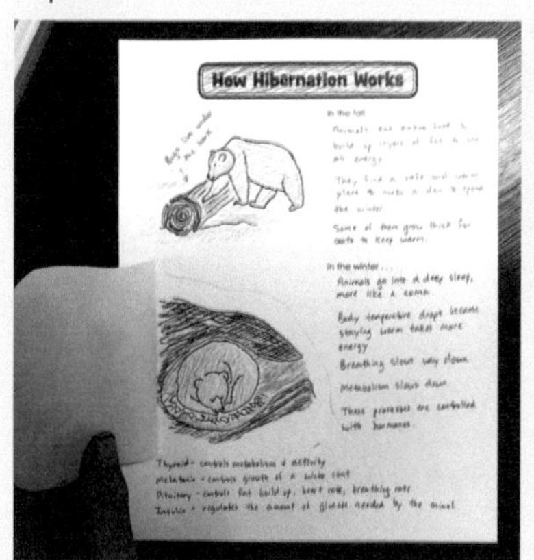

You will need to read websites, books, or watch videos about hibernation in order to complete the sheet.

Add a cave flap over the top of the hibernating bear.

Writer's Workshop

Animals often compete with one another for food, mates, or space.

Pick any animal and learn how it competes in the wild.

Write a paragraph describing the competition behavior.

Conservation

Taking care of the natural world by using resources responsibly, composting or recycling, and putting trash where it belongs is how we make sure we leave the world better than we found it.

EXPERIMENT: Algae and Pollution

WARNING: This experiment uses chemicals that can be dangerous if handled incorrectly. Keep out of reach of small children, wear gloves when handling chemicals, and never mix two chemicals together.

For this activity, you will need:

- 2 or more large jars, one jar for each chemical tested plus one control
- A nearby pond
- Grease pencil or dry erase marker to label the jars
- Disposable gloves
- Electronic balance and a dish for the balance
- Bone meal fertilizer
- All purpose fertilizer (optional)
- Dishwasher or laundry detergent (optional)
- White vinegar (optional - simulates acid rain)
- Science Notebook

Pollution is when a **contaminant** is introduced to a natural environment. Contaminants could be fertilizers, chemicals, waste water, trash, heat, light, or noise. Pollution has negative effects on the living things in an ecosystem and

Bookworms

The Lorax by Dr. Seuss is funny and quirky but also has an ecological message when the Lorax "speaks for the trees."

On the Web

Leave No Trace is an outdoor ethic of seven principles of outdoor behavior people can embrace to preserve the wild. The National Park Service, Scouts USA,the Girls Scouts, and other organizations embrace and teach Leave No Trace. Look them up online and see what they are about.

Memorization Station

Pollution: a substance that has harmful effects in the environment

Contaminant: any physical, chemical, or biological substance that harms or poisons a natural environment

Natural resources: raw materials people harvest or use from the earth to make everything we need to survive and thrive

usually harms humans as well.

Algae grows naturally in ponds and is a normal part of the ecosystem of a pond. But certain pollutants can affect the growth of algae. Phosphorus is one of the most important nutrients for plant growth, but if too much phosphorus makes its way into a pond, river, or lake, it can cause the algae and other aquatic plants to explode in growth. They suffocate out other life forms like frogs and fish when the plants die off.

1. Visit a pond and collect two or more jars of pond water, making sure to get some green living things in the sample. Take the pond water home and let the jars sit in a sunny spot to get the algae growing for 3 to 7 days.
2. Label one of the jars "control." Label each of the other jars with the chemical you intend to test. The control jar will be left untouched so you can compare its growth to the other jars.
3. Write down a hypothesis for each type of chemical you test. Do you think it will cause the algae to grow more, grow less, or have no effect? Keep track of your experiment in your Science Notebook.
4. Wear gloves when handling chemicals. Using an electronic balance and a dish to hold the chemicals, measure out 6 grams of bone meal and add it to a jar. Repeat with any other chemicals you want to test. Do not mix the chemicals with each other or inside the jars!
5. Leave the jars in a sunny place for 1-2 weeks. Visually inspect the jars to see how the algae growth is affected by the pollutants.
6. Discuss the reasons we should be careful about where the chemicals and things we use end up after we are done with them.
7. Can you think of how to expand or redesign the experiment you performed to make it better or test new things?

EXPLORATION: Using Natural Resources

For this activity, you will need:

- "Natural Resources" from the Printable Pack
- Colored pencils or crayons
- Scissors
- Glue stick
- Science Notebook

Everything we need to live and enjoy life comes from the

Teaching Tip

The most important thing you can do to help your kids gain a respect and value for the natural world is to get out of the city and into the woods, desert, seaside, or mountains.

Nature is vast, majestic, and awe-inspiring. It will sell itself.

Writer's Workshop

Conservation efforts are often successful.

The Hawaiian Goose, also known as the nene, is endemic to the Hawaiian Islands. Hunting and introduced predators reduced the population to just 30 birds by 1952. It was brought back from the brink of extinction. The population is now at around 2,500 birds, which are protected in Hawaii.

Learn about a conservation success near your home. Write a short paper about what happened and how the animal or plant was saved.

Deep Thoughts

What do you think is the most important conservation issue today? Discuss your ideas and explain your reasoning.

Fabulous Fact

Extinction theory says that species can and do go extinct. At one time, scientists thought that all the living things that ever had existed or ever would exist were on Earth presently. Now we know that all species eventually go extinct and new ones take their places. We know that change is normal on Earth.

Bookworms

Stingers by Randy Wayne White is about the invasive lionfish that have invaded Florida and the Bahamas. It's also about pirate treasure and outlaws. Can Luke, Maribel, and Sabina navigate these treacherous waters?

This is the second in the Sharks Incorporated series for ages 8 to 12.

earth. We have to use its **natural resources** to survive and thrive. As we are using natural resources, it is important to avoid wanton destruction and unnecessary waste as we harvest and create the things we need.

1. Collect ten items from around your house. Discuss how each one is made from natural resources. Do your best to identify the materials that your items are made of. Where are each of these materials gathered?
2. Color the "Natural Resources" page. Cut out the title, the information box, and the squares.
3. Glue the title to a page in your Science Notebook.
4. Sort the items so they are next to the natural resource used to make the item, and then next to the source of that natural resource.
5. Glue the information rectangle and the image squares into the Science Notebook in order.

EXPLORATION: Introduced Species

For this activity, you will need:

- Science Notebook
- Internet to look up introduced or invasive species

Since the Age of Exploration, humans have been moving about the world more and more and carrying species from one place to another, sometimes on purpose and often accidentally. When a non-native species is introduced to a new environment it can have unexpected consequences. An **introduced species** that is highly successful and outcompetes native species is called an **invasive species**.

If you bring your boat to Idaho in the United States, you will be required to stop at a boat check station along the road. The officials there will examine your boat for signs of non-native species that you are inadvertently moving from one lake or river to another. Many lakes in the United States are now infected with zebra mussels which are native to fresh waters in Eurasia. They are an aggressive species that kills the native mussels, clogs power plants and water treatment plants, and eats all the native algae, preventing other species from getting the food they need. Stories like this of introduced species that have caused damage to ecosystems and human systems are common.

1. Head a page in your Science Notebook "Introduced Species" and add it to your table of contents.
2. Read online about introduced species or watch a video about introduced species.
3. Search for an invasive species near where you live and the impact it has had on the ecosystem. Find out what is being done to get rid of it or mitigate its effects. Take notes in your Science Notebook.

EXPLORATION: Extinction

For this activity, you will need:

- Book or video about extinction from the Library List or YouTube playlist for this unit
- "Extinction" from the Printable Pack
- Colored pencils or crayons

Extinction means there are no more of a particular species left on earth. Eventually, every species goes extinct as the conditions on the earth change. An estimated five billion different species have lived on the earth and, of those, more than 99% are now extinct. As one species goes extinct, others move in to fill the space that is left empty and new species emerge. Species go extinct when they can not get enough food, water, shelter, or space to survive. This can happen naturally as Earth's climate changes, new species competition moves in from another area, a drought or flood wipes out a habitat, a new disease wipes out a population, and so on. It can also happen because of human activities such as harvesting a forest, mining a mountain, spilling garbage onto a reef, overharvesting for food, or building a highway.

Ecologists are concerned about extinctions that are caused by humans. They study the reasons for past extinctions to

Memorization Station

Introduced species: an organism that is not native to a particular area

Invasive species: a non-native organism that is successful and outcompetes the native organisms in the new environment

Extinction: the end of a species

Endangered species: an organism that is threated with extinction

Fabulous Fact

Mass extinctions, where millions of species go extinct all at once, are rare in the history of the earth, but they do happen. A mass extinction event wiped out the dinosaurs.

Today, many biologists believe we are in the middle of a mass extinction event caused by humans. They call it the Holocene extinction.

Deep Thoughts

Biodiversity and the preservation of species is mostly a moral issue. Does it really impact anyone's life if one poisonous frog in the Amazon goes extinct?

Endangered blue poison dart frog

Do you care? Why or why not?

Additional Layer

Look up threatened and endangered species in your corner of the world. Make a map of the range of this species, previously and currently. Learn what factors have caused the decline of the species. Find out what is being done for conservation or recovery of the species. Put your information on a sheet of card stock and make a mini-poster that you can post on the wall until the end of this unit.

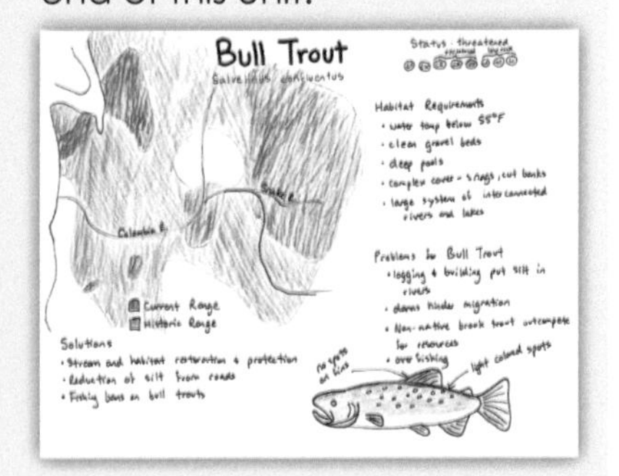

The bull trout is endangered in Idaho. It is illegal to harvest them.

Additional Layer

One way to keep trash out of landfills is to recycle what you can. Make a recycling system for your house.

Make sure you know where and how to recycle in your community.

Find out what happens to the things you toss in the recycling bin.

understand why current species may be struggling. Then they take steps to protect the **endangered species**. Usually, the most important step is simply protecting the habitat of the species from human use.

1. Read a book or watch a video about extinction. Pay attention to the causes and effects of extinction.
2. Color the "Extinction" worksheet. X out one animal from the food web.
3. Fill in the boxes at the bottom of the sheet with your ideas about what the removal of that species from the food web would do to the rest of the web.
4. As a group, discuss how the removal of one species affects all the others, even if they don't directly interact.
5. Put your "Extinction" worksheet in your Layers of Learning Notebook in the Science section.

EXPERIMENT: Decomposing Trash

For this activity, you will need:

- Trash from your house
- An outdoor area with dirt you can dig in
- Spade
- Craft sticks
- Permanent marker
- Science Notebook

Modern households produce lots and lots of trash. The trash often ends up in landfills. A big hole is dug and trash that was collected from households and stores is dumped into the hole. In well-run landfills, the trash is covered up every day with a layer of soil or wood chips to keep vermin down. Eventually, the landfill is full. Sometimes the land is then reused for development. But underneath, what has happened to all that trash? Some of it certainly decomposes, but does it all? How long does that take?

1. Do an experiment to see how long it takes to decompose some of your trash. Choose five different items of trash, things you often throw away. Bury them underground in a trench. Mark the places you've left them with labeled craft sticks.
2. Predict how decomposed you think each item will be at the end of the experiment, then leave them for 4-6 weeks. Longer is better. At the end of your time, dig them up and observe how much they have decomposed.

   Create a scale like this: 1 - no decomposition; 2 -

some holes, beginnings of decomposition; 3- lots of holes, about halfway to decomposition, 4 - mostly decomposed; 5 - completely decomposed.

3. Record your experiment set up and results in your Science Notebook.

Step 3: Show What You Know

During this unit, choose one of the assignments below to show what you have learned during the unit. Add this work to your Layers of Learning Notebook. You can also use this assignment to show your supervising teacher or your charter school as a sample of what you've been working on in your homeschool, if needed.

There are more ideas for writing assignments in the "Writer's Workshop" sidebars.

Coloring or Narration Page

For this activity, you will need:

- "Ecology" from the Printable Pack
- Writing or drawing utensils

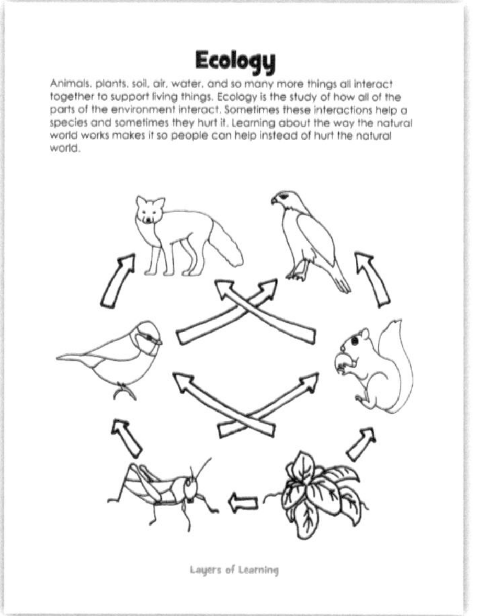

1. Depending on the age and ability of the child, choose either the "Ecology" coloring sheet or the "Ecology" narration page from the Printable Pack.
2. Younger kids can color the coloring sheet as you review some of the things you learned about during this unit. On the bottom of the coloring page, kids can write a sentence about what they learned. Very young children can explain their ideas orally while a parent writes for them.
3. Older kids can write about some of the concepts you learned on the narration page and color the picture as well.
4. Add this to the Science section of your Layers of Learning Notebook.

Additional Layer

Sustainability is living in such a way that people and businesses use fewer resources and destroy less of the environment while still having plentiful lives.

Learn about sustainable city and home design.

Make a poster that shows the problem and at least two sustainable solutions.

Bookworms

Consider the Octopus by Nora Raleigh Baskin and Gae Polisner is about two kids, JB and Sidney, who end up on an ocean research vessel in the Great Pacific Garbage Patch studying plastic pollution. JB does not want to be there and Sidney isn't supposed to be there.

Well-written and entertaining plot. Ages 10 and up.

Unit Trivia Questions

1. What is the difference between a biome and an ecosystem?

   A biome is a large geographical area with a similar climate. An ecosystem is defined by the person and includes all the living and non-living things in a particular place.

2. Name the four things every organism needs.

   Food, water, shelter, space

3. Organisms that make food for themselves from sunlight are called:

   a) Producers

   b) Consumers

   c) Decomposers

   d) Herbivores

   e) Carnivores

4. Fill in the food chain:

   Grass -> ______-> owl

   Answers vary (ie. mouse, frog, lizard, fish, snake)

5. True or false - There is more energy at the top of a food chain than at the bottom.

   False, the highest energy is at the bottom where the producers are getting energy from the sun.

6. In winter, animals can stay put, migrate, or ________.

   Hibernate

7. If all the members of a species die out, it is called:

   a) Extinction

   b) Depletion

   c) Instinct

   d) Extermination

Science Experiment Write-Up

For this activity, you will need:

- The "Experiment" write-up or "Experiment Report Template" from the Printable Pack

1. Choose one of the experiments you completed during this unit and create a careful and complete experiment write-up for it. Make sure you have included every specific detail of each step so your experiment could be repeated by anyone who wanted to try it.
2. Do a careful revision and edit of your write-up, taking it through the writing process, before you turn it in for grading.

Nature Documentary

For this activity, you will need:

- Camera, a phone camera works great
- A local natural environment
- Video editing software (Filmora and Adobe Premiere Rush are both free)
- Computer to edit the video on
- TV to show the video on

Having kids come up their own original ideas and use their creativity helps to really solidify the information. You can set a time limit on this process, perhaps they have two hours for the whole thing or two days of two hours each. Emphasize that a perfect finished product isn't the goal, the process of being creative and explaining their ideas is the goal. Young children will need a mentor or older sibling to assist with the whole process. This can be an individual or a group assignment. If it is group, make sure each child gets to be in front of the camera.

1. Write a script explaining a local food web, a local endangered animal, or a local ecosystem. Keep it to a page and a half at the most.
2. Practice reciting the major points of your script.
3. Film yourself in the natural environment (even if that environment is your back yard or a park) reciting your script. It is easier if you break your film up into short pieces so you don't have to remember the whole thing at once.
4. Edit your film to put all the pieces together and cut out mistakes. You can add music behind it and also clips of animals or plants in the natural setting.
5. Share your film with your family or friends.

S.O.S.

For this activity, you will need:

- A computer or a piece of paper and a writing utensil

S.O.S. stands for statement, opinion, support. The mentor gives a statement, the student writes his or her opinion, and then backs up the opinion with supporting facts.

1. Give your students one of these statements:
    a. Rainforest ecosystems are the most important.
    b. Honeybees are a keystone species.
    c. Nature is resilient and doesn't need human help.
2. The students should write their opinion about the statement and then back their opinion up with supporting facts. Make this closed book, so the student has to remember and draw upon her or his own ideas.

Big Book of Knowledge

For this activity, you will need:

- "Big Book of Knowledge: Ecology" printable from the Printable Pack, printed on card stock
- Writing or drawing utensils
- Big Book of Knowledge

1. Color, draw on, write on, or add to each of the Big Book of Knowledge pages you are using. Only add the printables if you learned these concepts during this unit. If there are other topics you focused on, feel free to add your own pages to your Big Book of Knowledge.

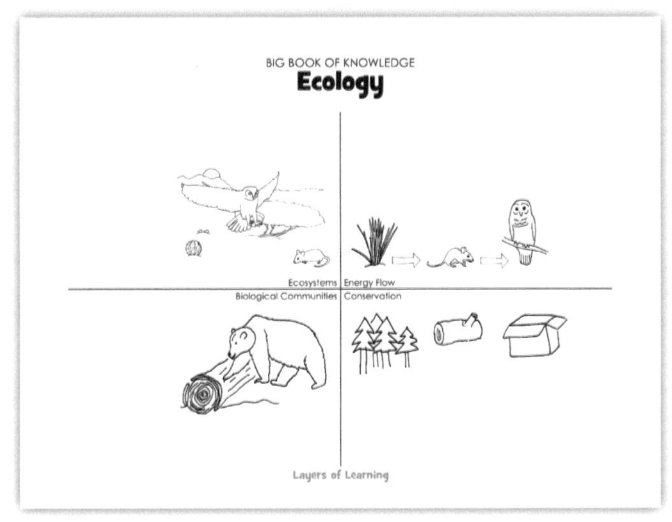

2. Use your Big Book of Knowledge regularly to help you review, quiz, or create games that will help you commit the things you've learned to memory.

Big Book of Knowledge

The Big Book of Knowledge is a book for you, the mentor, to use as a constant review of all of the things you're learning about. You can use it to quiz your kids or prepare tests or review games. Whenever you learn something in Layers of Learning that you want your kids to remember, add it to your Big Book of Knowledge.

Assemble your Big Book of Knowledge in a binder or with binder rings. Divide it into sections for each subject.

In the Printable Pack for this unit you will find a "Big Book of Knowledge" sheet. You can add this sheet to others you collect or create yourself as you progress through the Layers of Learning curriculum. Customize the Big Book of Knowledge to your family by adding facts and topics that you enjoyed exploring as you were learning.

Visit Layers of Learning online to find more information on how to assemble and use your own Big Book of Knowledge.

You will also find cover and section pages to print along with creative games to play with your Big Book of Knowledge to keep school, even the tests, fun!

HUMAN BODY

Unit Overview

Key Concepts:
- The major systems of the body are the skeletal, integumentary (skin), muscular, circulatory, respiratory, digestive, and nervous.
- It is important to know the correct anatomical names for the major organs and parts of our bodies.
- Understanding how some of our organs and body systems work helps us to take care better care of them and be healthier.

Vocabulary:
- Skeleton
- Joint
- Muscle
- Autonomic nervous system
- Skin
- Heart
- Blood vessel
- Arteries
- Veins
- Lungs
- Diaphragm
- Cellular respiration
- Glucose
- Mouth
- Stomach
- Intestine
- Neuron
- Brain

Scientific Skills:
- Memorizing the anatomical names of body parts is an important skill for doctors, anatomists, and physiologists.
- Dissection skills are essential for understanding organs and body systems.

Our bodies are made up of cells which form tissues, which form organs, which form organ systems, which work together to make the whole body function. The body tries to maintain perfect balance keeping the temperature, oxygen, sugars, and other chemicals in homeostasis.

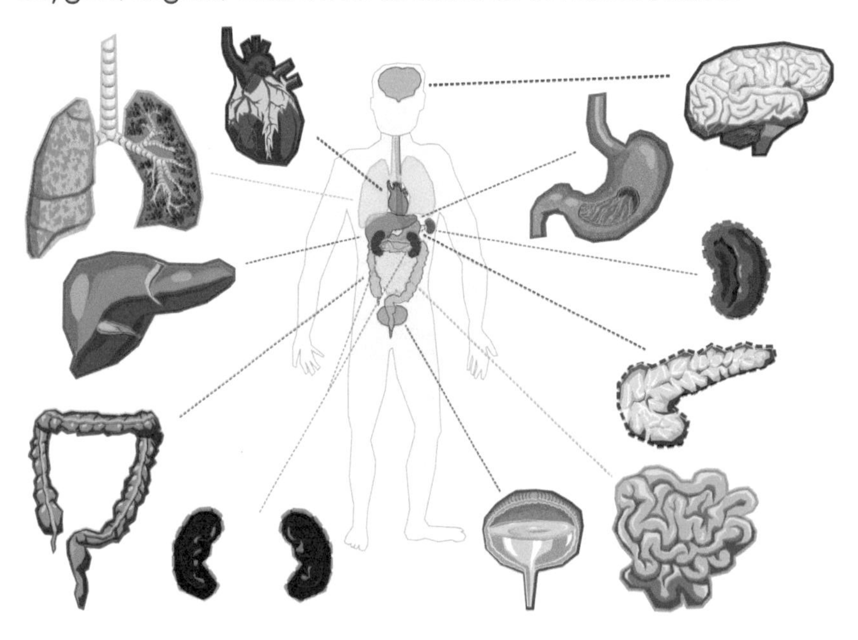

The body's shape is determined by the skeleton, made of bone and cartilage. The skeleton is surrounded by muscle and fat tissues that help us move, insulate and cushion us, and provide energy. The muscles and fat are, in turn, surrounded by skin that covers and protects the whole body.

The heart, veins, and arteries make up a circulatory system that provides the transportation for nutrients, water, hormones, and wastes around the body. Humans continually need oxygen, which cannot be stored in the body. The oxygen is provided through the respiratory system.

The body gets energy and necessary nutrients from food that passes through the digestive system, changing the food into a form that can be used by the cells.

The nervous system includes the brain, nerves, and sensory organs like eyes and ears. The nervous system controls the body and gives it information about the environment.

All of the body systems work together to keep the body functioning and alive. The more we understand about our bodies, the better we can care for them.

Step I: Library List

Choose books from your library that go with this topic. Here's a list of some favorites and also a list of search terms so you can utilize what your library offers. Read the books with your kids and/or assign them some to read independently. It is from these books your kids will learn most of the facts they need from this unit.

Search for: human body, human anatomy, physiology, skeletal system, muscular system, human eye, circulatory system, digestive system, nervous system

WARNING: Most anatomy books for all ages include information on the reproductve system. Pre-read and guide your children as needed.

☻ ☻ ☻*Encyclopedia of Science* from DK. Read "Teeth and Jaws" through "Brains" on pages 344-361.

☻ ☻ ☻*Kingfisher Science Encyclopedia*. Read "Body Organization" through "Waste Disposal" on pages 98-131. This is a lot of reading, so it's okay to pick your favorite sections or just browse through and read the best bits.

☻ ☻ ☻*The Usborne Science Encyclopedia*. Read "The Skeleton" through "Digestion" on pages 346-355 and "The respiratory system" through "The nose and tongue" on pages 358-375.

☻ ☻ ☻*Human Anatomy Coloring Book* by Margaret Matt and Joe Zieman. This is an inexpensive coloring book of all of the major human systems. Buy one for each child.

☻ ☻ ☻*DK: The Human Body Book.* A thorough full-color resource that covers all of the body systems as well as health. Also recommended in the Health unit.

☻ ☻ ☻*Human Anatomy in Full-Color by John Green.* Simple, clear, labeled illustrations that showcase the body. This is a great resource for helping you label body systems.

☻*Neurology: The Amazing Central Nervous System* by April Chloe Terrazas. The author takes super advanced subjects, like neuroscience, and breaks them down into language a child can understand.

☻*My Amazing Body Machine* by Robert Winston. Full of colorful illustrations and photos, this is about all the body systems on a young child's level.

☻*The Magic School Bus Inside the Human Body* by Joanna Cole and Bruce Degen. Also look for *The Magic School Bus Explores the Senses.*

Family School Levels

The colored smilies in this unit help you choose the correct levels of books and activities for your child.

☻ = Ages 6-9
☻ = Ages 10-13
☻ = Ages 14-18

On the Web

For videos, web pages, games, and more to add to this unit, visit the Biology Resources at Layers-of-Learning.com.

You will find a link to video playlists, web links, and more.

Bookworms

If you're looking for a family read-aloud, we'd like to suggest this one.

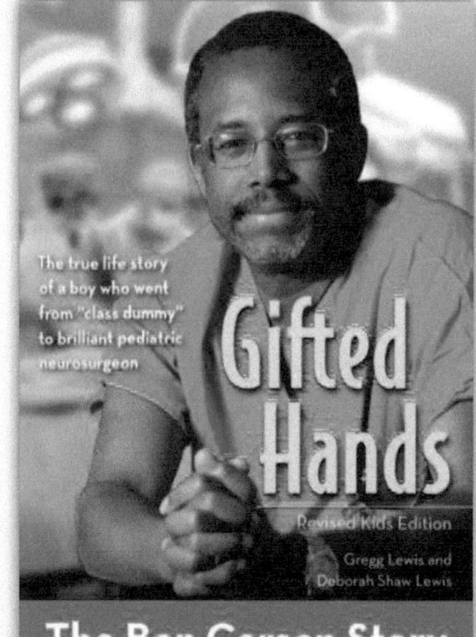

Gifted Hands (Kids Edition) by Gregg Lewis and Deborah Shaw Lewis is a biography of a boy from inner city America who overcame criticism, racism, and poverty to become a top pediatric brain surgeon.

Additional Layers

Homeostasis is the body keeping everything on an even keel. The levels of salts in the blood stay the same, the amount of water in the body stays stable, the amount of sugar in the blood stays even, and our body temperature stays steady.

Each of the parts of our bodies that need to be regulated have organs that work to regulate them in concert with other organisms.

Look up a YouTube video or two on homeostasis to learn how it works and why its so important.

Memorization Station

Skeleton: a system of hard organs, such as bones or cartilage, that provides support to an organism

Bookworms

Elizabeth Blackwell by Joanne Landers Henry is part of the excellent Childhood of Famous Americans series.

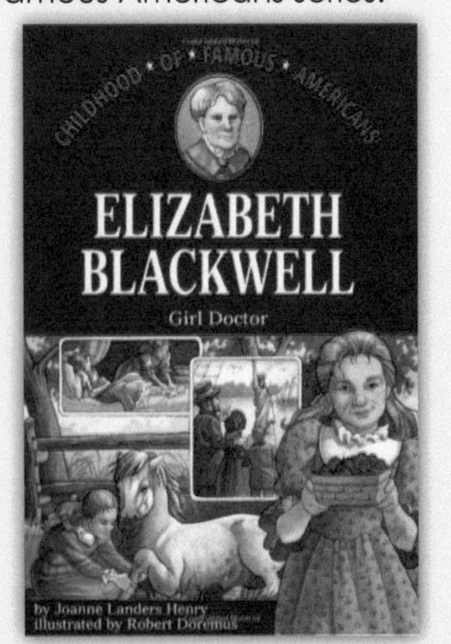

It is about the first woman doctor, focusing on her childhood.

☺ ☻*Ultimate Bodypedia* by Patricia Daniels, Christina Wilsdon, and Jen Agresta. Full of color and facts that appeal to kids. Includes simple experiments and health tips.

☺ ☻*My First Human Body Book* by Patricia J. Wynne. Filled with black and white line drawings and a fair amount of text, this is meant to be an instructional coloring book.

☺ ☻*The Fantastic Body: What Makes You Tick & How You Get Sick* by Howard Bennett. Written by a pediatrician, this book includes full-color photos and illustrations plus lots of information on body systems, injuries, and disease. It is hefty at 256 pages. If you are choosing to purchase one book, this is it.

☺ ☻*The Brain* by Seymour Simon. Full of colorful photos, this book does not shy away from tough vocabulary like synapses and dendrite. Worth the effort. Look for *The Heart, Bones, Eyes and Ears,* and *Lungs* by the same author.

☻*Science Comics: The Digestive System* by Jason Viola. Comic book-style science in a fun and accessible way. Talks about mucus, glands, digestive juices, and more!

☺ ☻*The Way We Work: Getting To Know the Amazing Human Body* by David Macaulay. Illustrated in the author's signature style, this is thorough and accurate plus easy to understand.

☻*The Complete Human Body* by Dr. Alice Roberts. This is a coffee table-style book from DK with full-color images, diagrams, and photos of the various systems of the human body. An accurate and thorough look at anatomy.

☻*The Body: A Guide For Occupants* by Bill Bryson. A popular level book about anatomy and physiology, including how and why the body does what it does. This book takes a negative view of the concept of an intelligent creator, so some parents may want to pass.

☻*Gut* by Giulia Enders. All about the digestive tract, this book goes into fascinating detail about glands, juices, cells, and more, starting with poop. Well written.

☻*How the Body Works* by DK. This is a fully illustrated and colorful book that details not just what the systems of the body are but also how they work. Great for teens and as a reference.

☻*The Digestive System, The Respiratory System, The Circulatory System, The Endocrine System, The Nervous System, The Muscular System, Osmoregulation*, and *Homeostatic Loops* from Bozeman Science on YouTube. Use these videos as lectures for your high schooler.

Step 2: Explore

Choose hands-on explorations from this section to work on as a family. They should be appealing activities that will create mental hooks so your kids remember the information in the unit. Save the rest of the explorations for the next time you do this unit in four years when your kids are older. You can also read the sidebars together and explore some little rabbit trails.

This unit includes printables. See the introduction for instructions on retrieving your Printable Pack.

Skeleton & Muscles

EXPLORATION: Skeleton
For this activity, you will need:

- Book or video about the skeleton
- "Skeletal System" from the Printable Pack
- Card stock - white
- Scissors
- Glue stick
- "Label the Skeleton" from the Printable Pack
- Pencil and colored pencils

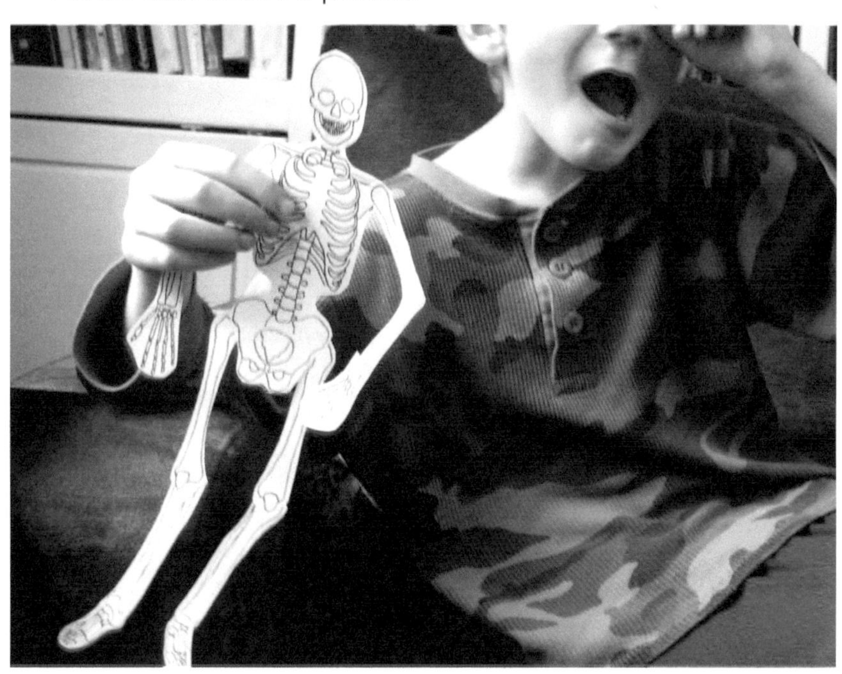

To give the body structure and to protect it from damage, humans have a **skeleton** made of 206 hard bones. Long bones are long and rounded, like a column. They make up most of the bones of the arms and legs. Short bones aren't any longer than they are wide. The bones in wrists and ankles are short bones. Flat bones provide structure and

Additional Layer

Read a book about bones to your little ones, then have them craft a skeleton out of pieces of pasta.

Additional Layer

It takes lots and lots of calcium to make strong bones. This is especially important for growing children who are building bones and pregnant mothers who are building the bodies of their children.

Discuss the kinds of foods that have calcium in them and why it is important to eat calcium.

Here is a list of some calcium rich foods:

- Broccoli
- Collard Greens
- Soybeans
- Oranges
- Cheese
- Yogurt
- Milk
- Fortified cereals

protection. They include the bones of the head, the shoulders, the rib cage, and the pelvis. Finally, there are irregular bones that provide very specific functions. These include the vertebrae, the tiny bones inside the ear, and the bones of the face.

1. Read a book of watch a video about the skeletal system.
2. Cut out the skeleton from the "Skeletal System" printable. Glue it together, putting the bones in the correct places as you go. Refer back to your book or video if you're not sure.
3. Learn the names of the bones that are printed on the skeleton. Play a memory game where someone calls out a bone and each person has to touch that bone on his or her body.
4. Practice learning the names of the bones for several days until you feel confident, then use the "Label the Skeleton" printable as a quiz. Color it if you like.

Additional Layer

Play "Simon Says" skeleton-style.

The game is that if you preface your instructions with "Simon says . . . " then the instructions must be followed. But if you leave that off then the instructions must not be followed. The goal is to catch someone out by making them follow an instruction when they aren't supposed to.

In this game you will say something like, "Simon says touch your ulna with your metatarsals."

Use the game to practice the names of the bones.

☻☻☻EXPLORATION: Bones

For this activity, you will need:

- Beef marrow pipe bones from a butcher
- Magnifying glass
- Science Notebook
- Pencil and colored pencils
- Vinegar
- Beaker or jar
- Plastic wrap or a lid to cover the beaker

Bones provide structure to hold up our bodies, protect our internal organs from injury, and work with our muscles and ligaments to help us move. To do all these things, bones have to be tough and rigid. They are made of collagen fibers, minerals, and water. Calcium is the main mineral that makes bones hard.

Additional Layer

Learn the first aid for a broken bone.

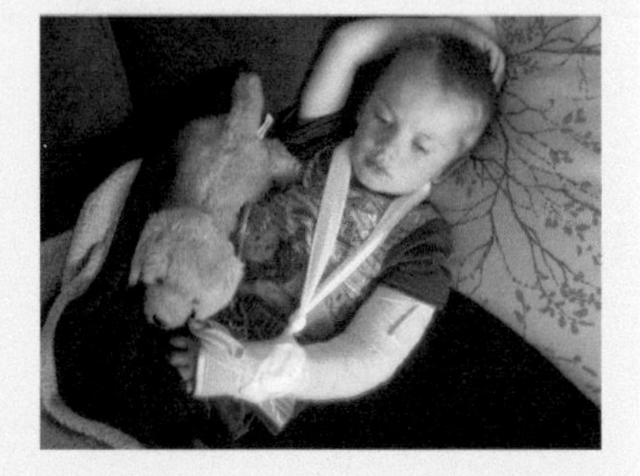

1. Stop any bleeding.
2. Immobilize the injured area (never move someone with a neck or back injury).
3. Apply a cold compress to the area.
4. Treat the patient for shock. Keep them warm and calm and elevate the feet.
5. Get the patient to a doctor immediately. Call for an ambulance if you suspect a neck or back injury.

1. Examine the section of bone with a magnifying glass.
2. Outside the bone will be very hard and rigid. The outside of bones is covered with a membrane called the periosteum. It may or may not still be present on your bone sample. You may also see some blood vessels attached to the outside of the bone.

   Inside, there is a soft section. This is the marrow. In the larger bones of our bodies—ribs, sternum, vertebrae, and pelvis—new blood and lymph cells are made.

   If you have the end of a long bone, you may also be able to see cartilage protecting the end of the bone.

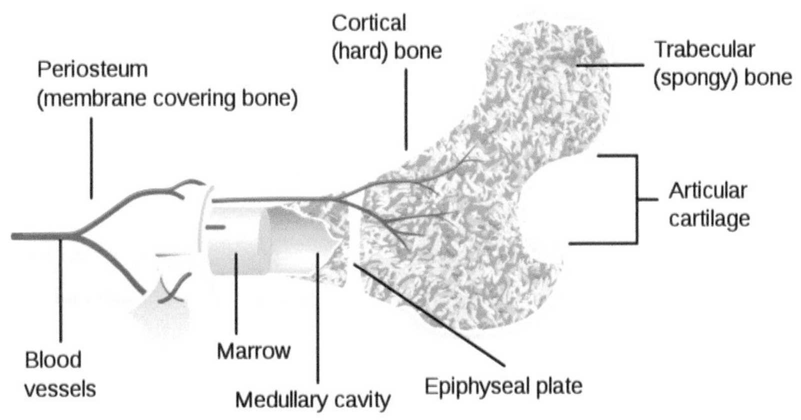

Image by Pbroks13, CC by SA 3.0, Wikimedia

3. Draw a diagram of the structure of the bone in your Science Notebook. Make sure to label it clearly and add the page description to your table of contents.
4. Pour 300 ml of vinegar into a beaker. Place a piece of bone into the beaker with the vinegar. You may want to cover the container because of the strong smell of vinegar. Let the bone sit in the vinegar for several days. Check the bone each day.
5. The bone will become soft and rubbery because vinegar dissolves calcium and other minerals.

EXPERIMENT: Three Types of Joints

For this activity, you will need:

- "Three Types of Joints" from the Printable Pack
- Pencil and colored pencils
- Craft sticks
- Rubber bands
- Plastic Easter egg or one section of an egg carton
- Play dough or clay
- Wooden blocks
- Kitchen sponge
- Scissors
- Craft foam

Memorization Station

Memorize the major bones of the body using "Label the Skeleton" from the Printable Pack.

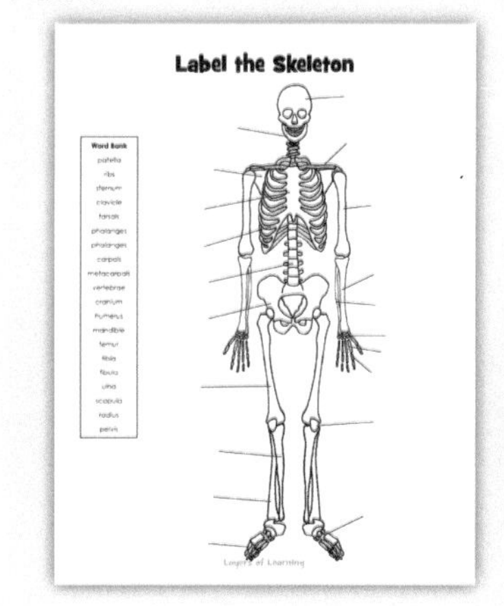

Bookworms

Give Me Back My Bones! by Kim Norman is a picture book story of a pirate who is searching for his bones that are scattered on the sea floor. It uses the correct names of each of the bones as they are reassembled into a skeleton. Ages up to 8.

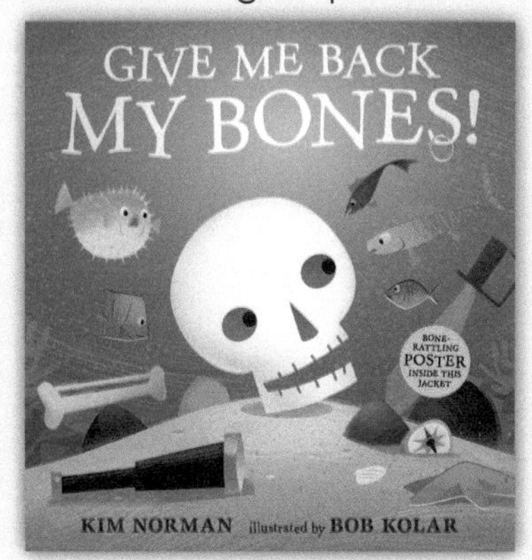

Writer's Workshop

Write about something fascinating you learned about bones. Write with enthusiasm so your reader will find it just as fascinating as you did.

Memorization Station

Joint: the point at which two parts of a skeleton are joined together

Fabulous Fact

An anatomist is a scientist who studies the physical structures of living things, often humans.

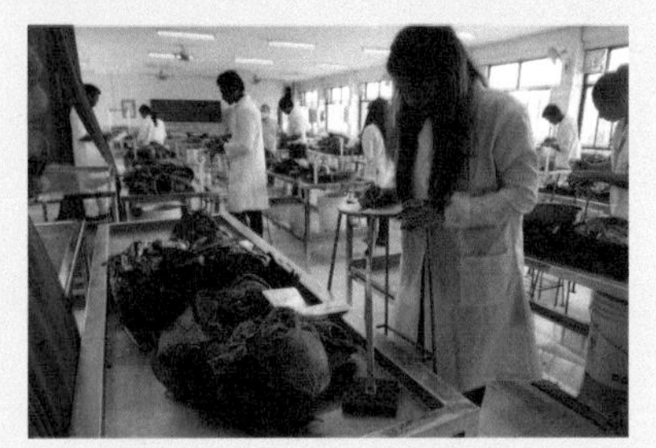

These are students in an anatomy lab in Pune, India. Image by Raghav93 CC by SA 4.0, Wikimedia

A physiologist studies the way the human body works, including organ functions, hormones, metabolic processes, and immunity.

Famous Folks

Gunther von Hagens is a German anatomist who discovered a revolutionary process for preserving anatomical specimens called plastination. In plastination, the water and fat are replaced by plastics, which completely prevents decay.

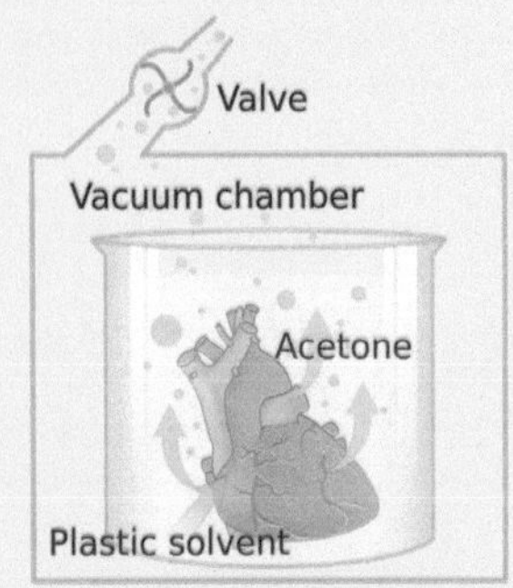

Forced Impregnation

Joints are where two bones connect. We are going to learn about three types of joints—fibrous, cartilaginous, and synovial.

Fibrous joints do not allow any movement between the bones. The bones are knit together with tough collagen fibers. The bones of the cranium and teeth are of this type.

Cartilaginous joints are connected together only with cartilage. This type of joint allows a little movement but not much. The cartilage does function as shock absorption. Vertebrae in the spinal column are cartilaginous joints.

Synovial joints allow lots of movement. They are constructed with ligaments, cartilage cushions, and bones. The ball and socket joint of your hip and the hinge joint of your knee are both synovial joints.

1. Color the "Three Types of Joints" sheet. Draw a line from the description to the picture of the joint that matches it. Use the description above to check the answers. When you're done, add it to the Science section of your Layers of Learning Notebook.

   Find examples of these types of joints on your body. Which kind do you have the most of? **(synovial)**

2. The cranium is made up of many bones jointed together

like a puzzle. Cut a piece of craft foam like a puzzle. The pieces stay together much better than they would if they had straight edges. Cut another puzzle out of craft foam with straight sides. Compare your two puzzles. The jagged edges and the collagen fibers holding the skull together make rigid immovable joints to protect your brain.

3. Make a model section of vertebrae. Cut a kitchen sponge in half. Put the two halves, one on top of the other, between two blocks of wood. Wrap two rubber bands around the whole thing.

   Squeeze and flex the model. The wood blocks are like the vertebrae bones. The sponges are like the cartilage and fluid. The rubber bands are like the ligaments.

4. A hinge joint is one kind of synovial joint. Attach two craft sticks together with a rubber band in an X pattern. The joint will be able to move back and forth with some resistance from the rubber band. The sticks are like bones and the rubber bands are like ligaments, the tough bands that connect one bone to another.

5. A ball and socket joint is another kind of synovial joint. Roll some clay into a ball and put it in one half of a plastic Easter egg. Push a craft stick into the clay. Then roll the clay ball around inside the Easter egg using the stick as a handle. The Easter egg is like the socket, the stick and clay are like the large long bone that sits inside the socket.

EXPLORATION: Knee Joint

For this activity, you will need:

- "Diagram of a Knee" and "Diagram of a Knee Answers" from the Printable Pack
- Scissors
- Glue stick
- Colored pencils or crayons
- Video about knee bones and ligaments from this unit's YouTube playlist

The knee is a hinge joint that can mainly move back and forth but only in one direction. It allows for a small amount of twisting as well. The ligaments of the knee hold the bones together firmly, while allowing for flexing and movement.

1. Cut the squares off the bottom of the "Diagram of a Knee" and cut them apart. The diagram is a front view of the knee when it is bent at 90 degrees.

2. Glue the squares onto the sheet in the dashed boxes.

Deep Thoughts

This Rembrandt painting is called *The Anatomy Lesson of Dr. Nicolaes Tulp*. The painting documented a real dissection. The cadaver (dead body) was a criminal, Aris Kindt, who had been executed the same day.

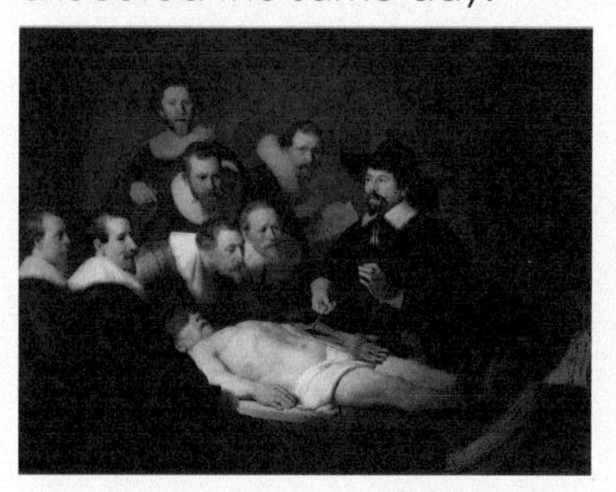

When it was painted in 1632, cutting up human bodies was extremely controversial. The city of Amsterdam allowed only one dissection per year.

But without examining and understanding the anatomy of the body, no progress could be made in learning how to heal the body.

Experiments done on human bodies, dead or alive, are still controversial and probably always will be.

What do you think is a good standard to decide what kinds of experiments are okay and which are not?

Writer's Workshop

Write about something amazing that your body does, something you are grateful for.

Set a timer and write for ten minutes straight without stopping. It's okay if you repeat yourself or write nonsense, but focus on the wonderful things about the body.

Bookworms

The Midwife's Apprentice and *Matilda Bone* by Karen Cushman are historical novels set in the medical world of the Middle Ages.

Brat is a destitute orphan who gets taken in by the local cranky midwife and has to decide whether to stay.

Matilda was raised by a priest to shun worldliness, but then she is thrust into the world of Blood and Bone Alley, home of leeches, barber-surgeons, and apothecaries. Can she learn to love the world?

Ages 9 and up.

Memorization Station

Muscle: a bundle of fibrous tissues that has the ability to expand and contract, producing movement

Autonomic nervous system: the part of the nervous system that automatically controls the muscles of internal organs and glands

Use the "Diagram of a Knee Answers" as a guide. Color the bones yellow, the cartilage blue, and the ligaments pink. Add it to the Science section of your Layers of Learning Notebook.

3. Watch a video about the knee joint.

☻☻☻EXPERIMENT: Three Types of Muscles

For this activity, you will need:

- Book or video about muscles
- Prepared microscope slide of muscle tissue (optional)
- Microscope (optional)
- "Three Types of Muscles" from the Printable Pack
- Pencil and colored pencils

The muscular system allows the body to move, stand upright, and circulate blood through the body. There are three types of **muscles**—skeletal, cardiac, and smooth.

Skeletal (striated) muscles are made of muscle fibers which can contract or shorten, making the whole muscle contract and causing movement. The skeletal muscles are voluntary; you can mostly control them.

Cardiac muscles and smooth muscles are both involuntary, they are controlled by the **autonomic nervous system**. Cardiac muscles are in the heart, and smooth muscles are in the digestive system and in blood vessels.

1. Watch a video or read a book about muscles.
2. Look at a prepared muscle tissue slide under a microscope. If you don't have a microscope or a muscle tissue slide, look up microscopic images of muscle tissue online and use those.
3. Use the "Three Types of Muscles" sheet to sketch what you see in the microscope. Color the sheet.
4. Fill in the blanks on the sheet, then add it to the Science section of your Layers of Learning Notebook. **The answers in order are: body, movement, heart, internal organs.**

☻☻☻EXPLORATION: Muscle Model

For this activity, you will need:

- "Tendons and Ligaments" from the Printable Pack
- Cardboard, like from a cereal box
- 2 balloons
- Metal paper fastener (brad)

- String
- Scissors

Skeletal muscles are attached to bone with ligaments. Ligaments are made of the same material and function the same way that tendons do except they connect muscles to bone instead of bone to bone. Most skeletal muscles are arranged in pairs so that when one muscle contracts, another relaxes. In this activity, the bones are represented by cardboard pieces, the muscles by balloons, and the tendons by string.

1. Bend and extend your arm at the elbow joint. Feel the muscles in your upper arm, your bicep and tricep. When one is flexed, the other is relaxed.
2. Cut two pieces of cardboard—5 cm by 20 cm. Shape them with scissors to look like long bones. In the first bone, punch two holes in one end and one centered hole in the other end. In the second bone punch three holes in a row in one end.
3. Connect the two bones together with a metal paper fastener going through the center holes.
4. Punch two holes, one in each end, of two balloons. Tie the balloons from the holes in one end of the first balloon to the end of the second balloon. When the joint is straight, the balloons should be taut but not stretched.
5. As you bend the joint, the balloons will contract (get shorter) or relax (get longer) with the movement. Only one is stretched at a time. One extends the bone, the other pulls it back.

Skin

EXPLORATION: Skin Booklet
For this activity, you will need:

- Book or video about human skin
- "Human Skin" from the Printable Pack

Deep Thoughts

Some people donate their bodies to science after they die so they can be used for medical research. Would you do that? Why or why not?

Writer's Workshop

What if you had to remember to make your heart beat or actively make your stomach digest food? What if you had to spend mental effort on remembering to breath at the right rate to keep yourself from passing out?

Write about a day where your involuntary muscles suddenly became voluntary. What happens to you?

Fabulous Fact

Muscles are made of bundles of tiny muscle fibers. Each muscle fiber expands and contracts with a set of chemical reactions.

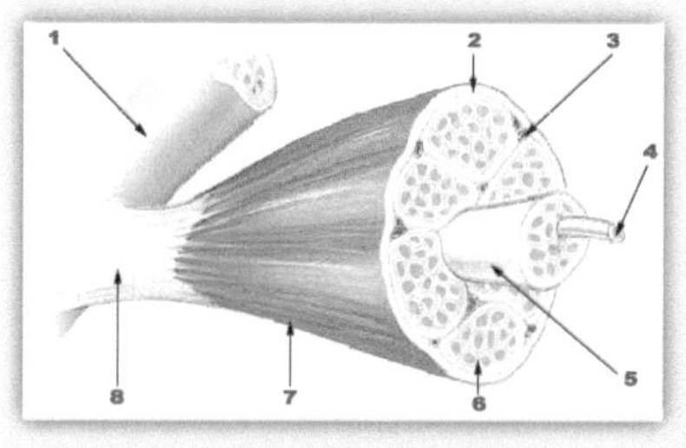

1. Bone	5. Fascicle
2. Perimysium	6. Endomysium
3. Blood vessel	7. Epimysium
4. Muscle fiber	8. Tendon

Fabulous Fact

About 1,000 different species of bacteria live on your skin. It's normal and natural and probably beneficial, although that is an area still being researched.

Memorization Station

Skin: the layer of soft tissue that forms an outer covering over the whole body

Fabulous Fact

Until recently it was thought that the strateum corneum, the very top layer of skin cells, was dead. But now we know it's alive.

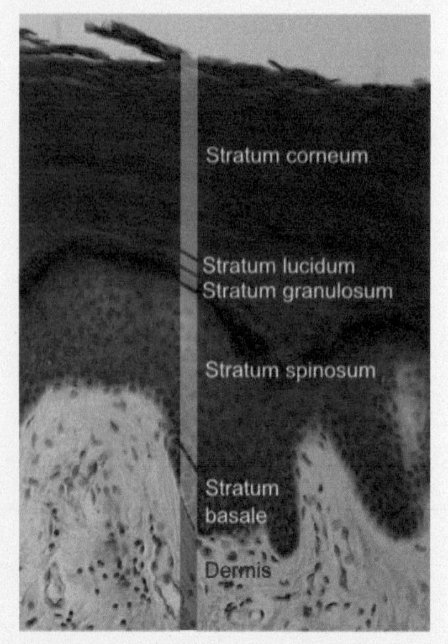

This layer actively protects the lower skin layers against shear and impacts. It also regulates how much moisture gets in and out of the skin, fosters certain microbes while blocking others, and selectively allows some things through the skin while blocking toxins and allergens.

Fabulous Fact

The scientific name of skin is epidermis. "Epi" is a Greek root that means over or upon. "Dermis" means skin. Below the epidermis is the dermis.

- Microscope (optional)
- Prepared microscope slide of human skin (optional)
- Video or book about fingerprints
- Colored pencils
- Pen
- Scissors
- Stapler
- Ink pad
- Magnifying glass

Skin is the layer of soft tissue that covers the entire body. It protects the body, helps regulate the internal environment, allows us to feel sensations of touch, cold, heat, and pain, and protects us from disease causing pathogens.

1. Cut apart the "Human Skin" printable on the solid lines. Assemble the pages into a book. There are page numbers so you can assemble them in order. Staple the pages together at the staple marks on the first page.

2. Watch a video or read a book about skin. Using page 1 of your book, color the top layer of skin to match your skin color. This epidermis protects your body from damage and from pathogens and regulates your temperature.

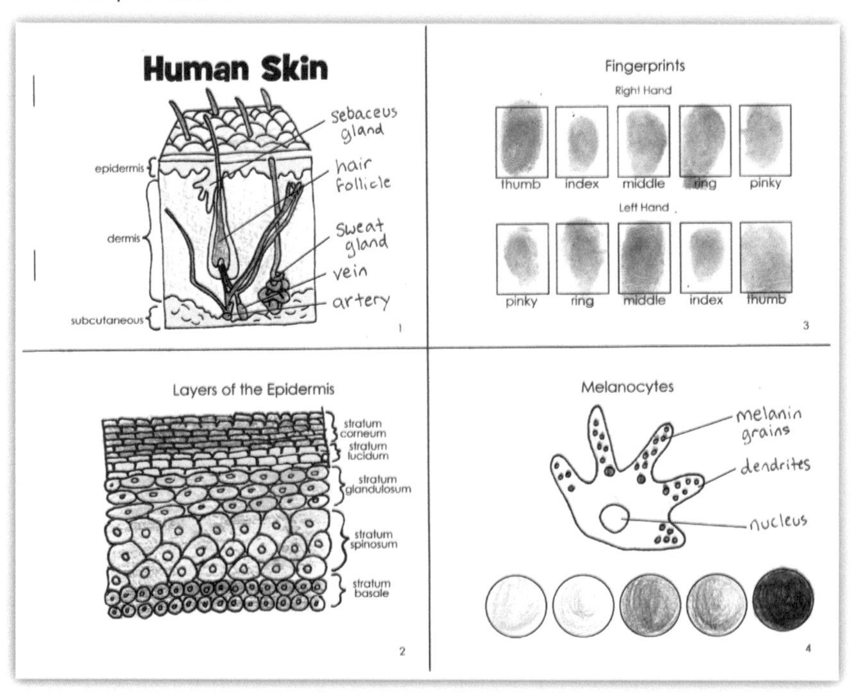

Write "sebaceous gland" on the top labeling line. Color it yellow along with the lower layer of epidermis. Sebaceous glands secrete oily or waxy substances called sebum. Sebum lubricates skin and hair and keeps it soft and shining.

Color the dermis tissue pink. Label the next labeling

line "hair follicle" and color the hair follicles to match your hair color. Each hair follicle is made up of twenty different types of cells. The length the hair grows and whether it is curly or straight is all controlled by hormones, proteins, and immune cells. The shape of the hair follicle determines if your hair is coarse or fine.

Label the next line "sweat gland" and color the sweat gland darker pink. Sweat glands secrete a watery substance to cool the body. They are controlled by nerve fibers and fed by capillaries. The sweat itself is made from blood plasma.

Label the next line "vein" and color it blue. Label the last line "artery" and color it red. The skin, made up of living cells, is fed by small blood vessels that extend through the subcutaneous and dermis layers, but not into the epidermis. The epidermis is fed though diffusion.

Color the subcutaneous layer yellow. This layer isn't part of the skin, but attaches the skin to the underlying bone and muscle. About 50% of this layer is fat that feeds the skin and helps keep the body warm and padded.

3. Look at a human skin cross section sample under a microscope. Zoom in on the top layer of skin. You will see very thin flaky-looking cells at the very top, then a layer of rounder cells just beneath. This is the epidermis.

   Using page 2 of your book, color each layer of the epidermis a different color. The skin cells at the lowest part of the epidermis are new. As the cells age, they move up the layers. At the top layer, they die and flake off your body. But all the layers are alive and doing vital jobs for your body.

4. Your fingertips have patterns of raised ridges that leave behind fingerprints when you touch a smooth surface. The pattern of raised ridges is unique to each person and can be used as an identifying mark.

   Use an ink pad to carefully record your fingerprints on page 3 of your booklet. Examine your fingerprints closely with a magnifying glass.

   Watch a video or read a book about fingerprints. After you have learned more about fingerprints, go back and examine yours again. Can you spot any loops, arches, or whorls in your fingerprints?

5. Melanocytes are cells located in the bottom layer of the epidermis, the stratum basale, as well as a few other places in the body. They produce melanin, which is the main cause behind different skin colors. People with very

Additional Layer

Melanocytes are also part of the immune system. They can absorb and digest microbes and viruses, and produce antigens to mark viruses for destruction.

Since they are placed in the upper level of skin, melanocytes are vital for protecting your body from pathogens trying to enter from outside.

Additional Layer

Melanin in skin naturally blocks much of the harmful UV rays from the sun, but it can't block all of them.

UVA rays, in particular, can penetrate deeply into the dermis layer of the skin, damaging nerves, blood vessels, and immune cells. The deep penetration of UVA rays causes skin to wrinkle and lose elasticity over time. It is what makes skin look old.

UV rays can also cause cancer if the sun damage is severe or if an individual is susceptible. Sunscreen prevents the damage while still allowing your skin to produce Vitamin D, a vitamin essential for a healthy immune system and one you only get from exposure to sunlight.

Tanning beds produce the same UV rays that the sun does and cause just as much damage.

Having naturally dark skin doesn't protect you from skin cancer either.

Learn more about how sunscreen works.

Famous Folks

Most women who die in childbirth do so because they are bleeding out, or hemorrhaging. James Blundell, a British obstetrician, was the first to use a blood transfusion to save a mother's life.

Not all of Blundell's patients responded well to the blood transfusions. No one knew why until in 1901 when Karl Landsteiner, an Austrian doctor discovered that blood comes in different "types."

Memorization Station

Heart: a hollow muscular organ that acts as a pump to move blood around the body

Blood vessel: a tubular structure that carries blood through the tissues and organs of the body

Famous Folks

Andreas Vasalius was the first to dissect and study a human heart, among other organs.

He wrote an anatomy book in 1538, the first of its kind.

light skin have little melanin. People with very dark skin have the most melanin.

The amount of melanin your melanocytes produce is controlled by hormones. Exposure to sun increases the production of melanin as well, darkening your natural skin pigment even further.

Label the melanocyte on page 4 of your booklet, from top to bottom—melanin grains, dendrites, nucleus. Color the melanocyte a light color. Then, color in the melanin grains with dark brown. Color the circles at the bottom of page 4 with different skin tones. Skin tones are caused by melanin production.

Circulation

EXPERIMENT: Heart Dissection

WARNING: This activity uses sharp tools and includes powerful chemicals. Children should be taught to use tools correctly and to never eat during this lab. Thoroughly wash hands and all surfaces after the experiment.

For this activity, you will need:

- "Heart Diagram" and "Heart Diagram Instructions" from the Printable Pack
- Colored pencils or crayons
- Sheep heart dissection kit
- Disposable gloves

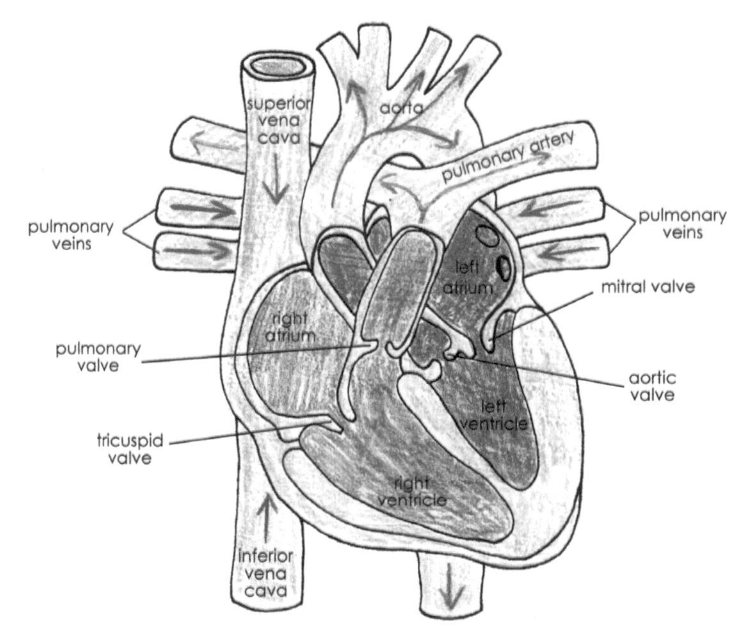

The **heart** works with the **blood vessels** to provide constant blood flow throughout the body. Blood flows from the heart to the lungs where it picks up oxygen, comes back to the heart, then flows out to the rest of the body, delivering oxygen and nutrients to the cells.

The heart is a bundle of muscles that has to beat all the time, never resting, so the heart muscles are special. They are made of cardiac muscle, strong fibers bound to one another in a lattice work and then swirling around the heart in complex patterns.

1. Color the "Heart Diagram" using the "Heart Diagram Instructions."
2. Dissect a sheep heart with instructions from the kit you purchased.

EXPERIMENT: Blood Model

For this activity, you will need:

- Microscope (optional)
- Prepared microscope slide of blood, or make your own slide with a prick of your finger (optional)
- Science Notebook
- Corn syrup
- Mini marshmallows
- Red candies
- Decorating sprinkles
- Beaker or medium bowl
- "Blood Clotting" from the Printable Pack
- Colored pencils

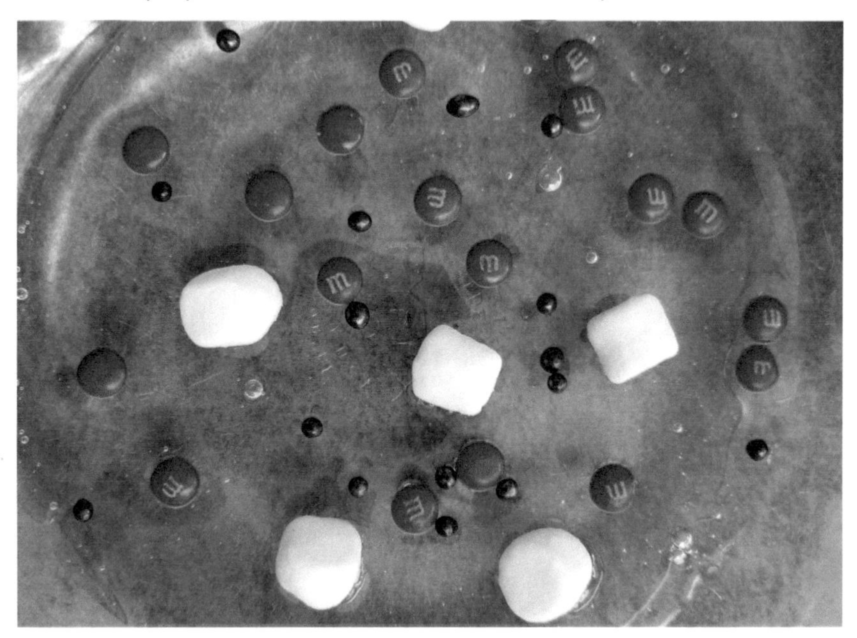

Blood delivers necessary substances (like oxygen) from the lungs, nutrients from the digestive system, and hormones and proteins from glands and cells all around the body. Blood also takes waste from the cells and transports it to the lungs, the kidneys, and the skin to get rid of it.

1. Look at a microscope slide of blood. If you don't have a microscope or slide, look up microscopic images of

Bookworms

Florence Nightingale: Lady With the Lamp by Trina Robbins and Anne Timmons is a graphic-style biography of the woman who created the profession of nursing during the horrible conditions of the Crimean War. Ages 9 to 12.

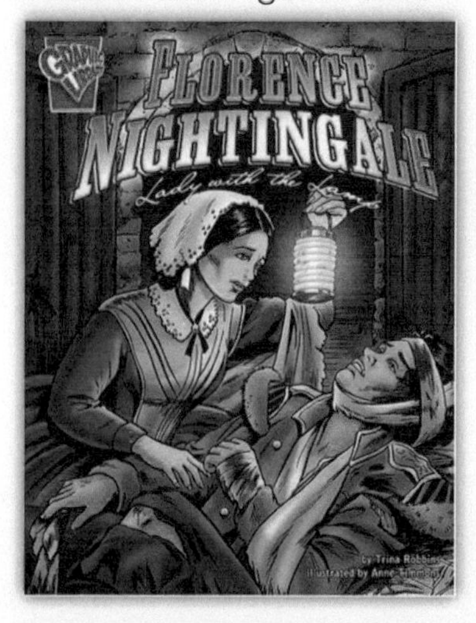

There are dozens of excellent biographies of this famous woman, from picture books to fat tomes.

Famous Folks

William Harvey was an English physician. In 1628, he published a short book explaining for the first time how blood circulates through the body.

Learn about his life.

blood online. Under the microscope you can see that blood is made up of many cells. You can see hundreds of round red blood cells. These carry oxygen. You can also see a few slightly larger cells (stained purple). These are the white blood cells. Their primary job is to protect the body from infection by bacteria or viruses. You may be able to see tiny purple dots. These are platelets; they cause blood to clot when there is a wound.

2. Zoom in as far as you can with the microscope and then draw a sketch of what you see in your Science Notebook. Label the red blood cells, white blood cells, and platelets. Write a few sentences describing what blood is made of.
3. Add 300 ml corn syrup to a beaker. Corn syrup represents the plasma. Plasma is the fluid which carries all the other parts of the blood around. It is made mostly of water.
4. Add 6 mini marshmallows. These are the white blood cells, which seek out and destroy bacteria and viruses that try to infect our bodies.
5. Add about 20 red hots. These are the red blood cells that carry oxygen and nutrients to the cells.
6. Add about 20 sprinkles. These are the platelets that help the blood clot when an injury occurs.
7. Read and color the "Blood Clotting" sheet from the Printable Pack. Start by coloring the key at the top. Color the images below according to the key. Add it to the Science section of your Layers of Learning Notebook.

Bookworms

Tiny Stitches: Life of Medical Pioneer Vivien Thomas by Gwendolyn Hooks is a picture book biography of the man who pioneered the surgical cure for infant heart defects.

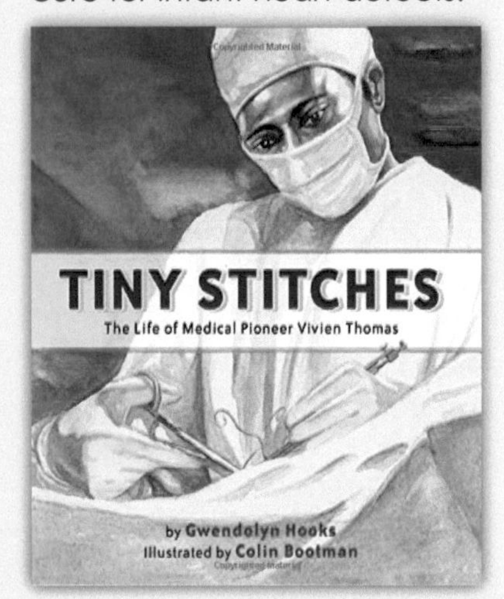

For ages 6 to 12 years.

Deep Thoughts

Discuss the four principles of medical ethics.

1. Respect patient autonomy
2. Do no harm
3. Promote the patient's good
4. Follow the law and be just to all patients

Apply these principles to a controversial medical issue and decide how the ethics would be applied.

- Covid-19 handling
- Abortion
- Euthanasia
- Gene therapy
- Organ donation
- Medicinal cannabis
- Transgender surgery

Do you think these ethical guidelines are correct and sufficient to protect people?

EXPLORATION: Arteries and Veins

For this activity, you will need:

- "Circulatory System" and "Circulatory System Answers" from the Printable Pack
- "Arteries and Veins" from the Printable Pack
- Colored pencils
- Stopwatch or timer
- Book or video about veins and arteries

Arteries move blood from the heart to the body. The arteries are full of oxygen-rich blood. **Veins** move blood from the body to the heart. Veins have oxygen-depleted blood and are full of waste to remove from the body. Arteries are muscular and, like the heart, help blood to pump through the body. Veins also have muscles but not as many. Instead, veins have valves that prevent the blood from flowing backward as it is returned to the heart.

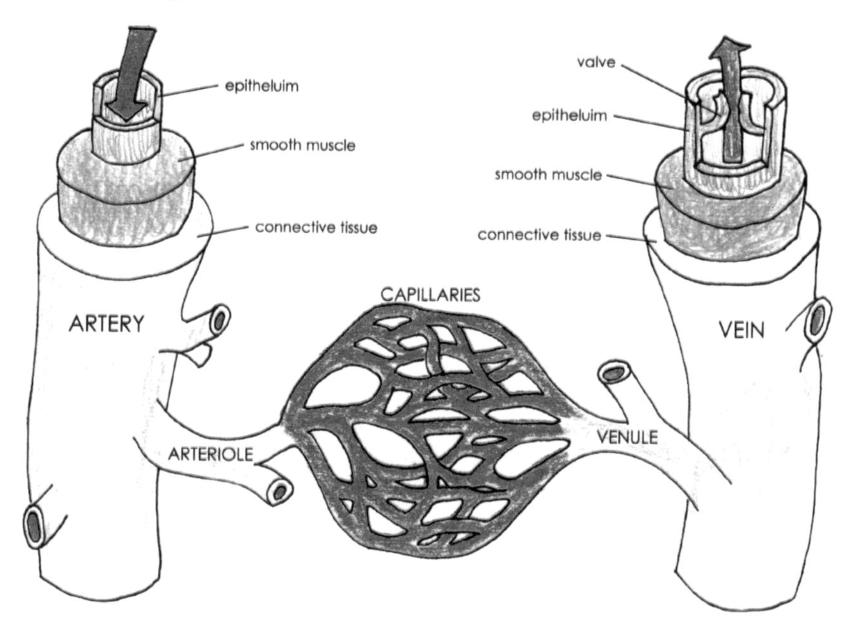

1. Color the "Circulatory System" printable. The left side of the body diagram should be colored blue for veins. The right side of the body diagram should be colored red for arteries. Color the heart red. Of course, in reality, arteries and veins are on both sides of the body. The diagram is drawn this way for simplicity.
2. Label the major arteries and veins using the word bank and the "Circulatory System Answers" sheet.
3. Feel for your jugular vein in your neck. You can feel blood pulsing through on its way back from your head. Count the number of pulses in 15 seconds. Multiply by four to get your beats per minute. Repeat this three times and then average your results. When you are resting, your heart rate should be between 60 and 100 beats per minute for anyone above the age of 10 and between 70 and 115 for children over the age of 5.
4. Do 15 jumping jacks or another vigorous exercise and then check your heart rate again. Your heart rate increases when you exercise or if you are stressed or ill.
5. Watch a video or read a book about arteries and veins.
6. On the "Arteries and Veins" printable, color the connective tissue light peach. This tissue is really mostly transparent. The veins under your skin in your wrists can look blue because of the red blood combined with the overlying fat tissue, giving a blue appearance.

   Color the smooth muscle pink. Both veins and arteries have smooth muscles in their walls, but arteries are thicker.

Memorization Station

Arteries: muscular-walled tubes that convey blood from the heart to the body

Veins: a tube with valves that returns deoxygenated blood from the body to the heart

Additional Layer

Kidneys are the filter of the body. All of the blood in your body passes through them several times a day. The kidneys sift through the blood and take out any unneeded salts, vitamins, urea, and excess water.

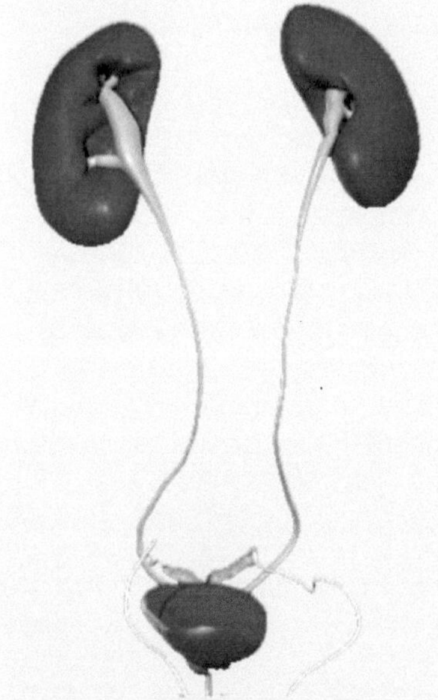

Above, you see two kidneys with ureter tubes leading to the bladder, then a short urethra leading outside the body.

Their job is to keep all the little bits our blood carries around in balance.

The kidneys also create a couple of vital hormones.

Learn more about the kidneys and their function.

Fabulous Fact

In most charts of blood vessels, veins are blue and arteries are red, but that's just to help us keep them straight. In your body, your blood is red even when it's deoxygenated. It may change shades, but it isn't blue.

So why do your veins look blue? Fat under your skin only allows blue light to penetrate it, so this is what is reflected back to your eyes.

Deep Thoughts

Until the 1940s and1950s, cigarettes and cigars were thought of as benign or even healthy. Then an uptick in the number of smokers in the early 20th century meant an epidemic of lung cancer swept the globe. It wasn't until decades later that doctors were universally convinced of the dangers of tobacco. There are still about 1.5 million lung cancer deaths caused by cigarettes per year worldwide.

Why does it take so long to convince people and cultures to change their perceptions about health?

Why do people keep doing things they know are risky or unhealthy?

Which bad health habits do you find it hard to break? Why?

Besides cigarettes, what do you think are the unhealthiest habits in your country? Look up research and see if you are right.

Color the epithelium light brown. Epithelium is any tissue that lines organs or body cavities. Your skin is one kind of epithelium and the lining of your veins and arteries is another.

Color the valve in the vein light brown. Valves prevent the blood from flowing backward.

Color the arrow leaving the vein blue to represent blood flowing toward the heart. Color the arrow entering the artery red to represent blood flowing away from the heart.

Color the arteriole and venule light peach. Arterioles and venules are between arteries and capillaries and between veins and capillaries in size.

Color the capillaries red. Capillaries are so tiny that you can't see them with the naked eye. They have a wall that is only one cell thick. Water, nutrients, oxygen, and wastes can travel right through the capillary wall and to the cells. The capillaries also connect the artery system to the vein system.

Respiration

EXPLORATION: Model Lung
For this activity, you will need:

- Video or book about lungs
- Narrow-necked plastic bottle
- 2 large balloons
- Scissors
- "Lungs" from the Printable Pack
- Science Notebook

1. Watch a video or read a book about the **lungs**.
2. Title a page in your Science Notebook "Lungs" and add it to your table of contents.
3. Cut the bottom off a plastic soda pop bottle. Cut the neck off a large balloon and stretch it over the cut bottom of the bottle.
4. Poke a second balloon into the neck of the bottle, stretching the neck of the balloon over the neck of the bottle.
5. Push and pull on the bottom balloon. This balloon represents the **diaphragm**, a sheet of skeletal muscle that sits just beneath the lungs. Observe what happens to the balloon inside the bottle.
6. Draw a diagram of the model lungs you made in your Science Notebook and write down an explanation of how the diaphragm works in your own words.
7. Air is drawn into your lungs when the diaphragm contracts. You can control your breathing with your diaphragm. Try it. Take a deep breath in, hold it for 30 seconds, then let the breath out. Could you feel your diaphragm?
8. Color the "Lungs" printable using the directions on the sheet. Glue the diagram into your Science Notebook.

☻EXPERIMENT: Cellular Respiration

WARNING: This exploration includes chemistry concepts that you may not have learned yet. If so, just absorb the chemistry and focus on the process.

For this activity, you will need:

- Three 150 ml beakers or small bowls
- Baker's yeast
- Water
- Sugar or honey
- Salt
- Electronic balance and a small dish for measuring
- Grease pencil or masking tape and marker for labeling
- Video about cellular respiration from the YouTube playlist
- "Cellular Respiration" and "Cellular Respiration Notes" from the Printable Pack
- Colored pencils
- Science Notebook
- Glue stick

Cellular respiration is the exchange of nutrients and oxygen for energy, water, and carbon dioxide inside the cells. The materials and wastes get carried by the blood

Writer's Workshop

Make a poster comparing human anatomy to an animal of your choice.

Draw a digram of six different human organs and write a paragraph describing each one.

Draw a diagram of six analogous organs in the animal of your choice and write a paragraph for each one, explaining how it is the same and how it is different from the human organs.

Put your information together on a poster. Present the poster to an audience.

If you want to grade the project, do so based on:

- Content - is the information thorough and accurate?
- Visual appeal - is the poster attractive and neat?
- Presentation - did you speak clearly and knowledgeably about your subject?

Memorization Station

Lungs: a pair of organs in the body that supply the body with oxygen and remove carbon dioxide

Diaphragm: a large muscle that separates the thorax from abdomen and contracts to enlarge the thorax, inflating the lungs

Fabulous Fact

The energy used inside your cells is ATP, or adenosine triphosphate. When it is broken apart it releases energy.

and expelled by the lungs and kidneys. The reactions happening in cellular respiration are why your body needs oxygen and the respiratory system. All cells respire whether they are the cells in your body or yeast cells.

1. Title a page in your Science Notebook "Cellular Respiration" and add it to your table of contents. Draw and record your experiment in your Notebook.
2. Label your beakers "sugar," "salt," and "control."
3. Add 100 ml warm tap water to each beaker. Add 1.5 g bakers yeast to each dish.
4. Add 3 g sugar to the first dish. Add 3 g salt to the second dish. Add nothing additional to the third dish.
5. After ten minutes, observe. In the first beaker, with added sugar, the yeast will have multiplied and begun to give off carbon dioxide gas, as evidence by the bubbles. The second and third dishes have not much growth, if any. Sugar is the food source for the yeast. Human cells also use sugar (**glucose**) to provide energy. Yeast and human cells both give off carbon dioxide.

$$C_6H_{12}O_6 + O_2 \rightarrow CO_2 + H_2O + ATP$$

glucose + oxygen → carbon dioxide + water + energy

6. Watch a video about cellular respiration. As you watch, take notes on the "Cellular Respiration" printable. You may need to watch the video more than once and pause it from time to time.

   We have provided a "Cellular Respiration Notes" page of completed notes to use as a reference, but don't copy ours; make your own notes!

7. Glue your notes into your Science Notebook.

Writer's Workshop

Choose any organ in the body and make an illustrated fact sheet about it.

Use one side of a plain white piece of paper and fill it with facts, lists, diagrams, and labels about that one organ.

Memorization Station

Cellular respiration: the process where cells combine oxygen with glucose to produce ATP, which the body uses for energy

Glucose: a simple sugar that is the main source of energy for the body

Famous Folks

Your body needs all of its organs. If an organ fails, you will die. Dr. Richard H. Lawler performed the first organ transplant, a kidney, in 1950.

The surgery was semi-successful. The woman lived another 5 years, but her body rejected the foreign tissue. Lawler was heavily criticized at the time because of his experimentation, but in 1970 he was nominated for a Nobel Prize.

Digestion

EXPLORATION: Digestive System Model

For this activity, you will need:

- Video or book about the digestive system
- "Digestive System" and "Digestive System Answers" from the Printable Pack
- Colored pencils
- Pen or pencil
- Banana
- Bowl
- Potato masher or fork
- Sandwich-size zipper bag
- Orange juice or lemon juice
- Nylon stocking - knee high
- Scissors

The digestive system breaks food down into molecules of glucose. It starts with chewing and saliva in the **mouth**, grinding and mixing with harsh acids in the **stomach**, and more digestive juices provided by the gall bladder, liver, and pancreas mixing with food in the small **intestine.** There, most of the nutrition from food gets absorbed into our blood. The waste products exit through the rectum and out of the body.

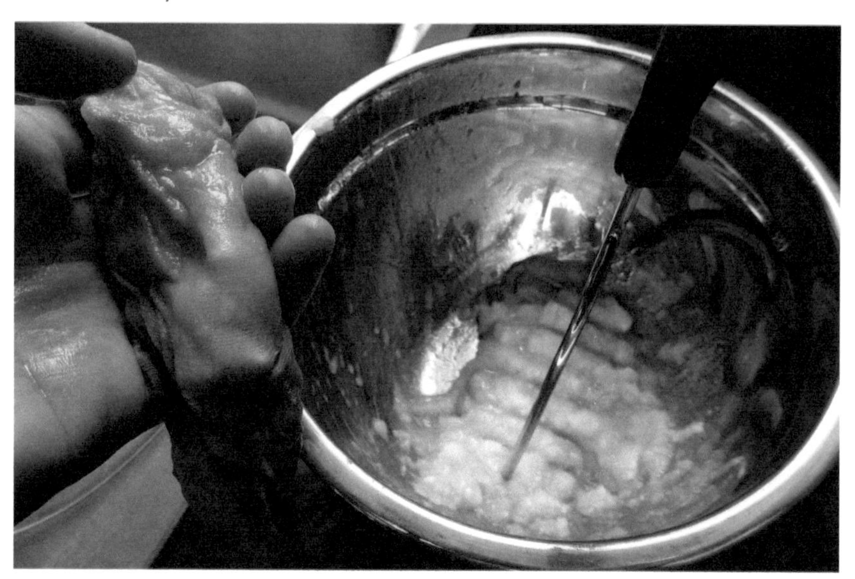

1. Read a book or watch a video about the digestive system.
2. Label and color the "Digestive System" using the "Digestive System Answers" as a guide.
3. Mash up a banana in a bowl with a potato masher or a fork. This represents your teeth chewing up your food.
4. Add a splash of orange juice and mash some more. This represents your saliva beginning to break down the starch in foods.

Bookworms

Gross and Ghastly: Human Body by Kev Payne is about the disgusting things your body does. It also includes activities, interactive quizzes, and puzzles. For ages 6 to 9.

Memorization Station

Mouth: opening in the face through which food and water are taken in

Stomach: internal organ where food the bulk of digestion occurs

Intestine: a long tube-shaped organ that completes digestion and absorbs nutrients and energy from food into the body

Fabulous Fact

Did you know that the intestines have nearly as many neurons as the brain?

New research is investigating how your gut can affect your mood and overall health.

5. Dump the banana mush into a zipper bag. Add another splash of orange juice. Squeeze out the air, seal it, and mash the banana even further with your hands. This represents the stomach with its acids digesting the starches and beginning digestion of the proteins.
6. Cut the toe off a nylon stocking. Over the top of a sink or basin, dump the "stomach" contents into the end of the stocking. Use your hands to squeeze the banana mush from one end of the stocking to the other. This represents the food moving through your intestines. The liquids and digested nutrients move through the intestine walls and into the blood stream. The remains are squeezed out the end of the anus and out of the body.
7. Take turns using your "Digestive System" diagrams to explain the parts of the digestive system to each other, then add them to the Science section of your Layers of Learning Notebook.

EXPERIMENT: Digestion Starts In Your Mouth

For this activity, you will need:

- Mirror
- Bright light - flashlight or lamp
- Science Notebook
- Colored pencils
- Soda crackers
- Napkins or paper towel
- Iodine with dropper

Digestion is the process of turning any kind of food into glucose. Glucose is the form of energy that cells can actually use. Glucose is a type of sugar. The mouth digests food mechanically by grinding it up with teeth. It also digests food chemically with saliva.

1. Use a bright light and a mirror to examine the inside of your mouth. Lift your tongue and find the openings of the salivary glands. Look at your teeth. How are the ones in front and the ones the in back shaped? Look at the texture on your tongue. Look at the roof of your mouth and the back of your throat.
2. Draw a detailed diagram of the inside of your mouth in your Science Notebook. Label your tongue and the salivary gland underneath it. Label your flat molar teeth in the back of your mouth and the sharp incisor teeth in the front of your mouth. Label the palate, also known as the roof of your mouth. Label your uvula, the dangly flesh that hangs down in the middle of your throat. Label your tonsils on each side of the opening to your throat.
3. Look up any of the mouth parts you are curious about or aren't sure of their function.

Additional Layer

Hormones are molecules that signal cell behaviors and functions throughout the body. They control everything your body does, from how it stores fat and how tall you grow to how big your muscles get and what your mood is.

There are cells all over your body that secrete hormones. It happens in your skin, in your small intestine, in your muscles, and in your kidneys. But there are special organs called glands whose main job is to produce various types of hormones. These glands make up the endocrine system.

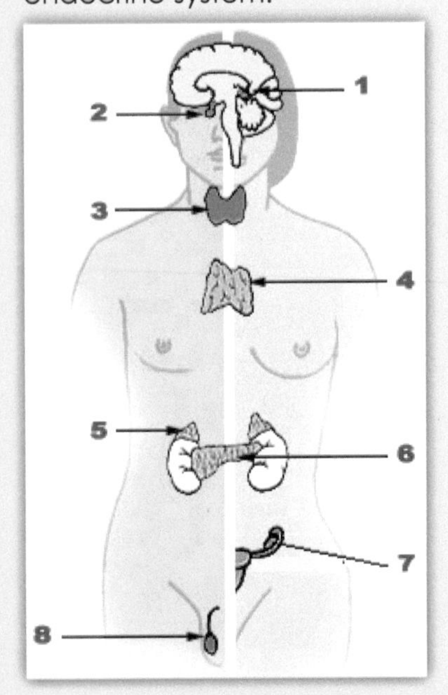

Here are the major endocrine glands, numbered to match the image above:

1. Pineal gland
2. Pituitary gland
3. Thyroid gland
4. Thymus gland
5. Adrenal gland
6. Pancreas
7. Ovary (female)
8. Testis (male)

Learn more about the endocrine system.

4. Chew a soda cracker 25 times and spit the chewed bit out onto a napkin.
5. Place a whole undamaged cracker next to it.
6. Test the two crackers for starch content by dropping a bit of iodine onto each one.

   In the presence of starch, the dark brown iodine turns black. Has the chewed cracker changed? Digestion of starches, like the flour and sugar in the cracker, begins in your mouth.

Brain & Nerves

EXPLORATION: Neurons
For this activity, you will need:

- Video about neurons
- Clay
- Colored markers
- Science Notebook

A **neuron** is a nerve cell. It has the ability to transport electricity from one end of the cell to the other. Neurons form chains and networks for passing electrical signals around the body. The electrical signals tell muscles to relax and contract, signal pain or gentler sensations, detect light, and tell glands to produce secretions.

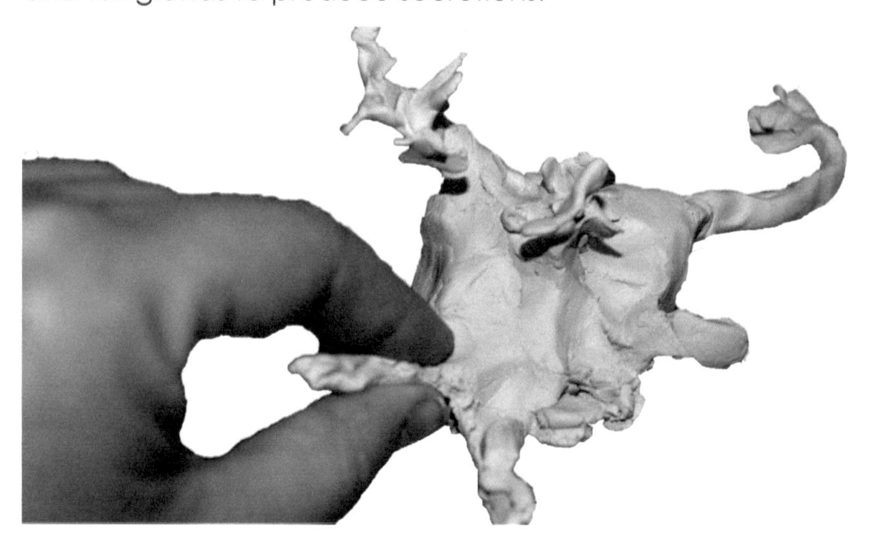

1. Watch a video about neurons.
2. Make a model neuron out of clay. Neurons have a central cell body with a nucleus, long arms extending out from the cell body called dendrites, and a long thin axon which ends in a branched terminal.

   Electrical signals enter through the dendrites, get processed in the cell body, and then exit through the

Memorization Station

Neuron: a type of cell that receives and sends messages between the brain and body; also called a nerve cell

Additional Layer

Complete "Label the Neuron" from the Printable Pack using the answer sheet.

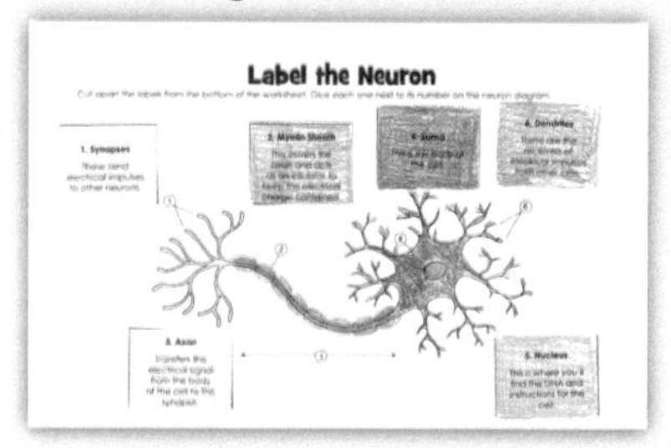

Pair the worksheet with a book or video for a complete lesson. You could also add this to the "Neuron" exploration if you want to learn a bit more.

Additional Layer

The brain and nervous system allow us to sense the world through sight, touch, smell, hearing, and taste.

Set up an obstacle course:

- Put on a blindfold while having to walk around obstacles
- Put on nose plugs while having to follow the smell of orange peels
- Put on thick gloves while having to feel the sandpaper path
- Put in ear plugs while following the sound of the music
- Identify the flavor of cookie without tasting it before you can proceed

Fabulous Fact

Just like you are left-handed or right-handed, you are also left-eyed or right-eyed.

Extend your arms out in front of your face and make a triangle with your fingers. Focus on an object about 6 meters away.

Now, without moving your head or hands, close one eye and then the other. One of your eyes will still be able to see the object through the triangle of your fingers while the other will not.

The eye that can still see the object is your dominant eye.

Additional Layer

Read a book or watch a video about the ear, then label and color the ear diagram from the Printable Pack.

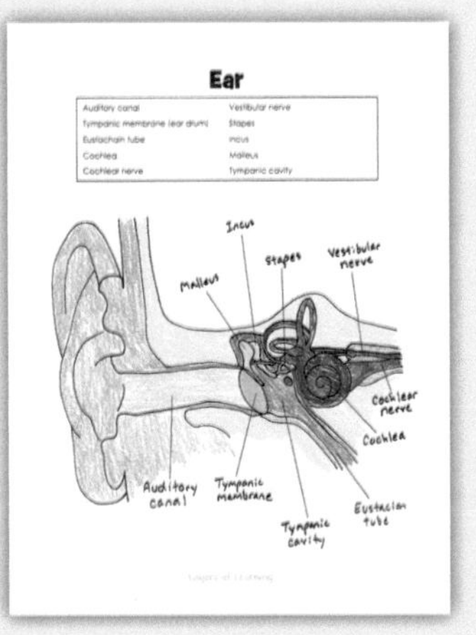

axon to the next neuron cell.

3. Draw ten dots, each a different color, down the left side of a page in your Science Notebook and ten more down the right side.
4. Draw a line from each dot on the left to each dot on the right. This represents neurons making connections with other neurons. In reality, its much more complicated, with each neuron making thousands of connections. Neurons connect with other neurons as you move, think, walk, run, catch a ball, read, watch TV, and listen and speak. The more you learn and experience, the more connections your neurons make.
5. Draw a diagram of a neuron in your Science Notebook and label it. Add a title to your page and add it to your table of contents.

EXPLORATION: Brain Cap & Dissection

WARNING: This activity uses sharp tools and includes powerful chemicals. Children should be taught to use tools correctly and to never eat during this lab. Thoroughly wash hands and all surfaces after the experiment.

For this activity, you will need:

- "Brain" from the Printable Pack
- White swim cap
- Colored permanent markers
- Sheep brain dissection kit (for middle grades and up)
- Safety goggles
- Disposable gloves

The **brain** is made up of billions of neurons connected together in complex patterns. Different regions of the brain focus on different jobs. For example, the brain stem controls involuntary functions like digestion.

1. Make a brain map with a white swimming cap and permanent markers. Use the "Brain" sheet as a guide.
2. Read a book or watch a video about the brain.
3. If you are learning with only younger kids, stop here. For middle schoolers and up, go on to the brain dissection.
4. Dissect a sheep brain using a dissection kit with its guide. We highly recommend watching a sheep brain dissection video as you go.

EXPLORATION: Eye Dissection

WARNING: This activity uses sharp tools and includes powerful chemicals. Children should be taught to use tools correctly and to never eat during this lab. Thoroughly wash hands and all surfaces after the experiment.

For this activity, you will need:

- "Eye" and "Eye Answers" from the Printable Pack
- Colored pencils
- Cow eye dissection kit
- Safety goggles
- Disposable gloves
- Video or book about eyes.

Eyes are part of the nervous system. They are sensory organs that can detect light and transmit the light information as neural signals to the brain.

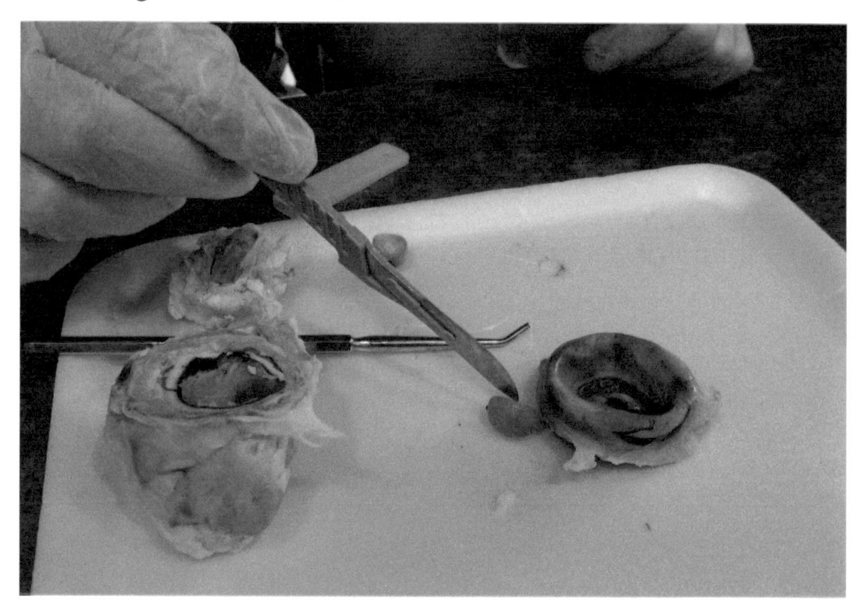

1. Color and label the "Eye" diagram using the "Eye

Fabulous Fact

There are two ways that messages are sent around the body. One is through the nervous system and the other is through hormones. Hormones and the nervous system often work together.

Famous Folks

Dr. James Tour is an American nanotechnologist. He builds carbon-based things atom by atom.

Photo by Jeff Fitlow, CC by SA 4.0, Wikimedia

He built a "bridge" using graphene from one side of a severed spinal cord in an adult rat to the other side of the severed spinal cord. The nano bridge made it possible for the nerves to regrow themselves and the rat recovered full mobility.

Research like this gives hope to people who are paralyzed or blind because of nerve damage.

Memorization Station

Brain: organ inside the head, made up of billions of nerve cells, that controls all of the body functions

Answers" as a guide.

2. Dissect a cow eyeball with the instructions in the kit you purchased.

3. Watch a video or read a book about how eyes work.

Step 3: Show What You Know

During this unit, choose one of the assignments below to show what you have learned during the unit. Add this work to your Layers of Learning Notebook. You can also use this assignment to show your supervising teacher or your charter school as a sample of what you've been working on in your homeschool, if needed.

There are more ideas for writing assignments in the "Writer's Workshop" sidebars.

Coloring or Narration Page

For this activity, you will need:

- "Human Body" from the Printable Pack
- Writing or drawing utensils

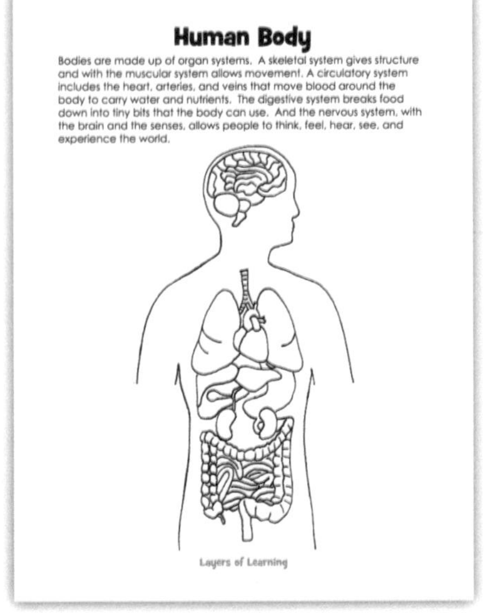
Human Body

Bodies are made up of organ systems. A skeletal system gives structure and with the muscular system allows movement. A circulatory system includes the heart, arteries, and veins that move blood around the body to carry water and nutrients. The digestive system breaks food down into tiny bits that the body can use. And the nervous system, with the brain and the senses, allows people to think, feel, hear, see, and experience the world.

Layers of Learning

1. Depending on the age and ability of the child, choose either the "Human Body" coloring sheet or the "Human Body" narration page from the Printable Pack.

2. Younger kids can color the coloring sheet as you review some of the things you learned about during this unit. Discuss the parts of the body, internally and externally and how important our organs are. On the bottom of the coloring page, kids can write a sentence about what they learned. Very young children can explain their ideas orally while a parent writes for them.

3. Older kids can write about some of the concepts you learned on the narration page and color the picture as well.

4. Add this to the Science section of your Layers of Learning Notebook.

Bookworms

Eye: How It Works by David Macaulay is an intricately drawn and carefully explained book about human eyes, how they take in light, how the images are interpreted by our brains, and more. Ages 6 and up.

Deep Thoughts

Physicist Michio Kaku once said:

> *[The brain] is the most complex object in the known universe. But the brain only uses 20 watts of power. It would require a nuclear power plant to energize a computer the size of a city block to mimic your brain, and your brain does it with just 20 watts.*

What do you think he means? Is the human brain really the most complex thing in the universe?

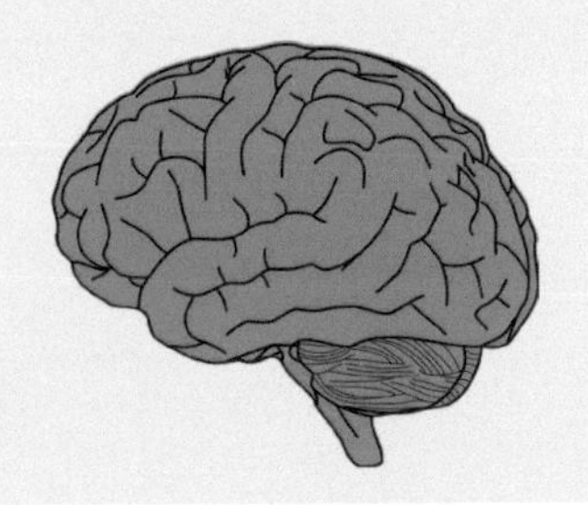

Science Experiment Write-Up

For this activity, you will need:

- The "Experiment" write-up or "Experiment Report Template" from the Printable Pack

1. Choose one of the experiments you completed during this unit and create a careful and complete experiment write-up for it. Make sure you have included every specific detail of each step so your experiment could be repeated by anyone who wanted to try it.
2. Do a careful revision and edit of your write-up, taking it through the writing process, before you turn it in for grading.

EXPLORATION: My Human Body

For this activity, you will need:

- Large sheet of white butcher paper
- Black permanent marker
- Pencil
- Construction paper in various colors
- Scissors
- Tape
- Human body book that shows anatomy diagrams

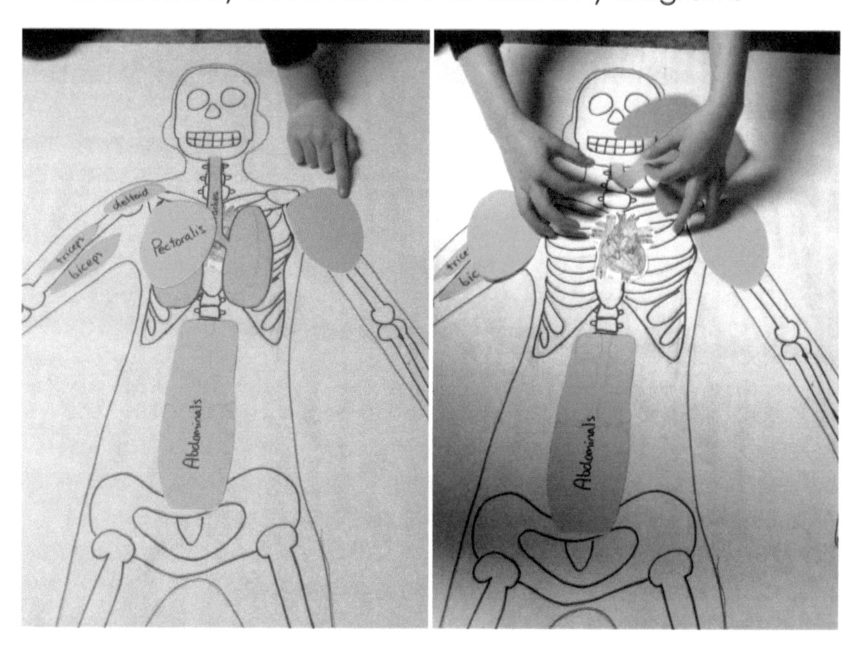

This is an excellent activity for covering all of the body systems with one big project. You may need to spend several days on this project. On the first day, make the body outline and draw on the bones, then read a book (or section from your human body book) about the skeletal system. On the second day, add the muscles and read a

Famous Folks

Insulin is a hormone that turns your food into sugar and controls the amount of sugar in your blood stream. If the hormone is out of balance or no longer functions, you die.

In 1921, Canadian doctor Frederick Banting and his assistant, Charles Best, isolated insulin from the pancreas of a dog and found that injecting the insulin into a diabetic dog cured the diabetes. People with diabetes can live long healthy lives today if they take insulin.

Additional Layer

Find out what it takes to become a brain surgeon.

- How many years of college and training are required?
- What universities and programs offer this training?
- How much does it cost?
- What is the average salary of a brain surgeon?
- Would you want to be responsible for the welfare of someone else's brain during surgery?

Unit Trivia Questions

1. Point to your femur.

   Thigh bone, repeat for as many bones as your children have learned

2. True or false - The heart is an autonomic muscle.

   True. The heart beats automatically without you having to think about it.

3. Which is not a part of skin?

   a) Alveoli

   b) Sebaceous gland

   c) Epidermis

   d) Melanocytes

   e) Hair follicle

4. Name the four main components of the blood.

   Platelets, red blood cells, white blood cells, plasma

5. Arteries carry blood _______ the heart and veins carry blood _______ the heart.

   away from, toward

6. What is the big muscle that separates your abdomen from your thorax and helps you breathe?

   Diaphragm

7. True or false - Human cells respire by taking in oxygen and food energy and releasing carbon dioxide.

   True. It is for the chemical reaction inside cells that produces energy that we need to breathe in oxygen and breathe out carbon dioxide.

8. Human cells use a simple sugar called ________ to get energy.

   glucose

book about muscles. On the third day, add the heart and read a book about the circulatory system, and so on.

Have your child do most of the drawing himself. It will be far from perfect, but children get much more from doing than just watching. The results will be charming if not accurate.

1. Have the child lay down on a piece of butcher paper and trace around his or her body with a black permanent marker. Verbally name the exterior parts of the anatomy like "hand, head, arms, torso" and so on.
2. Use an anatomy book to have the child draw in her own bones on the body outline. Start with a pencil and then trace over with a permanent marker. Write the names of the major bones right on the drawing.
3. Use an anatomy book to draw several of the major body muscles on pink construction paper. Cut them out and then tape them to the body so they can be lifted like a flap.
4. Repeat this process using different colors of construction paper, markers, and tape to make the circulatory system, respiratory system, digestive system, and brain.
5. ☻ Older teens can also add parts of the endocrine system and nervous system. You may want to include detailed "blow up" drawings of nerves, skin, or muscle cells, depending on what you learned during this unit.
6. Name as many organs of the body as you can from memory.

☺ ☺ ☻ **Writer's Workshop**

For this activity, you will need:

- A computer or a piece of paper and a writing utensil

Choose from one of the ideas below or write about something else you learned during this unit. Each of these prompts corresponds with one of the units from the Layers of Learning Writer's Workshop curriculum, so you may choose to coordinate the assignment with the monthly unit you are learning about in Writer's Workshop.

- **Sentences, Paragraphs, & Narrations:** Read a page about a body system from a science encyclopedia and then rewrite the information in your own words.
- **Descriptions & Instructions:** Write instructions telling your brain how to get information to the muscles of your finger. Include the way signals move through nerves.

- **Fanciful Stories**: Make a character sketch for a human organ as though it is a person with a personality.
- **Poetry:** Write a Clerihew poem about a famous doctor or scientist who made advances in medical knowledge.
- **True Stories:** Find out how blood circulation was discovered. Write an account of the discovery.
- **Reports & Essays:** Pick your favorite diagram you colored during this unit. Write a report to go with it.
- **Letters:** Pick a career in the medical field. Find out what education and training it takes to get that job. Write up a mock resume as through you have the skills and are applying.
- **Persuasive Writing:** The first anatomists had to secretly steal or buy dead bodies to dissect because it was illegal. Write a persuasive essay to convince the authorities that dissections should be made legal.

☻☻☻Big Book of Knowledge

For this activity, you will need:

- "Big Book of Knowledge: Human Body" printable from the Printable Pack, printed on card stock
- Writing or drawing utensils
- Big Book of Knowledge

1. Color, draw on, write on, or add to each of the Big Book of Knowledge pages you are using. Only add the printables if you learned these concepts during this unit. If there are other topics you focused on, feel free to add your own pages to your Big Book of Knowledge.

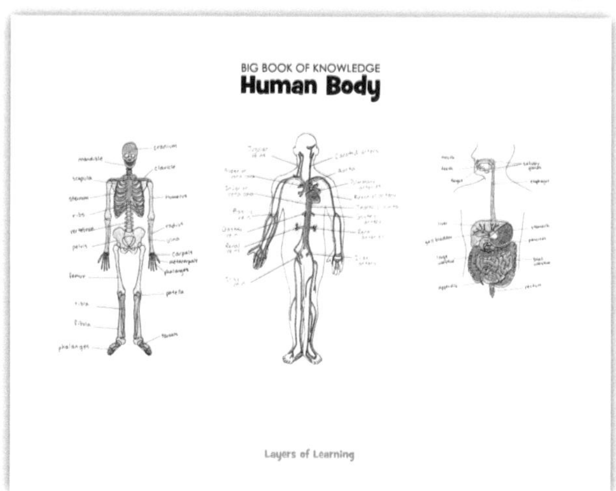

2. Use your Big Book of Knowledge regularly to help you review, quiz, or create games that will help you commit the things you've learned to memory.

Big Book of Knowledge

The Big Book of Knowledge is a book for you, the mentor, to use as a constant review of all of the things you're learning about. You can use it to quiz your kids or prepare tests or review games. Whenever you learn something in Layers of Learning that you want your kids to remember, add it to your Big Book of Knowledge.

Assemble your Big Book of Knowledge in a binder or with binder rings. Divide it into sections for each subject.

In the Printable Pack for this unit you will find a "Big Book of Knowledge" sheet. You can add this sheet to others you collect or create yourself as you progress through the Layers of Learning curriculum. Customize the Big Book of Knowledge to your family by adding facts and topics that you enjoyed exploring as you were learning.

Visit Layers of Learning online to find more information on how to assemble and use your own Big Book of Knowledge.

You will also find cover and section pages to print along with creative games to play with your Big Book of Knowledge to keep school, even the tests, fun!

Unit Overview

Key Concepts:
- Our immune systems help us fight off sickness and disease within our bodies.
- Good nutrition and exercise helps our bodies to be strong and healthy.
- Taking care of our bodies also includes taking safety precautions, having good hygiene, getting adequate sleep, and maintaining good mental and emotional health.
- Our reproductive systems allow us to have offspring. It's important to learn about the process of reproduction and how to be healthy and free of disease.

Vocabulary:
- Germ
- Virus
- Vector
- Immune system
- White blood cells
- Vaccine
- Macronutrients
- Carbohydrates
- Proteins
- Fats
- Micronutrients
- Essential micronutrients
- Drug
- Addiction
- Blastocyst
- Embryo
- Fetus

Theories, Laws, & Hypotheses:
- Germ Theory

HEALTH

Being healthy means being free from illness, but it also means feeling strong, sleeping well, and being able to work and play. The immune system is miraculous in its ability to fight off pathogens. Even so, there will be times when we experience illness. Even people who take very good care of themselves can become ill or hurt. When we do, it's important that we have the resources to get well again.

A healthy lifestyle involves making conscious choices and adopting habits that promote physical, mental, and emotional well-being. This includes:

- A balanced diet
- Regular exercise
- Adequate rest
- Hydration
- Stress management
- Limiting screen time
- Healthy relationships
- Regular health check-ups
- Personal hygiene

Part of health is also understanding our bodies—nutrition, safety, cardiovascular health, our reproductive systems, and more. Learning how to care for our bodies and respect other people's bodies is an important step toward living a happy, healthy lifestyle. It's important to note that a healthy lifestyle is a lifelong journey.

Step I: Library List

Choose books from your library that go with this topic. Here's a list of some favorites and also a list of search terms so you can utilize what your library offers. Read the books with your kids and/or assign them some to read independently. It is from these books your kids will learn most of the facts they need from this unit.

Search for: body, immune system, disease, germs, vaccines, nutrition, hygiene, dental health, doctor visits, safety, mental health, stress, sleep, reproductive system, birth, pregnancy

WARNING: These resources may contain various viewpoints about disease, medicine, vaccines, mental health, and sex education. In addition, there may be resources with nudity. Please preview and make decisions accordingly.

☺ ☺ ☻ *Encyclopedia of Science* from DK. Read "Nutrition" on page 342, and "Human Reproduction" on page 368.

☺ ☺ ☻ *Kingfisher Science Encyclopedia*. Read "Food and Nutrition" on page 126, "Reproduction" and "Growth and Development" on page 132-134, and "Bacteria and Viruses" through "Medical Technology" on pages 136-142.

☺ ☺ ☻ *The Usborne Science Encyclopedia*. Read "Food and Diet" on page 356, "Reproduction" and "Growing and changing" on pages 376-379, and "Fighting disease" and "Medicine" on pages 386-391.

☺ ☺ ☻ *DK: The Human Body Book*. A thorough full-color resource that covers all of the body systems as well as health. Also recommended in the Human Body unit.

☺ *Why Do We Eat?* by Stephanie Turnbull. All about the nutrients our bodies need, with cartoon-style illustrations.

☺ *Germs Make Me Sick* by Melvin Berger. Talks about bacteria and viruses and how they can make us sick. Touches on antibodies and natural defenses we have to get well again.

☺ *A Germ's Journey* by Thom Rooke, M.D. A story about how one boy sneezes and spreads his germs. What happens next? Discusses the immune system.

☺ *Do Not Lick This Book* by Idan Ben-Barak and Julian Frost. Uses microscope pictures and the story of a microbe named Min to explain germs and the immune system.

☺ *The Tooth Book* by Edward Miller. An engaging picture book about taking good care of your teeth, visiting the dentist, and losing teeth.

Family School Levels

The colored smilies in this unit help you choose the correct levels of books and activities for your child.

☺ = Ages 6-9
☺ = Ages 10-13
☻ = Ages 14-18

On the Web

For videos, web pages, games, and more to add to this unit, visit the Biology Resources at Layers-of-Learning.com.

You will find a link to video playlists, web links, and more.

Bookworms

If you're looking for a family read-aloud, we'd like to suggest this one.

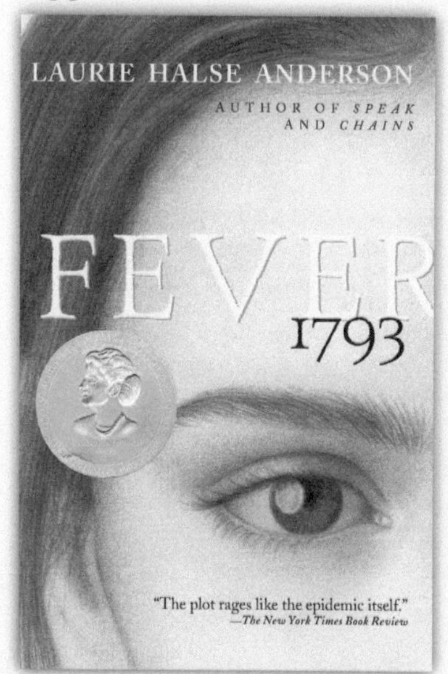

Fever 1793 by Laurie Halse Anderson is about a girl who dreams of turning her coffeehouse into the best in Philadelphia. Then fever strikes and the churchyards begin to fill with victims.

Famous Folks

Martinus Beijernick, a Dutch chemical engineer, was the first to demonstrate the existence of viruses in 1898.

He crushed infected tobacco leaves and was unable to see bacteria through the microscope, but was able to infect other tobacco plants with the juice.

Bookworms

Teach Your Dragon Good Hygiene by Steve Herman explains, in rhyme, why we need to be clean for social and health reasons.

There are 64 Dragon books in all, many of them about physical and emotional health, safety, coping, and more.

It's Not the Stork! by Robie H. Harris. This book is about how babies are made, including penises, vaginas, eggs, and sperm. However, the focus of the book is on bodies and babies, not sex. Pre-read and be ready to discuss.

Microbes: Discover an Unseen World by Christine Burillo-Kirch. Fairly detailed information on all kinds of microbes, how some are pathogens and what the body does to fight them off. Includes projects about bacteria.

Are You What You Eat? by DK. Colorful images and lots of fact boxes alongside the main text. Good basic information. It starts out with a page about evolution (for no real reason); skip this if you like.

Good Pictures Bad Pictures by Kristen A. Jenson. An excellent book to help your children navigate the modern world of screens by understanding the damaging effects of pornography. Read it aloud with your child - many times.

The Care and Keeping of You by Valorie Schaefer. There are a number of versions, some for girls and some for boys, that help them navigate puberty.

The Girl's Body Book by Kelli Dunham RN BSN. Covers everything from periods and bras to social pressure and making money. For 7 and up.

The Boy's Body Book by Kelli Dunham RN BSN. Discusses all those puberty changes plus hygiene, social pressure, consent and boundaries, plus more. For 7 and up.

Asking About Sex and Growing Up by Joanna Cole. Done in a question-and-answer style, this book remains neutral and factual, making it appropriate for all sorts of belief systems. We recommend you read it with your child because of topics like abortion, homosexuality, and masturbation being included.

Plagues: The Microscopic Battlefield by Falynn Koch. This is a graphic-style science book that delves into plagues, what causes them, and how modern science has learned to help out the immune system.

An Elegant Defense by Matt Richtel. This book is about the latest research on the immune system and where the research is headed now. Entertainingly written.

Deadliest Enemy by Michael T. Osterholm and Mark Olshaker. Written by an epidemiologist, this is about the deadliest diseases out there and how they are a threat to humans. Discusses what we are doing and what we *should* do to prepare for pandemics.

The Immune System, Viruses, Proteins, Lipids, Carbohydrates and *The Reproductive System* from Bozeman Science on YouTube. Use these videos as lectures for your high schooler. Have your student take notes.

Step 2: Explore

Choose hands-on explorations from this section to work on as a family. They should be appealing activities that will create mental hooks so your kids remember the information in the unit. Save the rest of the explorations for the next time you do this unit in four years when your kids are older. You can also read the sidebars together and explore some little rabbit trails.

This unit includes printables. See the introduction for instructions on retrieving your Printable Pack.

EXPLORATION: Health Log

For this activity, you will need:

- "Health Log" from the Printable Pack (at least one per student)

Begin keeping a health log to track how you are doing at keeping a healthy lifestyle.

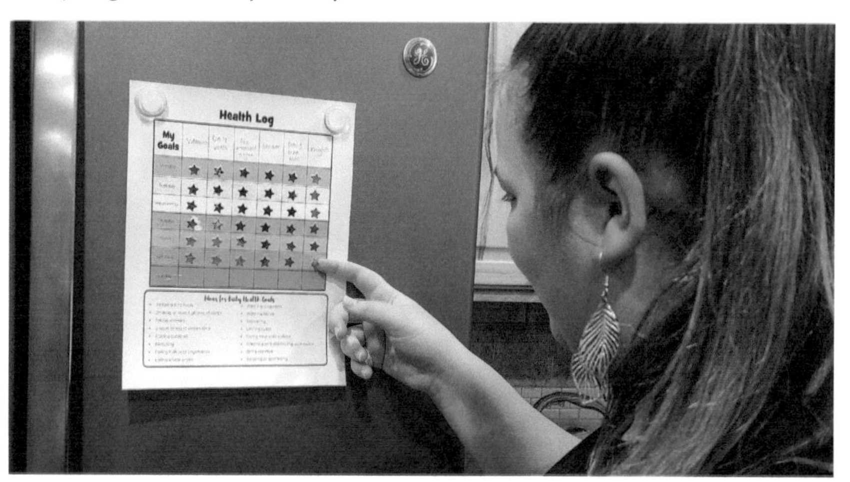

1. Begin by setting some health goals and recording them in each of the boxes along the edge. Hang it on the fridge, on a mirror, or near your bed—wherever you will see it daily.
2. Each day, mark the things you successfully completed that day.
3. At the end of the week (or month or unit), discuss how you did together and decide on some healthy habits you would like to keep maintaining.

Additional Layer

Bacteria are measured in micrometers. A micrometer is 1000 times smaller than a millimeter. The symbol for micrometers is μm.

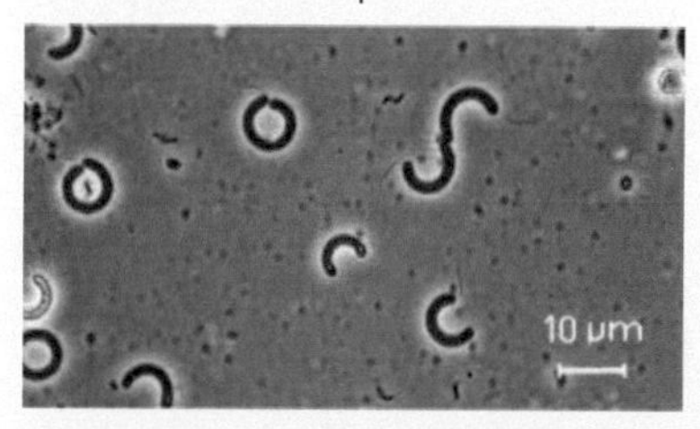

In your microscope view of bacteria, you can measure how big each bacterium is.

1. Determine the diameter of your field of view on the highest magnification.
2. Estimate how many bacteria would fit end to end across the diameter of your field of view.
3. Divide the diameter by the number of bacteria.

Additional Layer

Dirty kitchens, counters, rags, bathrooms, floors, laundry, and other household surroundings harbor bacteria.

Come up with a routine and reward system for keeping the house clean.

You may want to use nutrient agar and swabs to test dirty kitchen rags, dirty sinks, dirty laundry, and other surfaces.

Immune System

☺ ☺ ☻EXPERIMENT: Germs All Around Us

WARNING: Colonized bacteria can be harmful. Wash hands thoroughly. Never eat or drink anything while doing lab work. Sterilize countertops and tools with bleach or alcohol. When finished, seal bacteria colonies and petri dishes in plastic bags and dispose of them in the trash.

For this activity, you will need:

- Nutrient agar in a petri dish (from a science supplier)
- Clean cotton swabs
- Soap and a sink for washing
- Science Notebook

Most microorganisms, from bacteria to protists, are harmless to humans. Many thousands of these species are actually helpful. However, there are also hundreds of **germs** that can make us very sick. By far, the most effective way to avoid becoming infected with bad bacteria, viruses, or other microorganisms is to wash hands with warm soapy water.

1. Brush a clean cotton swab over unwashed hands and smear the cotton swab over nutrient agar. Label the petri dish with the date and the source of the sample.
2. Wash your hands with soap and water for at least 20 seconds, scrubbing between your fingers and under your finger nails. Take another cotton swab sample from your hands and place it on a second petri dish. Label it and cover your dishes with lids.
3. Put both petri dishes in a warm place. Check both petri dishes every day for a week. Count the number of bacteria colonies. Each colony will appear as a different dot or splotch on the petri dish. Each type of bacteria

Memorization Station

Germ: a microorganism, especially one that causes disease

Virus: an infectious microbe consisting of a segment of DNA or RNA surrounded by a protein coat

Vector: an organism that transmits a pathogen, disease, or parasite from one organism to another

Additional Layer

Germ theory states that it is germs that cause disease.

Germs are small organisms, too small to see with the naked eye, but their growth and reproduction within their hosts can cause sickness. This includes bacteria, viruses, and other pathogens. Before this was adopted, most people believed that sicknesses came from bad air or imbalances in the body.

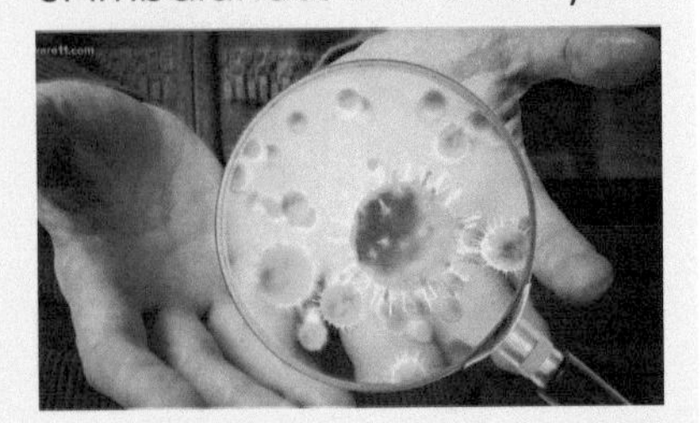

Additional Layer

Demonstrate how germs spread. Go outside and pour a bit of glitter into one person's hand. Shake hands, give high-5s, and interact, observing how the glitter spreads. Talk about how even invisible things, like germs, can spread just as easily. This doesn't mean we should be afraid to touch people. It just means we should wash our hands!

will be a different color and/or shape of colony. How effective do you think hand washing is at stopping harmful germs?

4. Dispose of the petri dishes by placing them in a plastic bag, sealing it, and throwing it in the trash. Be careful to clean all surfaces and hands thoroughly afterward.
5. Record your observations and findings in your Science Notebook.

EXPERIMENT: Viruses

For this activity, you will need:

- Video or book about viruses
- "Diagram of a Virus" from the Printable Pack
- Egg carton
- String or yarn
- Scissors
- Clay or play dough
- Push pins or metal brads

Viruses are so small that you can't see them under a light microscope. They can only be observed directly through an electron microscope because they are about 100 times smaller than a bacteria cell. Viruses must replicate inside of a host cell. They do not have organelles and they do not metabolize. Since they are not cells, the smallest unit of life, they are not considered life by most biologists.

Viruses are spread by **vectors**, like mosquitoes or rats, in droplets from a cough or sneeze, from feces to mouth when people don't wash their hands properly, or by direct touching of body fluids.

Viruses are known for their high adaptability. Some mutate and change so fast that new strains come out every year. Every new emergent virus can successfully infect many

Additional Layer

If you have older kids, take the Germs All Around Us Experiment further in at least one of these ways:

- Use more nutrient agar to test bacterial growth on numerous surfaces of your home.
- Use a toothpick to gently touch one of your bacterial colonies and smear it on to a microscope slide. Add one drop of methylene blue and top it with a cover slip. Examine it under the highest magnification on your microscope and draw a sketch of what you see.
- Place 3 antibacterial disks (from a science supplier) on your colonized petri dish near a bacterial colony. Observe what happens over two days. Watch a video about how antibiotics work.

Fabulous Fact

You can strengthen your immune system by:

- Getting enough sleep
- Reducing stress by being outside in the sun and fresh air, playing, laughing, and having happy relationships with your family
- Eating a variety of healthy whole foods
- Getting immunizations
- Having a clean, but not sterile, house
- Washing your hands before you eat
- Going easy on the sugar and junk food
- Exercising a little bit every day

Bookworms

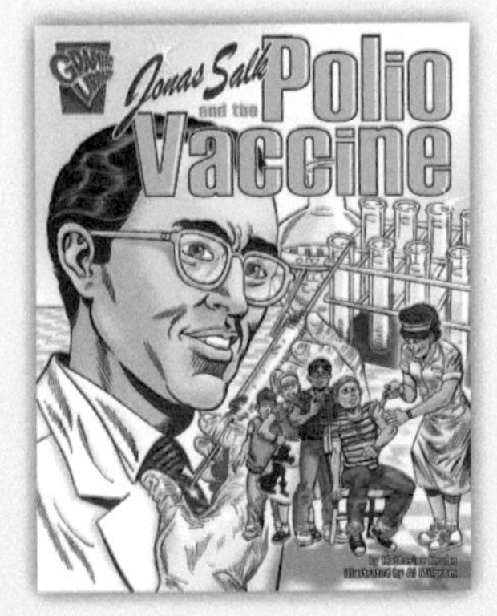

Jonas Salk and the Polio Vaccine by Katherine Krohn is a graphic-style biography of the man who defeated the plague of polio in the 20th century.

Fabulous Fact

For decades, doctors believed that the appendix was a vestigial organ that had no actual function.

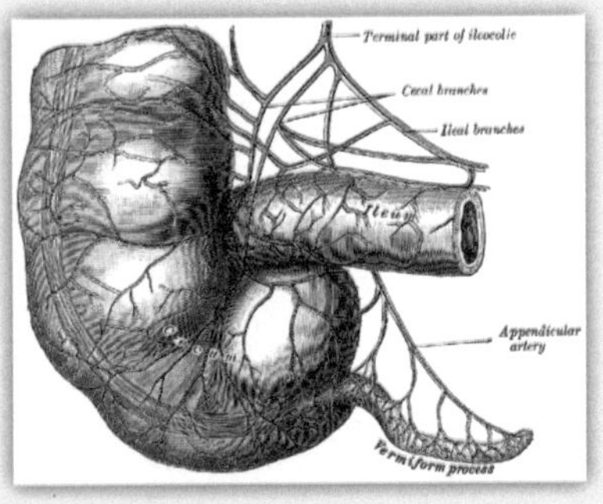

Then, in 2007, after the vital role of beneficial gut bacteria was discovered, scientists at Duke University proposed that it is actually a haven of beneficial bacteria.

The human body does function just fine without the appendix, however.

Memorization Station

Immune system: a network of organs, cells, and proteins that defends the body against infection

people because no one has immunity. The success of a virus depends on how easily it is transmitted from person to person and on how successful it is at getting past the **immune system**, an ability called virulence.

You can protect yourself from viral infections by washing your hands and getting vaccinated. You can protect other people by staying home when you are sick.

1. Watch a video or read a book about viruses to get the basics of what viruses are and how they replicate inside a host cell.
2. Color the "Diagram of a Virus" sheet and add it to the Science section of your Layers of Learning Notebook.
3. Make a virus model. First, use a piece of string or yarn to represent DNA or RNA. Surround this genetic material with an egg carton cup, representing the capsid. Add a layers of clay around the egg carton cup to represent the lipid layers stolen from the host cell. Finally, add push pins or metal brads to the outside of the clay to represent the protein markers on the surface.

EXPLORATION: Parts of the Immune System

For this activity, you will need:

- Video or book about the Immune System
- "Parts of the Immune System" and "Parts of the Immune System Flaps" from the Printable Pack
- Colored pencils or crayons
- Scissors
- Glue stick
- Internet

Your body has all kinds of things that help you fight disease.

- Your skin has melanocytes that attack anything trying to invade through the pores of the skin.
- The bacteria in your intestines teaches white blood cells what belongs in your body and what doesn't.
- Your lungs, nose, throat, and stomach all produce mucus that prevents most types of bacteria from getting to the cell walls of these organs.
- Saliva in your mouth contains antibacterial chemicals.
- Colonies of helpful bacteria take up space so invaders can't infiltrate.
- Bones produce white blood cells that remember and attack bacterial and viral invaders.
- Your spleen and other lymph nodes synthesize antibodies which are attached to bacteria and viruses so they can be removed from the body.

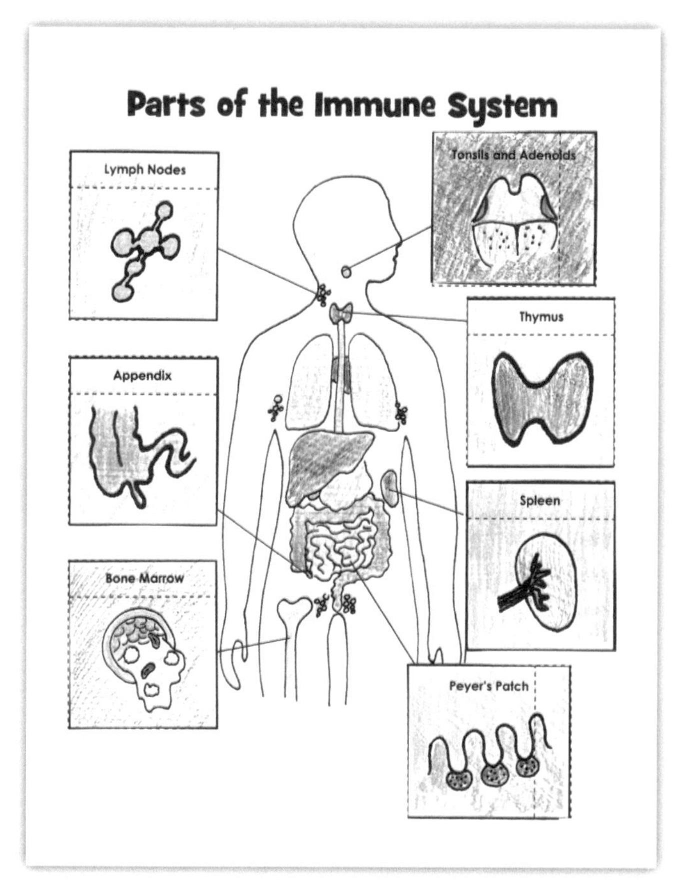

1. Watch a video or read a book about the immune system.
2. Color the body on the "Parts of the Immune System," then color the organs on the "Parts of the Immune System Flaps." Cut the flaps out on the solid lines, fold them along the dashed lines, then glue the narrow part of the folded flap to the correct places on the "Parts of the Immune System."
3. Go online and look up each of the organs one by one. Under each flap, write a brief description of how each organ functions in the immune system. The printable includes some sample answers.

☻ ☻ EXPLORATION: Six Types of Pathogens

For this activity, you will need:

- "Six Types of Pathogens" from the Printable Pack (4 pgs.)
- Colored pencils
- Glue stick

Pathogens are things that cause disease. There are six different major types: parasites, protozoa, fungi, bacteria, viruses, and prions. Viruses and prions are not alive, while the other four are.

Additional Layer

Invite a group of friends over to play some yard games or have a field day. A big part of health is building an active life that rewards you with fun, friends, and feeling good.

Writer's Workshop

Look up several online menus for restaurants you like and write down the calorie count and other health information about some meals they offer. Most, but not all, restaurants now provide this information online. You might see things listed like the amount of saturated fat or sodium in each meal.

Make a chart that shows an unhealthy option, a healthy option, and your favorite dish from each restaurant you look up. Write a paragraph for each restaurant that compares the options in price, health, and taste.

Fabulous Fact

Did you know that laughing is actually good for your heart? It increases your blood flow by about 20%!

1. On the front cover, page 1 of the "Six Types of Pathogens" booklet, color the six pathogens. Make the background behind the parasites, protozoa, fungus, and bacteria green because these are alive. Make the background behind the virus and prion red because these are not alive.
2. Fold all of the pages in half along the dashed lines. Glue the back of page 2 to the back of the cover page. Glue the rest of the pages to each other in order by the page numbers. Finish by gluing the back of page 7 to the back of the "How to Prevent Infections" page.
3. Read through the booklet together and color as you go. Discuss or look things up that interest you. Emphasize the importance of disease prevention.

   Note: We only had space to highlight a few prominent diseases. You may want to mention more that you know of. Also, the booklet mentions that HIV is spread through body fluids but does not mention sex or shared drug needles, You may want to discuss these risks with older children.

Fabulous Fact

The immune system can fight off nearly all pathogens, but it takes two weeks or so for the immune system to ramp up. For many diseases, this is too long. An already prepared immune system goes into action immediately.

You are exposed to pathogenic fungi, viruses, and bacteria all the time, maybe every day. But a healthy, well-trained immune system recognizes and kills the invaders before they can start their replication cycle that sends you to bed, the hospital, or the morgue.

Famous Folks

Dr. Edward Jenner proved that vaccination was a safer way of preventing smallpox than previous inoculation methods. In 1796, Jenner purposely gave cowpox, a mild form of the disease, to his gardener's boy, James Phipps. Watch the story on Vimeo: "James" from James Films.

☻☻☻EXPLORATION: Immune Cells

For this activity, you will need:

- 3 ping pong balls
- Permanent markers

Blood cells are made in bone marrow from stem cells. The immune cells in blood are the **white blood cells**. There are eleven different kinds of white blood cells, each doing a unique job in the body to prevent and fight off disease. We're going to learn about three of them: macrophages, B-cells, and killer T-cells. We're going to use ping pong balls to create three immune cell characters to help us remember what each of these three white blood cells do in our bodies.

Famous Folks

Jonas Salk developed the first polio vaccine during the outbreaks of the 1950s.

Learn more about him and the polio outbreaks.

1. Macrophages engulf and digest microbes, cancer cells, and foreign substances. They are found in every tissue of the body. Draw a big mouth on the first ball to remind you that macrophages "eat" the bad stuff in the body.

Add arms too, because it can spread out and surround foreign particles. Decorate the rest of it any way you like.

2. B-cells recognize antigens (foreign chemicals on the surface of bacteria or viruses). They manufacture antibodies that bind to the antigen and help the macrophages find the foreign particles. Some B-cells are assigned to remember each pathogen so antibodies can be immediately deployed. B-cells hang out mostly in the lymph nodes because the body herds foreign invaders into these nodes. Use a second ball to create a B-cell character with glasses and a big brain to remind you that it spots and remembers previous invaders.
3. When a body cell is damaged (like with cancer) or infected with a virus, it sends antigens to its surface. Killer T-cells can recognize these antigens and then destroy the compromised body cell, which sacrifices itself for the greater good. Decorate the third ball like a warrior and give it weapons. It will fight off invaders.
4. Pass the ping pong balls around and try to name the type of immune cell and its job each time you catch it.

Additional Layer

There is a virulent debate going on between doctors, officials, and parents about the safety of vaccines.

Each person should always investigate and research things they aren't sure about, like whether vaccines, or a particular vaccine, are safe. This is especially true when your child's health and safety are at stake.

Be sure to always read both sides of any debate to get the most complete picture. In the case of debates that hinge around science topics, making an effort to understand the technical terms will aid you immensely in coming to the best conclusions.

☻ ☻ EXPLORATION: How Vaccines Work

For this activity, you will need:

- Videos, books, or websites about some or all of these diseases: polio, smallpox, chicken pox, hepatitis, measles, mumps, rubella, tuberculosis, tetanus, diphtheria, pertussis, and meningitis
- "Types of Vaccines" from the Printable Pack
- Scissors
- Glue stick
- Science Notebook

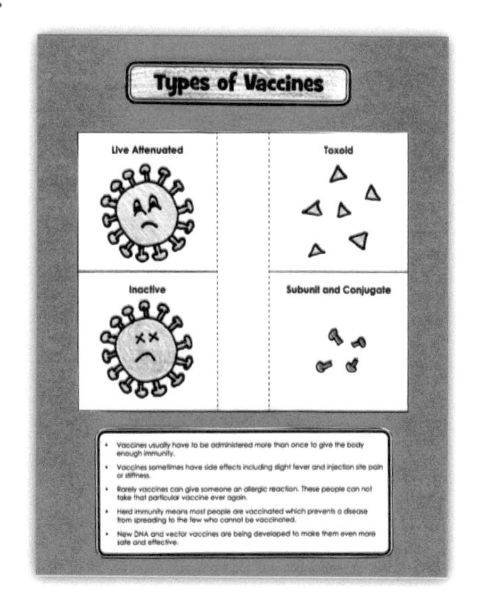

Vaccines teach your immune system by exposing your body to a weakened or killed virus, bacteria, microorganism, or toxin that the microorganism produces. The body recognizes the weakened invader as foreign, destroys it, and then remembers the threat so it can destroy any future wild and full-strength encounters rapidly. The effect is that your body can avoid getting sick with the pathogen even if you are exposed to it, because it can fight it off so quickly since it recognizes it.

Deep Thoughts

Should the government compel vaccination since it affects the health of not just one person, but the whole community?

Why or why not? Where exactly do personal rights end and the common good begin?

Are there alternatives to compulsion to gain compliance for vaccinations or for other things?

Memorization Station

White blood cells: these circulate through the bloodstream and fight viruses, bacteria, and other invaders that threaten your health

Vaccine: substance used to stimulate immunity to a particular infectious disease

Additional Layer

Bullying is the use of force or intimidation to dominate another person. It is wrong. We don't do it and we don't let others do it to people we care about.

Bullying can be physical, verbal, and/or emotional.

It can happen at home, school, church, online, or in the neighborhood. Bullying within the family is not normal or healthy. It is harmful and can have lasting impacts on family relationships and confidence.

Teach children:

- Every human being has value simply because they are a person.
- All people have feelings.
- All people have talents and skills and a reason for existing.
- All people need love and acceptance to become the best they can be.
- You don't have to be best friends with everyone, but you do have to be kind to everyone.
- To identify and express emotions in healthy ways instead of by lashing out or withdrawing.
- Be cautious about teasing. What you think is funny may be devastating to your target.
- To recognize bullying and tell the bully to stop.
- To ask for help if they are being made to feel inferior by anyone else.
- To understand that cyberbullying can be just as harmful as in person.

These lessons need to be repeated and practiced again and again and again all through a child's rearing.

1. Begin by watching videos about diseases that are commonly vaccinated for. Discuss the seriousness of these diseases. Not only do they involve pain and suffering, but they can be fatal or debilitating. Others weaken the immune system so the patient gets secondary infections like pneumonia.
2. Cut out all of the parts of the "Types of Vaccines" sheet on the solid lines. Glue the title to a page in your Science Notebook. Glue the four-part vaccine types section in the center so it creates four flaps. Glue the fact sheet at the bottom.
3. Read through the fact sheet together. Under each of the flaps, write a brief description of each type of vaccine.
   - Live Attenuated: Weakened pathogen that cannot be passed to other people or multiply in the body. (Tuberculosis, measles, mumps, rubella, typhoid)
   - Inactivated: Killed pathogen. (Pertussis, polio, rabies, hepatitis A, influenza)
   - Toxoid: An inactivated version of a toxin that a bacteria produces. (Tetanus, diphtheria)
   - Subunit: Uses part of the pathogen, like a coating or just the antigen. (Hepatitis B, influenza, human papillomavirus)

Nutrition & Exercise

EXPLORATION: Macronutrients Collage

For this activity, you will need:

- Video about macronutrients from this unit's YouTube playlist
- Pictures of food from magazines or the internet
- Scissors
- Glue stick
- Small poster board
- Marker

Macronutrients are types of foods that provide energy. There are three types: carbohydrates, proteins, and fats. All three groups are needed by the body to function optimally.

Carbohydrates are simple carbon chains of sugars or starches. You can find them in potatoes, wheat, rice, fruit, and vegetables. They are easy for our bodies to digest and break down, so they provide quick and accessible energy. Carbohydrates are healthy, but only when they are not heavily processed and stripped of nutrients.

Proteins are big molecules made of amino acids. They require more work for our bodies to digest, but the amino acids they contain are essential for life. You can find them in fish, eggs, milk, meat, beans, and nuts.

Fats are made of complex carbon and hydrogen groups. They are the most work for the body to digest and require a special process. They are found in some vegetables, fruits, nuts, and meats.

1. Watch a video about macronutrients—carbs, proteins, and fats—then discuss each of the macronutrients and what each does for your body.
2. Cut out pictures of healthy, whole foods, not processed junk foods. Create a collage of healthy foods. You can also use a marker to add your own words or pictures. Emphasize that all three food groups are necessary and healthy for most people.

EXPERIMENT: Micronutrients
For this activity, you will need:

- Video about micronutrients from this unit's YouTube playlist
- "Micronutrient Match" and "Micronutrient Match Answers" from the Printable Pack - for younger kids
- "Micronutrients Data Sheet" from the Printable Pack- for teens, 14 years and up
- Internet for research

Micronutrients are vitamins, minerals, and other substances that your body needs in smaller quantities. The body produces some micronutrients, like Vitamin D, but most are **essential**, which means they have to be eaten in food. No single food contains all that you need, so to stay healthy,

Bookworms

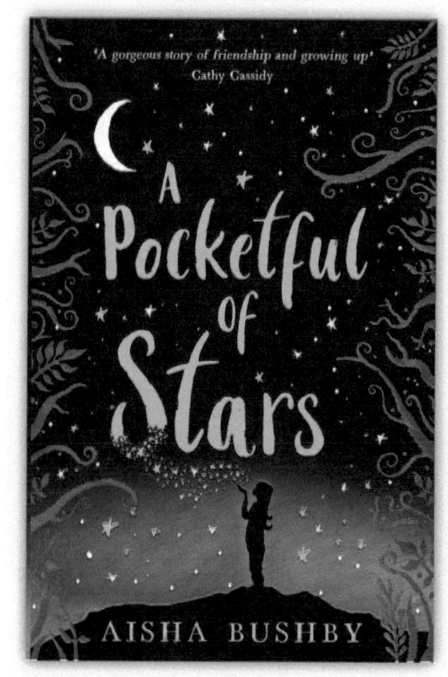

A Pocketful of Stars by Aisha Bushby is the story of a young girl who is dealing with her mom being in a coma. Along the way, Safiya also learns how to stand up to bullies and find her voice.

Memorization Station

Macronutrients: nutrients that are needed in large amounts and provide calories of energy

Carbohydrates: compounds that provide the body with glucose, which is converted to energy used to support bodily functions and physical activity

Proteins: large, complex molecules do most of the work in cells and are required for the structure, function, and regulation of the body's tissues and organs

Fats: nutrients in food that the body uses to build cell membranes, nerve tissue, and hormones; also used as fuel or stored

you need to eat a variety of foods. Whole foods contain many more micronutrients than processed foods.

1. Watch a video about micronutrients. Many foods in wealthy industrial nations are fortified with extra micronutrients to make sure people don't develop deficiencies even if they make poor food choices. Breakfast cereal, bread, and milk are commonly fortified, for example. We did not include fortified foods in the match up. So dairy, which is usually fortified with Vitamin D, is not matched to Vitamin D in the match up, because it is not naturally found in these foods.

   Processed foods often have half or more of their micronutrients stripped out or destroyed. That is the main reason they are not very healthy. That is where the phrase "empty calories" comes from.

2. Complete the "Micronutrient Match" together. First, discuss what micronutrients are and that we need to get most of them from the food we eat. Read each micronutrient and match it to the foods that contain it, using the answer sheet.

3. For teens, 14 and up, choose one micronutrient from the list below and fill out a "Micronutrient Data Sheet" by researching online.

   - Iron
   - Folate
   - Pyridoxine (B6)
   - Calcium
   - Vitamin A
   - Zinc
   - Vitamin C
   - Potassium
   - Vitamin D
   - Riboflavin (B2)
   - Vitamin K
   - Zinc

4. Have your teen explain his or her findings to the family.

EXPEDITION: Whole vs. Processed

For this activity, you will need:

- Internet connection to look up recipes
- Your regular grocery store or market
- "Whole Food Scavenger Hunt" from the Printable Pack
- Scissors
- Crayon or pencil
- Ingredients and cooking utensils to make a whole food recipe of your choice

Whole foods are raw ingredients that contain only one thing. Processed foods have ingredients lists. Oats are a whole food. Breakfast cereal is a processed food. Steak is a whole food. Hot dogs are a processed food. Peas are a whole food. Shepherd's pie from the frozen section is a processed food. Your diet should lean heavily toward whole foods.

Memorization Station

Micronutrients: vitamins, minerals, and other substances that your body needs in smaller quantities.

Essential micronutrients: substances your body needs but doesn't produce, so they must be eaten

Fabulous Fact

Processed food includes anything that has been cooked, canned, frozen, packaged, ground, or prepped in any way. When you cook in your kitchen, you are processing food.

Processed foods range from minimal processing to heavy processing. Minimal processing is not a problem, but heavy processing can be because of inferior ingredients coupled with added sugar and chemicals. Here's a list from least to most processed:

- Frozen fruits or veggies, pre-cut and packaged chicken, beef, or pork, roasted nuts
- Canned tomatoes, vegetables, tuna
- Jarred pasta sauce, salsa, condiments, yogurt, cake mixes
- Ready to eat foods like granola bars, breakfast cereal, crackers, deli meats, and cookies
- Pre-made meals like frozen pizza and boxed dinners

Most of the processed foods are found in the middle aisles of the store while most of the whole foods are found around the perimeter of the store.

1. Choose a processed food you usually buy, like maybe mayonnaise, salad dressing, or breakfast cereal. Find a recipe for it or a whole food substitute. Write down the ingredients.
2. Go to your regular grocery store. Together, read a label on a processed food. Many processed foods have extra chemicals added to preserve them, color them, add flavor, or put back in vitamins and minerals that have been stripped out by the processing. Compare several processed food labels. How many have extra sugar added? Compare different brands of the same food like mayonnaise, spaghetti sauce, or peanut butter. Do some brands seem healthier than others? Look for fewer ingredients and no added sugar.
3. Use the "Whole Food Scavenger Hunt" to find each of these types of whole foods in your grocery store. There are two copies of the scavenger hunt on each sheet of paper. Cut them apart and use one per child. Color or mark off each box as you find it.
4. Find the ingredients for the recipe you are going to make out of whole foods and purchase them.
5. Make your recipe from whole ingredients at home. Whole foods taste different than processed foods. Because processed foods contain extra salt and sugar, the taste of the real food is often masked. At first you may not like the whole food as well, but if you eat whole foods regularly, most people come to like the flavors even more than processed foods.

Additional Layer

Both salt and iodine are necessary for good health. Iodine is one of the few micronutrients that people in wealthy countries tend to become deficient in, so salt often has iodine added to it.

Investigate 3 or more types of salt to find out if each has iodine added. Put a spoonful of the salt into a dish. Add a few tablespoons of water, a tablespoon of vinegar, and a tablespoon of hydrogen peroxide. Stir. Pour in a few drops of laundry starch. If the salt has iodine, the solution will turn a purple/blue color.

The vinegar and hydrogen peroxide put the solution at the proper pH and change the iodide into iodine to help the reaction happen.

Fabulous Fact

Vitamin B is actually a group of related vitamins, each with its own chemical formula and function in the body. Vitamin B12 is the main vitamin vegetarians and vegans have trouble getting in their diets since it is found only in meat sources. They need to take supplements or eat foods fortified with B12.

Deficiency in B12 leads to anemia and nervous system damage including memory loss and psychosis.

Additional Layer

Gather toy or real foods and practice sorting them into food groups. You could also do the same activity using pictures of food from a grocery store advertisement.

Famous Folks

Lillian Wald was a nurse who noticed that many poor families did not have adequate access to healthcare. She created the first public health nursing program in the United States, then continued making programs for more health organizations, including the Red Cross. She wanted to help everyone, regardless of their ability to pay.

Writer's Workshop

Create a written menu for the healthy meals you are preparing and presenting.

EXPERIMENT: Planning Healthy Meals

For this activity, you will need:

- "A Healthy Meal" from the Printable Pack
- Ingredients to cook the meals planned

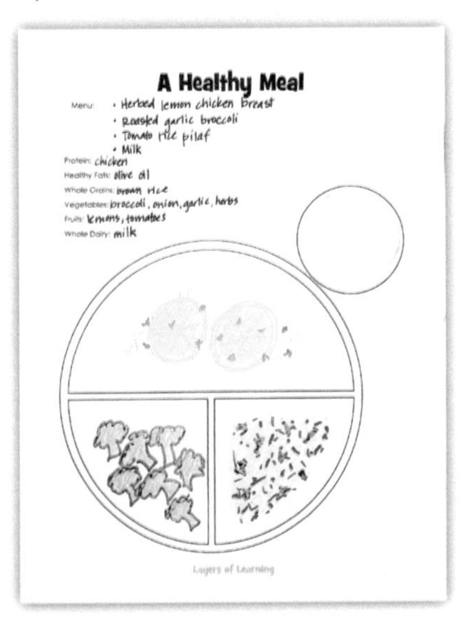

A Healthy Meal

Menu:
- Herbed lemon chicken breast
- Roasted garlic broccoli
- Tomato rice pilaf
- Milk

Protein: chicken
Healthy Fats: olive oil
Whole Grains: brown rice
Vegetables: broccoli, onion, garlic, herbs
Fruits: lemons, tomatoes
Whole Dairy: milk

Some of the most debilitating diseases are a result of poor eating. Heart disease, kidney disease, some types of cancer, and nutritional deficiencies like rickets, scurvy, and anemia are all caused by the foods we choose or fail to choose. Healthy meals include whole foods in all three macronutrient areas—carbohydrates, fats, and proteins. Fats should take up less space on your plate than carbohydrates or proteins and are normally mixed in with the other foods, like butter on your bread or salad dressing on your greens.

1. Use "A Healthy Meal" to have each child plan a meal for your family that includes all three macronutrient groups made from whole ingredients. The meal could be a breakfast, lunch, or dinner. There is space on the printable to write the menu, write which parts of the meal contain some important foods, and draw a picture of the meal on a plate.
2. Schedule out the meals and cook each one as a family over the next week.

EXPLORATION: Sugar

For this activity, you will need:

- Packaged foods from your cupboards
- Can of soda pop or fruit juice
- Electronic balance scale
- Medium cooking pot
- Spoon for stirring
- Stove or hot plate

Sugars occur naturally in all foods that have carbohydrates. This includes fruits, grains, vegetables, and dairy. When these foods are eaten in their whole forms, they are safe and healthy. Whole foods are complex and have fiber and nutrients. They digest slower and release sugars in a slow, controlled way.

Refined sugars like white sugar, brown sugar, corn syrup, high fructose corn syrup, fruit juice concentrate, dextrose, fructose, glucose, maltose, and sucrose have lots of calories, almost no nutrients, and enter the bloodstream quickly. This causes insulin spikes that affect body and brain chemistry. Sugar is linked to weight gain, diabetes, a suppressed immune system, mood swings, and depressed energy.

Sugar is also addictive. It is added to salsa, ketchup, breakfast cereal, salad dressing, crackers, granola bars, soup, bread, processed meats, and just about every other processed food. Sugary drinks not only contain lots of sugar, but the liquid form makes delivery to the bloodstream even faster.

1. Read the labels on some of the packaged foods you have in your cupboards. Look for the added sugar. Sugar is often added to mask the blah taste of foods that have had the nutrition stripped out of them.
2. Place your cooking pot on an electronic balance scale and record the weight. Add the soda pop and record the weight again. Calculate the weight of just the soda by subtracting the weight of the pot.
3. Bring the soda pop to a boil over the stove. As the water evaporates off, stir frequently to keep the liquid from burning. After about 10 or 15 minutes, you will have a thick syrup in the bottom of your pot. This is the sugar from your drink.
4. Let the pot cool, then weigh it again. Calculate the weight of the sugar. What percentage of the soda pop is sugar?
5. Discuss your sugar habits. For example, one important way to cut back on sugar is to never drink it. Cut out hot cocoa, soda pop, fruit juice, sports drinks, and other sugary drinks. Find some drinks to replace your former habits—ice cold water with a lemon wedge, sparkling water, or water infused with fruit.

EXPLORATION: Drugs and Addiction

For this activity, you will need:

- A video or book about addiction
- "How Does ______ Affect The Body?" from the Printable Pack (There is a female and male version. Choose whichever you want for your child.)
- "Effects of Drugs Parent Cheat Sheet" from the Printable Pack (2 pages)

Fabulous Fact

Refined sugar isn't ever really good for you, but a treat now and then is okay. Food is about more than our physical health, it also is part of our social life, celebrations, and family closeness.

A birthday party with a cake in an otherwise healthy week of eating will be just fine. A great deal of life is learning moderation and self control. Sugar is part of that.

Additional Layer

Take a tour of your home medicine cabinet. Talk about each drug, what it is for, and the effects of abuse. Show your children the dosage recommendations and the instructions on what to do in case of an overdose. Talk about following prescription drug directions exactly and potential side effects.

Also discuss other household products that we should never eat and teach your kids how to call poison control, just in case.

Additional Layer

For drugs that are not illegal, like alcohol or caffeine, discuss how they can be abused and how to avoid health and social problems due to abuse or addiction.

As adults, what rules do you impose on yourself regarding these substances? Explain them to your kids.

Memorization Station

Drug: any non-food substance that affects how the brain and body work

Addiction: a chronic, relapsing disorder that includes compulsive use of something despite adverse consequences

Writer's Workshop

Choose any drug, from ibuprofen to caffeine or cocaine to diphenhydramine, and research it. Make a poster about it, including how it affects peoples' bodies, how people obtain it, and potential side effects and consequences of consumption or overconsumption.

Fabulous Fact

People become addicted to lots of things even though they aren't good for them. Drugs, alcohol, tobacco, sugar, food, soda, gambling, pornography, screens, and shopping are just a few common addictions.

- Video of an addict talking about drug abuse

Drugs are substances that alter body chemistry or brain chemistry. Sometimes we use drugs to feel better. If we have a headache, we can take acetaminophen, ibuprofen, or aspirin to help ease the pain. Other drugs are taken to alter our emotions or energy levels. Alcohol makes most people feel mellow while caffeine gives a boost of energy. All drugs are dangerous if they are abused. Some drugs are so dangerous they should never be taken, others only with the direction of a doctor, and a few are okay to take in small quantities if we follow directions carefully.

Most countries control at least some drug substances. Some drugs are always illegal and others can only be administered by a physician. Children and teens who are still growing and developing and have smaller bodies need to be especially careful about the effects of drugs.

1. Watch a video or read a book about **addiction**. Drugs stimulate the reward centers in the brain, making you want to do them again and again even though they are harmful. Highlight to your kids that the most important way to prevent addiction is to have healthy relationships with family and friends. Discuss with your kids how to have healthy relationships.
2. Use the "How Does ________ Affect the Body" sheet. Have your child write in the name of a drug. Talk through how the drug makes a user feel when they take it and the short and long term health effects of the drug. Have your child write down information. Repeat for as many drugs as you want to highlight. You can briefly talk about other drugs as well.

   There is a completed "How Does ________ Affect the Body" in the Printable Pack so you can see an example.

   Parents can use the "Effects of Drugs Parent Cheat Sheet" for information on drugs. There are only 12 drugs or types of drugs detailed. Many more exist. We highly recommend you find out which drugs are the biggest problems among teens where you live and focus first on those. Information on how drugs feel and what they do to the body is easy to find from reputable health organizations online.
3. Either speak with or watch one or more videos of a former addict talking about their experience with drugs.
4. Come up with a family plan on how your kids should deal with it when they are confronted with drugs. The first rule for parents is never make your child more afraid of coming to you than they are of the effects of the

drugs. Role play some situations with both the actions of the child and the actions of the parent.

- You are at a party you aren't supposed to be at and there are lots of drugs and alcohol, making you uncomfortable.
- Someone at school/church/sports offers you a vape. You know they will tease you if you don't accept.
- Uncle ____ offers to let you take a sip of his whiskey.
- Aunt ____ is pressuring you to take a puff of her joint.
- Your date is drunk and wants to drive you home.
- You are at a concert with your friends and you start to feel strange. You think someone might have put something in your drink.
- Your friend's mom offers you brownies. Later you find out they were laced with marijuana.
- Your friends have started to drink alcohol and talk about their weekend parties and you feel left out.

5. Make sure your kids know the legal ramifications of getting caught with drugs beyond the health consequences. There can be consequences like jail time, fines, loss of driver's license, a permanent criminal record, inability to get a job, and so on.

☺ ☺ ☻EXPLORATION: Exercise

For this activity, you will need:

- Exercise equipment, as needed

Exercise is important for strong healthy bodies and robust mental health.

1. Make a plan to break up your school day with physical activity breaks.

- 10 minutes on the trampoline or with a jump rope
- Memorize spelling words while tossing bean bags
- Practice math facts while shooting baskets or running bases
- Take a half hour walk/ bike ride around the neighborhood
- For lunch, take a picnic to the playground
- Do a 10-minute circuit training routine every morning as part of your morning meeting time

Deep Thoughts

They say the worst drug is the first drug. What do you think this means?

Additional Layer

Begin with a fitness test. Complete each task and record the results.

- One-mile run - record the time
- Sit-ups - count how many are completed in one minute
- Trunk lift - lay on stomach and lift upper body off the floor, measuring how high the chin is lifted from the ground
- Push-ups - count how many are completed in one minute
- Sit-and-reach - measure the cm/inches beyond the feet

Set up an exercise program and do it consistently for at least one month.

At the end, have a second fitness test, record the results, and see if there are any improvements.

On the Web

On the Biology resource page, you'll find a link to a series of P.E. videos with skills you can teach your kids. You may want to set up a physical education routine as a family.

Caring For Your Body

Additional Layer

Another important part of safety is learning what to do in case of an emergency. Develop a family plan, a meeting place, and a regular time to practice what to do in case of a fire, an earthquake, a flood, a break-in, a snowstorm, or another disaster.

Additional Layer

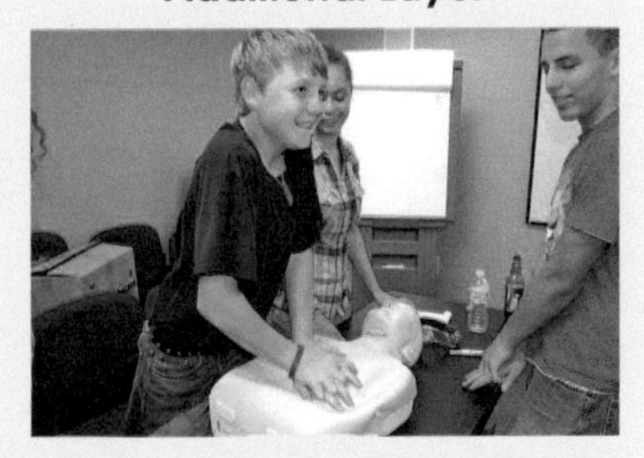

Take classes in first aid and CPR from your community so you can be prepared to help someone in need.

Additional Layer

Take time as a family to inventory and restock your first aid kits in your home and vehicles, replace the batteries in your smoke and carbon monoxide detectors, and take a home inventory. Make sure important documents are in a safety deposit box, fire safe, or backed up to the cloud.

EXPLORATION: Safety Gear

For this activity, you will need:

- Several melons
- Helmet
- High location
- Science Notebook
- Camera

1. Get several melons of the same type, like small watermelons or honeydew melons. Find a high spot to drop them from. Drop one with no protection. Then, drop one while it is wearing a bicycle helmet, a motorcycle helmet, or a hockey helmet.
2. Take pictures and record your results in your Science Notebook.
3. Discuss safety gear.
   a. How important is wearing a helmet to the safety of your head and brain?
   b. What does a seatbelt do for you? How about a life jacket?
   c. What safety gear is used for specific sports?
   d. What other safety gear can you think of that should be used and when?

EXPLORATION: Teeth

For this activity, you will need:

- White clay or play dough
- Pink construction paper or craft foam
- Scissors
- School glue or hot glue
- A book about teeth and brushing
- Boiled egg
- Dark colored sugary drink like cola or grape juice

Everybody has bacteria in their mouths and mostly this is good. The bacteria keep germs away and help the digestion process, but the bacteria also produce plaque when they eat the sugars in your mouth. Plaque build up causes your teeth to decay and your gums to become infected. Regular brushing removes the plaque.

1. Cut out a large oval shape from pink construction paper or craft foam. Fold it in half in the middle, then open it back up. Draw a tongue on the bottom half.
2. Use a book about teeth to learn about the teeth and why brushing is important.
3. Use the clay to model teeth to glue into the mouth. Add the correct number and shapes. Make some of them molars and some incisors.
4. Put a boiled egg into dark colored soda or grape juice overnight. Sugars give bacteria food to grow, they produce acid, which eats away at tooth enamel.
5. Brush the egg with toothpaste and a toothbrush to show how brushing can remove the build up of acidic plaque. 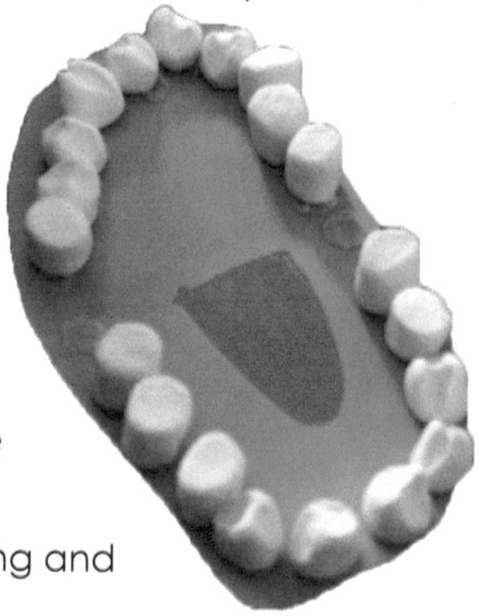
6. Take your kids to the sink and demonstrate good teeth brushing. Brush all the teeth, top, front, and back. Brush the roof of the mouth, the tongue, and the gums.
7. Demonstrate how to remove the plaque from between the teeth by flossing.
8. Have the kids practice brushing and flossing.

EXPERIMENT: Coughing and Sneezing
For this activity, you will need:

- Flour
- Poster board
- Construction paper
- Scissors
- Crayons or colored markers
- Paper tissue

When you cough or sneeze, water droplets from your mouth and nose travel further than you would imagine. If you are sick with a cold or flu, the germs travel with the water droplets. Another person can breath them in or touch a surface the germs landed on and get sick as well.

Writer's Workshop

Create a lap book all about teeth and oral health. Fill a file folder with diagrams, drawings, and captions all about teeth.

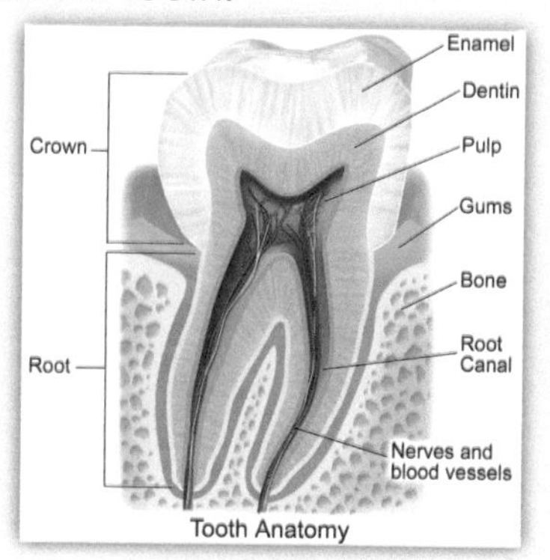

Tooth Anatomy

Famous Folks

William Kouwenhoven was the founder of modern CPR. He researched how electricity could be used to restart hearts. Learn more about him and CPR.

Deep Thoughts

Do you think all strangers are dangerous? What should you do if a stranger approaches you?

Discuss strangers, the buddy system, online predators, and ways to stay safe.

Additional Layer

This would also be a good time to make a list of emergency phone numbers and practice scenarios to make sure kids know how to call 9-1-1, grandparents or other close contacts, poison control, doctors, or others that could help in an emergency. Kids should also know parents' full names, their address, and other important information.

Writer's Workshop

Write a story from the point of view of a germ who is trying to survive but is constantly at risk from all the soaps, detergents, cleaners, and disinfectants. Will he be able to escape?

Additional Layer

Make a bath in the kitchen sink or tub and have your kids wash a doll or a pet. Take them through the steps of wetting, soaping, scrubbing everywhere, shampooing to the scalp, rinsing everything thoroughly, and then towel drying off. Talk about your expectations for keeping the bathroom tidy as well.

1. Go outside. Put flour on your child's hands. Have him or her fake cough or sneeze across the floured hands. Watch as the flour particles travel. That is what happens to a cough or sneeze. Wash your hands to show how hand washing removes the flour and it also removes germs.
2. Practice using a tissue, handkerchief, or the crook of an elbow to sneeze into. Practice turning your head away from other people when you need to cough or sneeze.
3. Make a coughing and sneezing poster. First, trace your child's upper body onto the poster. Let him or her color in the face, hair and clothes. Add long construction paper arms and hands. Glue a tissue or handkerchief to one of the hands.
4. Let your child practice moving the elbow or the handkerchief in front of the face when the poster child sneezes or coughs.

☺ ☺ ☻EXPERIMENT: Hand and Body Washing

For this activity, you will need:

- Potato, peeled and cut in two
- Two plastic zipper bags
- Soap and a sink with water
- Science Notebook

Washing regularly is important for health and for social reasons. Washing gets dirt and germs off our bodies so we don't get sick or get skin infections. We also wash so we look and smell clean.

Hands should be washed after using the bathroom, before eating, and any time the hands get dirty from work or play. Bathing or showering frequency depends on family and cultural norms as well as age. Right around the time kids hit puberty, most should start taking a daily shower or bath to get rid of the extra smells and oils that tend to accumulate.

1. Discuss that we wash our hands to get dirt & germs off.
2. Pass a peeled potato around to all the kids, let them handle it and grab it. Put it in a zipper bag.
3. Have everyone go wash hands with soap and water, scrubbing well. Pass the second potato around, letting everyone handle it and grasp it. Put it in the second zipper plastic bag. Label the bags.
4. Let the potatoes sit in a dark cupboard for a week. Check on them each day to see which potato is growing more colonies.
5. Record your results in your Science Notebook.

Teaching Tip

With preteens and teens, if you are experiencing bathing reluctance, know you're not alone. In private, gently discuss that as their body has matured, the extra hair, sweat, and oil being produced mean they smell bad and look bad. The smells come from bacteria which feed on sweat and oils. When you wash the bacteria and their products off your skin, you feel and smell and look better.

☻ ☻ ☻ **EXPLORATION: Looking, Smelling, and Feeling Good**

For this activity, you will need:

- Grooming supplies and products like nail clippers, tweezers, cotton swabs, hair products, brushes, deodorant, and so on - whatever you normally use in your house
- Foot baths, bath salts, foot brush, nail file, and towels
- Face cleansing products, masks, or creams
- Camera

Have a spa day together right in your home. As you take turns pampering each other, discuss hygiene and the simple ways we take care of the outsides of our bodies. You can take pictures just for fun.

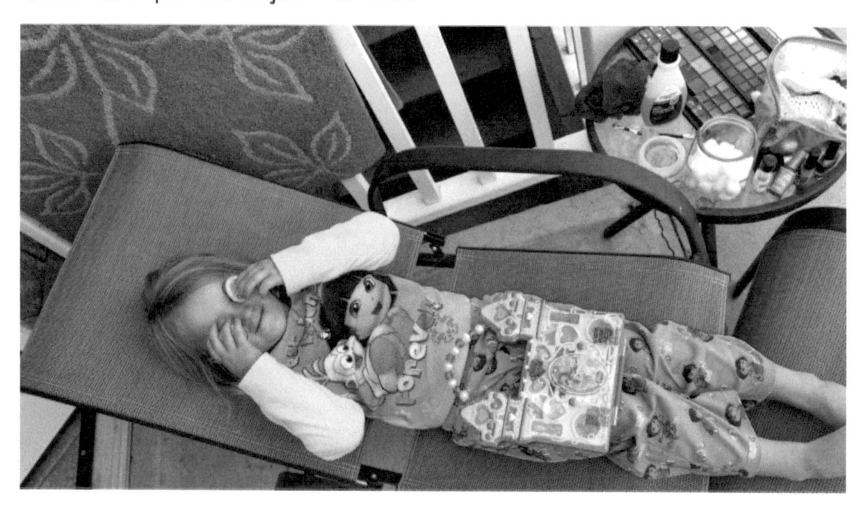

1. Make sure you have all the normal grooming supplies for your family. Teach the kids all together how to clip and file nails, how to tweeze unwanted hairs, how to clean ears, brush or comb hair, apply deodorant, and more.

Additional Layer

If acne is a problem, discuss how to care for it. See a doctor if the acne doesn't dissipate with over-the-counter medications.

Acne can be made worse by being dirty or eating the wrong foods, so eating healthy and face washing can help. But the cause is nearly always hormonal and not the fault of the person struggling with the problem.

Some teens notice that certain foods like sugar, milk, chocolate, or pizza cause break outs. That just means you're sensitive and should lay off those foods at least until the teen years are behind you.

Fabulous Fact

Having a positive mindset is an important part of health. This includes a positive attitude, self-confidence, resilience, a healthy body image, and a sense of gratitude and optimism.

Bookworms

Caleb and Kit by Beth Vrabel is about a 12-year-old boy named Caleb who has cystic fibrosis but doesn't want his diagnosis to define him.

Writer's Workshop

Create a digital slideshow or video presentation about one of these topics after you research it:

- Bicycle safety
- Water safety
- Internet safety
- Cyberbullying
- Online predators
- Physical abuse
- Sexual abuse
- Emotional abuse
- Stress
- Perfectionism
- Loss
- Empathy
- Anxiety
- Mood disorders
- Eating disorders
- Drug abuse
- Alcoholism
- Anxiety
- Depression
- Hygiene
- Mental illness

2. Discuss smelly feet. Soak your feet in a foot bath with bath salts, and clean them thoroughly, scrubbing and smoothing, including caring for your toenails.
3. Discuss how to care for the skin on your face, especially as kids approach puberty. Take turns cleansing and applying face masks or other products.

EXPLORATION: Sleep

For this activity, you will need:

- Book or video about sleep
- Science Notebook

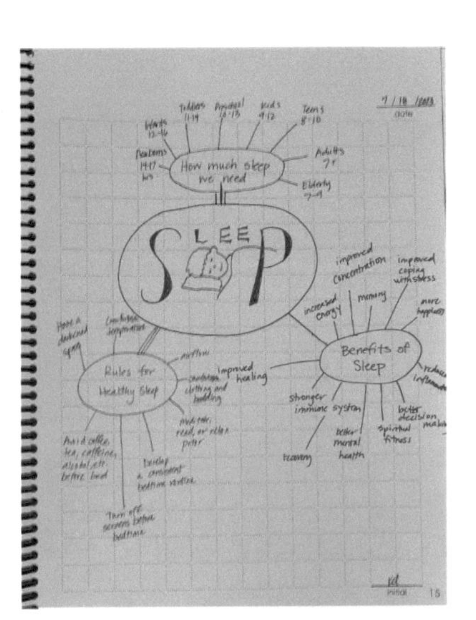

1. Learn about sleep together by reading a book or watching a video about the importance of sleep. Take notes in your Science Notebook as you read or watch.
2. Create a web diagram about the benefits of sleep. Include how much sleep people of different ages need, the benefits of sleep, and also a section with tips for healthy sleep habits.

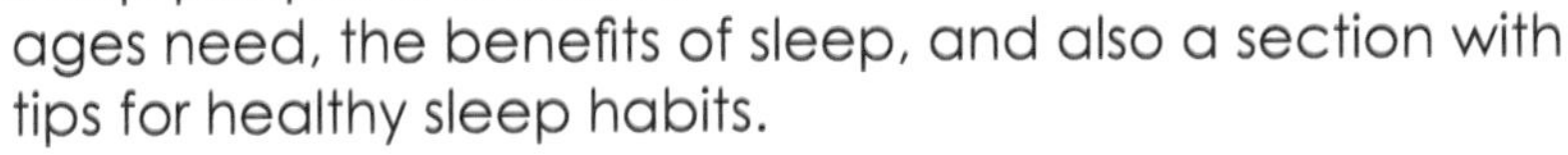

EXPLORATION: Mental and Emotional Health

For this activity, you will need:

- Books or videos about mental or emotional health
- "Emotional Regulation" from the Printable Pack
- Scissors
- Glue stick
- Colored pencils
- Science Notebook

Our mental health is just as important as our physical health. We can learn about how our moods, thoughts, and actions are all related. An important concept to teach kids is that they have the power to feel, to think, and to act, and that although these are related, they are ultimately in control of their thoughts and actions. When they are feeling certain things, they can learn to manage themselves regardless and become emotionally aware.

1. Read books or watch videos about any topics that have to do with mental or emotional health. Discuss what you read or view together. Can you spot ways you can work on mental and emotional health, resilience, and

emotional awareness?

2. Cut out the triangle on the "Emotional Regulation" sheet and discuss each color zone and what you think each one has to do with mental health.

   You may also want to talk about how sometimes we need to see mental health doctors just like we need doctors that treat our physical ailments. Fill in the circles at the bottom with faces that reflect each of the moods.

3. Color the triangle accordingly, fold along the dotted lines, and then glue it into your Science Notebook.

EXPEDITION: Visit a Health Professional

For this activity, you will need:

- Arrangements to visit a doctor, dentist, or another medical office

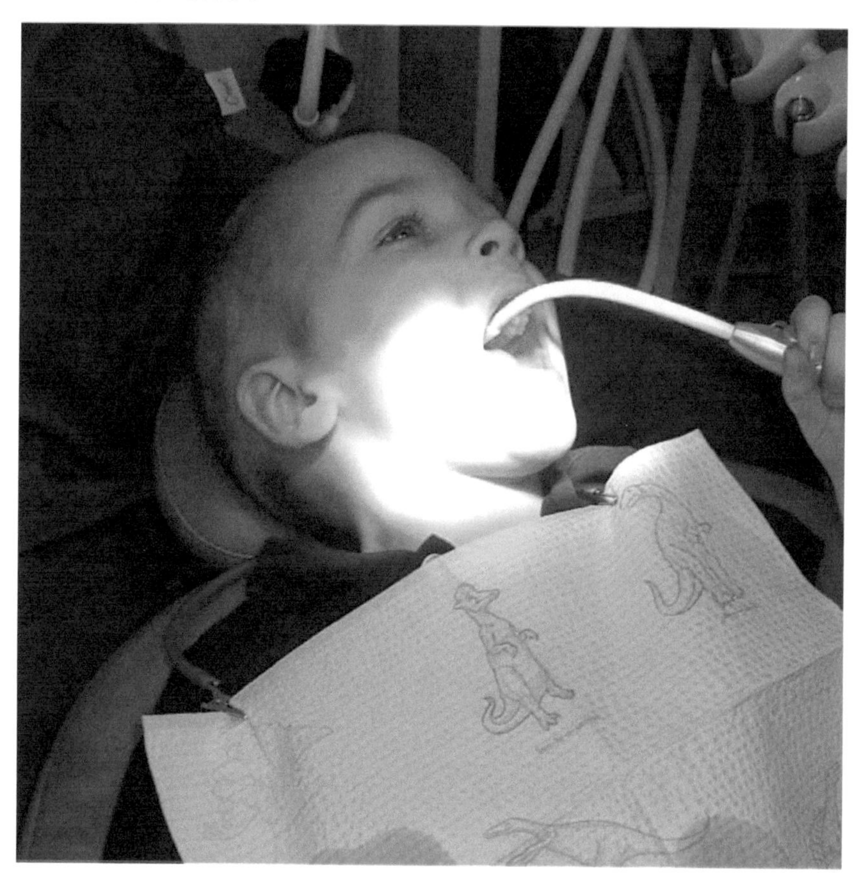

1. Make an appointment to visit a health professional (dentist, doctor, registered nurse, dietitian) to talk about hygiene, cleanliness, and overall health. Make sure you tell the doctor or dentist what you hope they will focus on. Prepare your kids to be respectful and to have a good question or two to ask.

Additional Layer

Discuss personal safety with your children.

- Never get into a car or go in the house of a stranger or anyone you have not previously arranged with your parents.
- If you ever feel unsafe or uncomfortable, you always have the right to leave - call parents to come get you.
- Know the phone numbers of parents and a trusted family friend or relative who you can call if you ever get into trouble.
- Don't go anywhere without telling parents or close friends where you are and who you are with.
- Don't ever disclose personal information like your name, age, or where you live to people online, even your gaming buddies.
- If you are attacked or grabbed, scream and try to run away. Go straight to any nearby adult to get help.
- No one should touch your private areas or any part of your body without your consent, even on the outside of your clothing.
- Don't keep bad secrets. Tell a parent, a teacher, or an adult friend.
- Don't hang out with people you can't trust to care about you.
- Adults don't need help from kids. If an adult asks you for help, don't do it. Leave immediately.
- Always go places with a buddy.

Practice role plays with your children for each of these situations.

Reproductive System

Teaching Tip

Reproduction is a topic that will be approached differently by different families. Parents are best equipped to know when and how to teach their kids these sensitive topics.

Here are a few suggestions many experts offer:

- Young children should learn the proper names for genitalia and be taught that they have a say over their own bodies and that others should not touch their genitals.
- By the time they are about 8 years old, kids should know the parts of their own reproductive system and be taught the body changes they will be experiencing with puberty.
- By about 12 years old, kids should also know about the reproductive system of the opposite sex. Both boys and girls need to understand erections, arousal, intercourse, and menstruation. In addition, its important to teach them about pornography, consent, and birth control.

As a parent, you can help your kids know that you are a safe person to ask questions of if you approach the topics comfortably and regularly instead of having an uncomfortable, one-sided talk about puberty. These discussions will help your kids understand the human body and reproduction, prepare them to know what to do if someone tries to violate them, and give you an opportunity to teach what your family believes about sexual intimacy and why.

EXPLORATION: Reproductive Systems

For this activity, you will need:

- Book or video about the reproductive system
- "Male Reproductive System" and "Female Reproductive System" from the Printable Pack, with "Answers"
- "How A Menstrual Cycle Works" from the Printable Pack
- Colored pencils

This lesson about male and female anatomy, periods, friendship, sex, and finding partners could take several days to complete.

Discuss these things frankly and without embarrassment. This is a body system like all the others. Invite your kids to ask questions and let them know they can ask you anything, anytime. Don't let this be a conversation you have once or just every four years when it comes up in the curriculum. These discussions need to be frequent, frank, and medically accurate.

1. Read a book or watch a video about the reproductive system.
2. Label the "Male Reproductive System." Discuss erections and ejaculation. Talk about arousal, what causes it, what it means, and how to control your actions in spite of your feelings.
3. Label the "Female Reproductive System." Discuss breast development as well. Talk about arousal, what causes it, what it means, and how to control your actions in spite of your feelings.
4. Color the "How A Menstrual Cycle Works" as you read through the sheet together. Discuss some of the symptoms beyond menstruation that women may experience when having a period—tender breasts, backaches, cramps, headaches, acne breakouts, low energy, joint pain, trouble sleeping, and diarrhea or constipation. Talk about how sometimes the symptoms can be extremely painful and debilitating and how to relieve some of the symptoms. Also, make sure girls know they should see a doctor if their periods are so bad they are causing major disruptions to their lives.

   Discuss premenstrual syndrome (PMS) which occurs in the days leading up to a period. Some women experience mood swings and food cravings during this time.

Discuss sanitary pads, tampons, period underwear, and menstrual discs and cups. Boys need to know about these things too.

5. Talk to pre-pubescent boys and girls about the body changes that will be coming soon such penis and testicle growth, body hair, muscle growth, and deepening voices for boys, and breast growth, pubic hair, and general body shape changes for girls. Talk about body odor, hygiene, and deodorant with all your children.
6. Discuss how to treat the opposite sex, people you are attracted to, and just people in general. Talk about how to be friends with boys or girls, how to handle crushes, and how to have conversations with both friends and romantic interests.
7. Discuss intercourse, sexual stimulation, and that the genitals feel good when they are touched in a way you like and by who you like and when you like. This is normal and right. Also discuss how emotional sex is and how it demands and creates deep intimacy. It's not only about physical pleasure.
8. Discuss respect, consent, and your family morals regarding dating, touching, and sexual intercourse. You may want to practice consent by having your kids always ask before they touch someone (including hugs, tickling, roughhousing) and never expecting them to submit to touches (including hugs, tickling, and roughhousing) that make them uncomfortable. Children who can ask and children who can assertively say yes or no will be teens and adults who respect their partner, do the things they want, and not just the things a persuasive or pressuring partner wants. Practice body respect by modeling it consistently in your home.
9. Discuss one committed partner vs multiple partners with children and how multiple partners and infidelity increase risks of harm emotionally and physically.

☻ ☻EXPLORATION: Pregnancy and Fetal Development
For this activity, you will need:

- "Pregnancy and Fetal Development" from the Printable Pack
- Colored pencils
- Video or book about fetal development
- Video or book about birth

Pregnancy happens if a woman's egg is fertilized by a

Deep Thoughts

Regardless of your moral or social beliefs about LGBTQ+ issues, take time to have a discussion about kindness, bullying, and the struggles all people have of fitting in.

- What is your point of view on these issues? Where does it stem from?
- Are there proper ways to debate these issues?
- Is either side 100% right or wrong?

Fabulous Fact

Birth control and family planning are the responsibility of both men and women. A child is the responsibility of both of its parents and deserves to be raised in a stable, loving home with two parents.

Birth control methods include:

- Abstinence
- Intrauterine devices
- Hormonal implants
- Contraceptive pills
- Emergency contraception pills
- Contraceptive patch
- Contraceptive shot
- Condoms
- Diaphragms
- Sponge with spermicide
- Cervical cap
- Natural rhythm method (does not work with teens with their irregular cycles)
- Tubal ligation for women (this is permanent)
- Vasectomy for men (this is also permanent)

Condoms, which come in male and female varieties, also prevent STIs, but they are not as foolproof against pregnancy as you might believe.

Memorization Station

Blastocyst: a rapidly dividing ball of cells

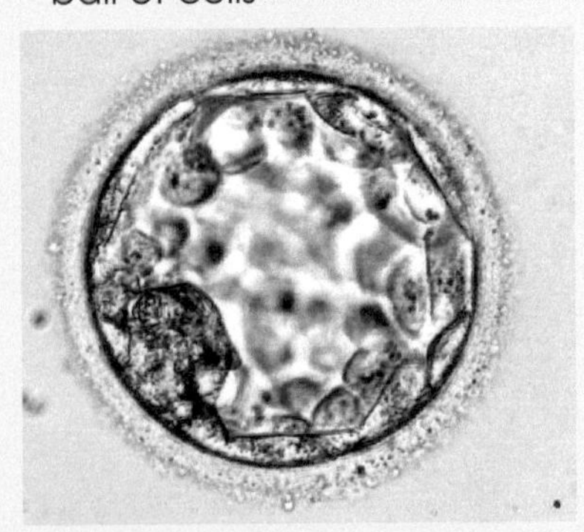

Embryo: earliest stages of development of a mammal before organs have begun to develop

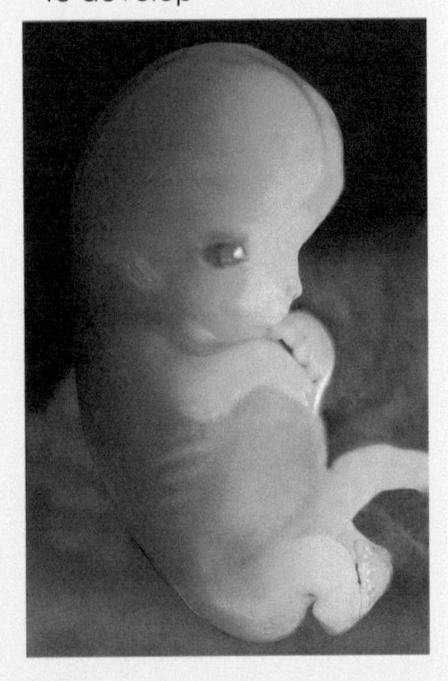

Fetus: after organs have begun to develop, offspring that grows within the uterus of a mammal

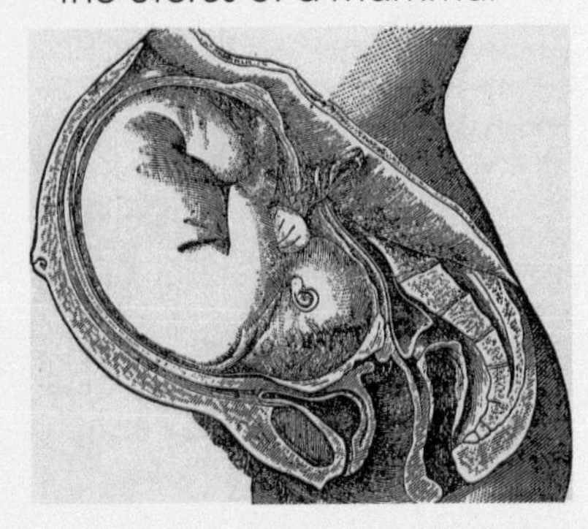

man's sperm during the time the woman is ovulating. Humans fertilize internally. The penis is inserted in the vagina where sperm is released. The sperm have to swim through the uterus and to the fallopian tubes. Fertilization normally happens inside the fallopian tube. The sperm enters the egg and immediately the egg's coating changes and prevents other sperm from entering. The baby's sex and genetic code is determined at the moment of fertilization. The sex of the baby is determined by the man's sperm. If the 26th chromosome is an X, the baby will be a girl. If it is a Y chromosome, the baby will be a boy.

Right after fertilization, the egg begins to rapidly divide; at this stage we call it a **blastocyst**. Next, it implants in the endometrium of the wall of the uterus. The endometrium thickens even further, then the blastocyst divides in two—half becomes the new baby and half becomes the placenta and umbilical cord. The baby is nourished from the mother through the placenta and umbilical cord until it is born. By three weeks after fertilization, the blastocyst has begun to develop a nervous system and is called an **embryo**. By ten weeks, the baby has a heartbeat, arms, legs, and a face. It is called a **fetus**, Latin for "offspring" or "hatching of young." The baby develops inside the uterus and inside an amniotic sac filled with fluid.

1. Read, color, and discuss "Pregnancy and Fetal Development" from the Printable Pack.
2. Watch a video or read a book about fetal development.
3. Watch a video or read a book about birth. Talk about the birth process. When it is time for the baby to be born the mother begins to have contractions, which are strong muscle cramps in the uterus. The contractions are preparing her body to push the baby out through the cervix and vagina. The vagina can stretch to fit the baby, but it's a tight squeeze. Both the contractions and the passage of the baby through the birth canal are painful.
4. Invite your kids to ask questions.

EXPLORATION: Sexually Transmitted Infections

For this activity, you will need:

- "Sexually Transmitted Infections" from the Printable Pack
- Internet connection
- Pencil or pen
- Video or book about sexually transmitted infections (or diseases)

Sexually transmitted infections (STIs) spread through direct contact with sexual body fluids like semen, blood, and mucus. Hands, sexual organs, and mouths can all spread these diseases. The diseases are very common and can be devastating. Some of them have no cure, so it is important to be careful not to contract one. Getting one sexually transmitted disease makes it more likely you will get another because it weakens your immune system. The only way to be absolutely sure you won't get an STI is to not have sex. If you have one healthy partner in a committed relationship then this also protects you from these diseases. If you do decide to have sex, using a condom usually (but not always) prevents infection.

1. Have your kids access a website like CDC.gov/std or another reputable health website that explains sexually transmitted infections. Use the "Sexually Transmitted Infections" worksheet from the Printable Pack and fill in the information about each disease.

   You can use the "Sexually Transmitted Infections Answers" if you need to.
2. Watch a video or read a book about each disease.
3. Make sure to discuss prevention of these diseases. All children need to understand vaccinations and condoms, even if you think they won't or shouldn't be sexually active outside of marriage.
4. Make sure kids know that if they are sexually active they should be tested for STIs at least once a year or if they suspect they may have been exposed. If you know you have an STI, it is important to inform any potential or recent partners.

Step 3: Show What You Know

During this unit, choose one of the assignments below to show what you have learned during the unit. Add this work to your Layers of Learning Notebook. You can also use this assignment to show your supervising teacher or your charter school as a sample of what you've been working on in your homeschool, if needed.

There are more ideas for writing assignments in the "Writer's Workshop" sidebars.

Coloring or Narration Page
For this activity, you will need:

- "Health" from the Printable Pack

Additional Layer

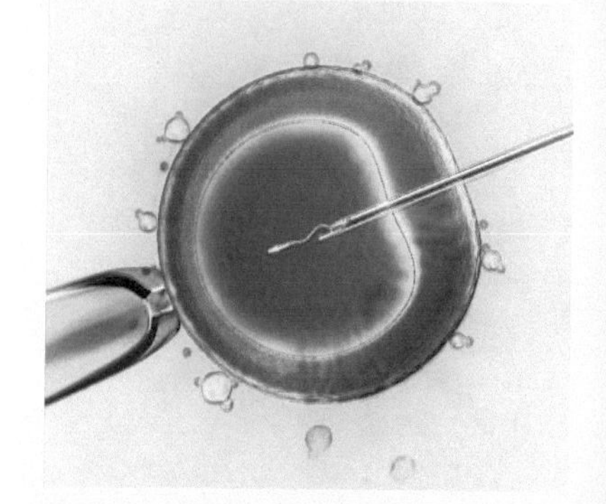

Depending on the ages of your children, you may also want to discuss adoption, in vitro fertilization, miscarriage, birth control methods, and abortion. These subjects should be broached by the time a child is around 13 years old, with more detail as they ask and as they grow older.

Famous Folks

Florence Nightingale is often called the Mother of Modern Nursing. She not only cared for wounded British soldiers during the Crimean War, but she also helped improve sanitation, reducing the death count by 66%. She also helped create a more formal health and nursing education program.

Unit Trivia Questions

1. True or false - Most microorganisms, like bacteria, are harmful to humans.

   False. Some are, but most are completely harmless.

2. What is the network of organs, cells, and proteins that defends your body against infection called?

   Your immune system

3. Where are blood cells made in your body?

   a) in your heart

   b) in your lungs

   c) in your bone marrow

   d) in your veins

4. True or false - Essential micronutrients are things your body produces.

   False. Your body needs them but doesn't produce them, so you need to eat them.

5. Give three examples of whole foods and three examples of processed foods.

   Answers will vary, but whole foods have just one ingredient, like tomatoes, while processed foods have lists of ingredients.

6. All drugs ______________.
   (choose all correct answers)

   a) are non-food substances

   b) are harmful

   c) are addictive

   d) affect your brain or body

7. Put these in order of development: embryo, fetus, blastocyst.

   Blastocyst, embryo, fetus

- Writing or drawing utensils

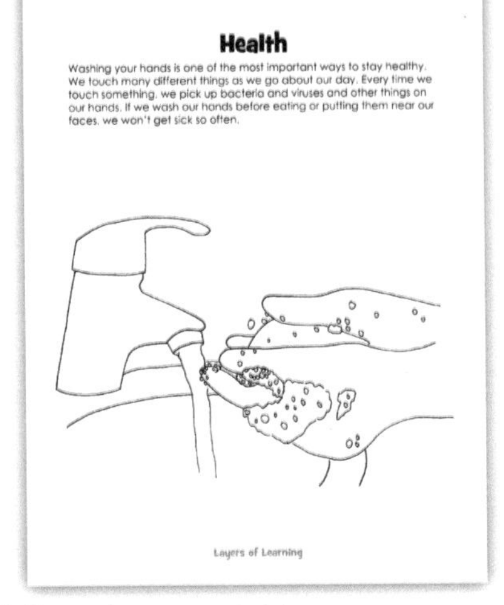

1. Depending on the age and ability of the child, choose either the "Health" coloring sheet or the "Health" narration page from the Printable Pack.
2. Younger kids can color the coloring sheet as you review some of the things you learned about during this unit. On the bottom of the coloring page, kids can write a sentence about what they learned. Very young children can explain their ideas orally while a parent writes for them.
3. Older kids can write about some of the concepts you learned on the narration page and color the picture as well.
4. Add this to the Science section of your Layers of Learning Notebook.

☻☻☻Science Experiment Write-Up

For this activity, you will need:

- The "Experiment" write-up or "Experiment Report Template" from the Printable Pack

1. Choose one of the experiments you completed during this unit and create a careful and complete experiment write-up for it. Make sure you have included every specific detail of each step so your experiment could be repeated by anyone who wanted to try it.
2. Do a careful revision and edit of your write-up, taking it through the writing process, before you turn it in for grading.

☻☻☻ Writer's Workshop

For this activity, you will need:

- A computer or a piece of paper and a writing utensil

Choose from one of the ideas below or write about something else you learned during this unit about history and why or how it is created. Each of these prompts corresponds with one of the units from the Layers of Learning Writer's Workshop curriculum, so you may choose to coordinate the assignment you choose with the monthly unit you are learning about in Writer's Workshop.

- **Sentences, Paragraphs, & Narrations:** Write a simple paragraph that describes basic things you can do to protect yourself from accidents. Include things like wearing a helmet and a seat belt.
- **Descriptions & Instructions:** Write out a recipe for a healthy dish. Include ingredients as well as a numbered set of instructions for making the dish.
- **Fanciful Stories**: Imagine a country that outlaws all processed sugars. Tell the story of what the people would do.
- **Poetry:** Write a poem about a baby who is growing in a mama's tummy. Use similes and metaphors.
- **True Stories:** Talk to someone who is dealing with a difficult illness. Interview the person thoroughly and write the person's true story about struggles, triumphs, hopes, and fears.
- **Reports and Essays:** Write a report about five ways to live a healthy life. Choose five specific areas and develop each one.
- **Letters:** Write a letter to yourself about the positive things you see in yourself. Fill it with affirmations and your strengths. Set it aside and read it whenever you're having a hard time or need a boost.
- **Persuasive Writing:** Create an anti-drug poster that convinces people of the harm of drugs, alcohol, or driving under the influence.

Big Book of Knowledge

For this activity, you will need:

- "Big Book of Knowledge: Health" printable from the Printable Pack, printed on card stock
- Writing or drawing utensils
- Big Book of Knowledge

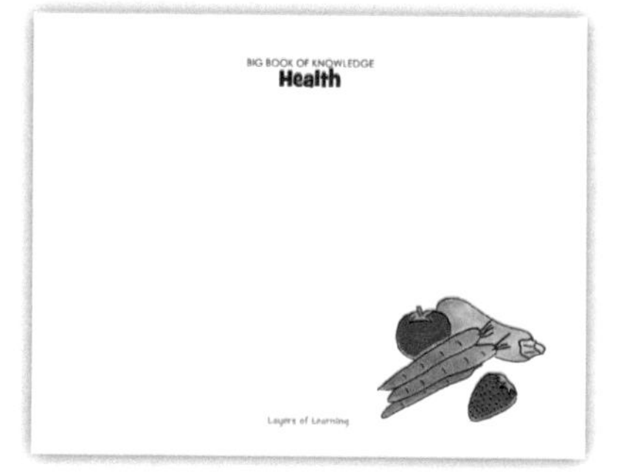

1. Color, draw on, write on, or add to each of the Big Book of Knowledge pages you are using. Only add the printables if you learned these concepts during this unit. If there are other topics you focused on, feel free to add your own pages to your Big Book of Knowledge.
2. Use your Big Book of Knowledge regularly to help you review, quiz, or create games that will help you commit the things you've learned to memory.

Big Book of Knowledge

The Big Book of Knowledge is a book for you, the mentor, to use as a constant review of all of the things you're learning about. You can use it to quiz your kids or prepare tests or review games. Whenever you learn something in Layers of Learning that you want your kids to remember, add it to your Big Book of Knowledge.

Assemble your Big Book of Knowledge in a binder or with binder rings. Divide it into sections for each subject.

In the Printable Pack for this unit you will find a "Big Book of Knowledge" sheet. You can add this sheet to others you collect or create yourself as you progress through the Layers of Learning curriculum. Customize the Big Book of Knowledge to your family by adding facts and topics that you enjoyed exploring as you were learning.

Visit Layers of Learning online to find more information on how to assemble and use your own Big Book of Knowledge.

You will also find cover and section pages to print along with creative games to play with your Big Book of Knowledge to keep school, even the tests, fun!

Glossary

Abdomen: the lower part of the body where the digestive organs are located 98

Adaptation: a trait that makes a species fit to survive in its environment 52

Addiction: a chronic, relapsing disorder that includes compulsive use of something despite adverse consequences 256

Adult: a fully grown and sexually mature individual of a species 134

Allele: alternative forms of a gene found at the same location on a chromosome, represented with letters like T or t 44

Allopatric speciation: a population becomes geographically isolated and diverges genetically or behaviorally so that it won't reproduce with the original species 60

Amphibian: four-limbed cold-blooded vertebrate that starts life with gills and then develops lungs 128

Anterior: the belly or the head (in fish) 100

Apex predator: an animal at the top of the food chain without natural predators of its own 191

Aquatic: an animal or plant that lives in the water 103

Arteries: muscular-walled tubes that convey blood from the heart to the body 227

Asexual reproduction: a single parent divides to produce offspring which are an exact copy of the parent 24

Autonomic nervous system: the part of the nervous system that automatically controls the muscles of internal organs and glands 220

Bacteria: single-celled organism lacking a nucleus 24

Beak: a bird's bony projections from the upper and lower mandibles of the face 159

Bilateral symmetry: a body that can be divided into two identical, but mirror image halves 98

Biodiversity: the variety of living things on the earth or in a specific ecosystem 195

Biology: the study of living things 9

Biome: a geographical region with dominant plant and animal populations due to the climate 184

Bird: warm-blooded vertebrate with feathers, wings, and a beak 155

Birth: emergence of the offspring from the mother's body 169

Blastocyst: a rapidly dividing ball of cells 266

Blood vessel: a tubular structure that carries blood through the tissues and organs of the body 224

Brain: organ inside the head, made up of billions of nerve cells, that controls all of the body functions 235

Bud: a compact growth on a plant that develops into a leaf, flower, or shoot 68

Camouflage: coloring or patterns meant to blend in with surroundings 164

Carbohydrates: compounds that provide the body with glucose, which is converted to energy used to support bodily functions and physical activity 251

Cell membrane: a layer of lipids and proteins that surrounds a cell 15

Cell: smallest living unit of an organism, consist of a cell wall or membrane surrounding cytoplasm and various organelles 15

Cellular respiration: the process where cells combine oxygen with glucose to produce ATP, which the body uses for energy 230

Centromere: the central region of a chromosome that appears during mitosis where the chromatids are held together in an X-shape 43

Cephalothorax: the fused head and thorax of spiders and some crustaceans 118

Chlorophyll: a green pigment present in all green plants that helps plant absorb light and photosynthesize 69

Chloroplasts: organelles in a plant cell that convert light into energy through photosynthesis 75

Chromatid: one of the two identical strands of a chromosome that separate during mitosis 43

Chromosome: a linear strand of DNA and proteins that is inside the nucleus of a cell and contains hereditary information 43

Classification: organizing things into groups based on shared characteristics 29

Cold-blooded: endotherm, an animal that has a body temperature that is regulated by the environment or varies with the environment 128

Colony: a community of animals or plants of one kind living closely together 104

Community: a group of species that live together in a specific place 195

Conifer: a tree that is cone-bearing, has evergreen needles, and grows in cool areas 81

Conservation: using natural resources responsibly so they are still available for the future 183

Consumer: an organism that eats other living things in order to get energy 191

Contaminant: any physical, chemical, or biological substance that harms or poisons a natural environment 204

Crepuscular: active at twilight 164

Cytosol: liquid inside a cell 15

Decomposer: an organism such as fungi, invertebrate, or bacteria that breaks down dead animals and plants 191

Diaphragm: a large muscle that separates the thorax from abdomen and contracts to enlarge the thorax, inflating the lungs 229

Dicot: flowering plant with embryonic leaf pairs 88

Distal: the terminal end of an appendage 100

Diurnal: active during the day 164

DNA: short for deoxyribonucleic acid, this is the code inside every living cell that directs the cell's activities 20, 39

Dominant: the trait that is expressed even if another genetic possibility is present. It is written with a capital letter, T 39

Dorsal: the back or spine 100

Drug: any non-food substance that affects how the brain and body work 256

Ecological niche: the role an organism plays in its ecosystem 187

Ecological succession: the process of a habitat recovering from a disturbance 201

Ecology: the science of how living things interact 183

Ecosystem: a community of plants and animals that interact with the non-living things in a specific place 184

Egg: the female reproductive cell in eukaryotic animals and plants 134

Embryo: an unborn offspring in the early stages of growth and development (2nd to 8th week after fertilization) 266

Embryo: earliest stages of development of a mammal before organs have begun to develop 169, 266

Endangered species: an organism that is threated with extinction 207

Endoskeleton: an internal skeleton 98

Energy flow: the movement of energy through an ecosystem as living things photosynthesize or eat 191

Energy: the ability to do work such as move, grow, or metabolize 191

Environmentalism: a political movement that seeks to protect nature though the force of government 183

Epidermis: the outer layer of cells that covers a plant (think of it as the "skin" of a plant) 75

Epigenetics: the study of changes in organisms caused by modification of gene expression rather than alteration of the genetic code itself 59

Essential micronutrients: substances your body needs but doesn't produce, so they must be eaten 252

Eukaryotic: large, complex cells with a nucleus, membrane bound organelles, and two strands of DNA 17, 69

Evolution: organisms change and diversify through successive generations due to environmental pressures 56

Exoskeleton: a rigid outer covering that protects the bodies of some animals 98

Extinction: the end of a species 207

Fallopian tube: one of a pair of tubes that carry the egg from the ovary to the uterus 169

Fats: nutrients in food that the body uses to build cell membranes, nerve tissue, and hormones; also used as fuel or stored 251

Feather: outer covering of a bird's skin that is made of hollow shafts fringed with barbules 160

Fern: flowerless plant with feathery fronds and spores 78

Fertilization: the union of a male reproductive cell, the sperm, with the female reproductive cell, the egg 134

Fetus: after organs have begun to develop, offspring that grows within the uterus of a mammal 169, 266

Fiddlehead: tightly coiled fronds of a young fern 78

Fish: aquatic vertebrate animal that has gills and no limbs with digits 128

Flower: seed-bearing part of a plant 68

Food chain: a series of organisms each dependent on the next for a food source 191

Food web: a system of interlocking food chains 192

Fruit: seed-bearing structure in flowering plants that is formed from the ovary after flowering 68

Fungi: plant-like organisms that do not make chlorophyll (singular = fungus) 89

Genes: a section of DNA that holds the code for a specific trait 39

Genetic mutation: a mistake in the copying of the genetic code or DNA that persists and is not corrected by the cell 49

Genetic variation: differences in genetic codes of individuals in a population 51

Genotype: the genes present in an individual 43

Germ: a microorganism, especially one that causes disease 244

Gills: paired organs on fish and some amphibians that have the ability to extract oxygen from the water 128

Glucose: a simple sugar that is the main source of energy for the body 230

Habitat: the place an organism lives, including its food, water, shelter, and space 188

Head: The front or upper part of an animal where the brain and sensory organs are concentrated 98

Heart: a hollow muscular organ that acts as a pump to move blood around the body 224

Heterozygous: having two different alleles of a particular gene 44

Homozygous: having two identical alleles of a particular gene 44

Hybridization: two species sexually reproduce and create a third viable offspring 60

Immune system: a network of organs, cells, and proteins that defends the body against infection 246

Inferior: lower down or further from the center of the body than another body part 100

Intestine: a long tube-shaped organ that completes digestion and absorbs nutrients and energy from food into the body 231

Introduced species: an organism that is not native to a particular area 207

Invasive species: a non-native organism that is successful and outcompetes the native organisms in the new environment 207

Invertebrates: animals without a backbone 98

Joint: the point at which two parts of a skeleton are joined together 218

Juvenile: a young and sexually immature individual of a species 134

Leaf: a flattened green outgrowth from the stem of a plant 68

Life: a living thing can grow, reproduce, use energy, and respond to its environment 15

Life cycle: stages of development that a living thing undergoes, including reproduction 134

Locus: the specific physical location of a gene or DNA sequence on a chromosome 43

Lungs: a pair of organs in the body that supply the body with oxygen and remove carbon dioxide 229

Macronutrients: nutrients that are needed in large amounts and provide calories of energy 251

Mammal: warm-blooded vertebrate with hair, milk production, and the birth of live young 164

Marine mammal: a mammal that spend most or all of its life in the sea and is dependent on the sea 174

Mesophyll: the inner tissue of a leaf 75

Metamorphosis: a profound body change from the immature stage to the adult 115

Micronutrients: vitamins, minerals, and other substances that your body needs in smaller quantities. 252

Microorganism: living things that are too small to see with the naked eye 24

Microscope: an instrument used to magnify and view very small objects 10

Migration: the long distance seasonal movement of animals from one location to another 117

Monocots: flowering plant with a single embryonic leaf 88

Moss: a small, flowerless, green plant that lacks true roots, grows in damp habitats, and reproduces with spores 81

Motile: able to move about 98

Mouth: opening in the face through which food and water are taken in 231

Multicellular organism: a living thing that is made up of more than one cell 21, 69

Muscle: a bundle of fibrous tissues that has the ability to expand and contract, producing movement 220

Natural resources: raw materials people harvest or use from the earth to make everything we need to survive and thrive 204

Natural selection: the process where living things that are better suited to the environment reproduce and pass on their traits to the next generation and those less suited die out or fail to pass on their genes 55

Neuron: a type of cell that receives and sends messages between the brain and body; also called a nerve cell 233

Nocturnal: active at night 164

Organelle: specialized structures inside a cell 15

Orifice: an opening in the body from the outside leading inside, like a mouth, ear, nostril, or anus 98

Ovary: female reproductive organ that produces eggs 169

Oviparous: Eggs are fertilized and develop outside the mother's body 132

Ovoviparous: eggs are fertilized inside the mother, but develop on their own with an egg sac 132

Phenotype: an observable trait in an individual, like being short or having stripes 43

Pheromones: substances secreted by a member of a species to communicate with another member of the same species 112

Photosynthesis: the process plants use to turn sunlight, water, and carbon dioxide into oxygen and energy (in the form of sugar) 73

Placenta: organ that nourishes the fetus in a pregnant female through an umbilical cord 169

Pollution: a substance that has harmful effects in the environment 204

Population: the individuals of one species in a particular location 201

Posterior: the back or tail (in fish) 100

Predator: an animal that hunts and eats other animals 132

Producer: an organism that makes its own food from sunlight energy 191

Prokaryotic cells: simple, small cells without a nucleus and one strand of DNA 17

Proteins: large, complex molecules do most of the work in cells and are required for the structure, function, and regulation of the body's tissues and organs 251

Protists: Single-celled eukaryotic creatures, also called protozoa 28

Proximal: where an appendage (leg, arm, tentacle) attaches to the body 100

Radial symmetry: a central axis around which the body is equally arranged 98

Recessive: a trait that is only expressed if it is the only one present, it is written with a lowercase letter, t 39

Reptile: air-breathing cold blooded vertebrate that has a scaly body 128

Roots: underground section of a plant body 68

Scientific method: the systematic observation, experimentation, and measurement of the natural world to find out why and how things happen 9

Seed: an undeveloped plant embryo 68

Segmented: a body formed of distinct parts 98

Sessile: fixed in one place, cannot move about 98

Sexual reproduction: gametes, or sex cells, from a male and a female parent combine to create a new, unique offspring 24

Simple plant: a plant that has no roots, stems, or leaves 78

Skeleton: a system of hard organs, such as bones or cartilage, that provides support to an organism 214

Skin: the layer of soft tissue that forms an outer covering over the whole body 222

Social insects: live in colonies, divide their labor, and are all descended from the same queen 112

Speciation: the emergence of a new species 60

Species: a specific kind of living thing that can exchange genes or reproduce with others of its kind 33, 60

Species fitness: the ability of an organism to thrive and pass its genetic code on to offspring 51

Spinal chord: a bundle of nerve fibers that runs the length of the body in animals and connects the body to the brain 128

Stem cell: has the ability to become any type of cell 21

Stem: the main stalk of a plant 68

Stomach: internal organ where food the bulk of digestion occurs 231

Stomata: microscopic pores that allow gas and water exchange in plants 75

Superior: higher or near the center of the body than another body part 100

Symbiotic: two different organisms living closely together and benefiting one another 104

Sympatric speciation: something other than geography divides a population and the two diverge into unique groups 60

Terrestrial: an animal or plant that lives on land 103

Thorax: central body cavity of an animal where the organs for circulation and respiration are located 98

Trait: an observable quality in a person, means the same thing as phenotype 39

Trophic level: the position an organism occupies in the food chain 194

Uterus: female organ where fertilization occurs and the offspring develop before birth 169

Vaccine: substance used to stimulate immunity to a particular infectious disease 249

Vascular bundle: part of the transport system in plants 75

Vector: an organism that transmits a pathogen, disease, or parasite from one organism to another 244

Veins: a tube with valves that returns deoxygenated blood from the body to the heart 227

Ventral: the belly 100

Vertebrate: an animal with a backbone and belonging to the phylum Chordata 128, 155

Virus: an infectious microbe consisting of a segment of DNA or RNA surrounded by a protein coat 244

Viviparous: eggs are fertilized and develop inside the mother's body 132

Warm-blooded: ectotherm, an animal that uses metabolic processes to keep its body temperature stable, regardless of the environment 128

White blood cells: these circulate through the bloodstream and fight viruses, bacteria, and other invaders that threaten your health 249

People

About the Authors

Michelle and Karen are sisters from Idaho, USA. They grew up playing in the woods and on the lakes of the northern Rockies. Karen is married with four children, two boys and two girls. Michelle is married with six kids, all boys.

Michelle has a BS in biology and Karen has a BA in education. Since the early 2000s, they have been homeschooling their kids and taking them to the lake as often as possible.

In 2008, at a family reunion (at the lake, of course), they were opining about all the things they wished they could have in a homeschool curriculum. Their mom suggested they write their own curriculum. They looked at each other in doubt, then thought, "Why not?" And Layers of Learning was born. Thanks, Mom.

Made in the USA
Columbia, SC
25 July 2024

38650834R40155